TERAHERTZ SPECTROSCOPY - A CUTTING EDGE TECHNOLOGY

Edited by **Jamal Uddin**

Terahertz Spectroscopy - A Cutting Edge Technology
http://dx.doi.org/10.5772/62805
Edited by Jamal Uddin

Contributors

Faustino Wahaia, Irmantas Kasalynas, Pedro L.Granja Granja, Catia Silva, Kapil Bhatt, C.C. Tripathi, Sandeep Kumar, Sandeep ., Shilpi Shriwastava, Kei Takeya, Kodo Kawase, Ci-Ling Pan, Sung-Liang Chen, L. Jay Guo, Maxime Bernier, Manabu Tsujimoto, Takanari Kashiwagi, Hidetoshi Minami, Kazuo Kadowaki, Yong-Duck Chung, Woo-Jung Lee, Ja-Yu Lu, Borwen You, Hiroshi Murakami, Xin Zhang, Zhuoyong Zhang, Oleksiy Shulika, Mykhailo Klymenko, Igor A. Sukhoivanov, Riccardo Piccoli, Roberto Morandotti, Luca Razzari, Andrea Rovere, Andrea Toma, Ronald Coutu, Jamal Uddin

Published by InTech

Janeza Trdine 9, 51000 Rijeka, Croatia

Publishing Process Manager Iva Lipovic
Technical Editor SPi Global
Cover InTech Design team

First published April, 2017
Printed in Croatia
Legal deposit, Croatia: National and University Library in Zagreb

Additional hard copies can be obtained from orders@intechopen.com

Terahertz Spectroscopy - A Cutting Edge Technology, Edited by Jamal Uddin
p. cm.
Print ISBN 978-953-51-3031-4
Online ISBN 978-953-51-3032-1

For EU product safety concerns:
IN TECH d.o.o., Prolaz Marije Krucifikse Kozulić 3, 51000 Rijeka, Croatia,
info@intechopen.com or visit our website at intechopen.com.

Meet the editor

Dr. Jamal Uddin received his BS and MS degrees from the Dhaka University and PhD degree in Chemistry from Osaka University. He did his postdoctoral research in Laser Photochemistry at Wayne State University, MI, and University of Maryland, MD. He is currently a professor of Coppin State University, Department of Natural Sciences, Baltimore, Maryland, and the founder and director of the Center for Nanotechnology. Dr. Uddin's research interest is in the areas of solar cell, laser photochemistry, nanotechnology, quantum dots, single-molecule imaging spectroscopy, terahertz spectroscopy, and metal enhance fluorescence. He is a member of the American Chemical Society (ACS) and the Inter-American Photochemical Society (I-APS). He is a corresponding secretary of the American Association of University Professors (AAUP), Coppin State University, Maryland. Dr. Uddin has received Wilson H. Elkins Professorship award in FY 2012–2014 for the STEM research in Nanotechnology. He was the recipient of 2012 and 2016 Daily Record Innovator of the year who has had a positive effect and tremendous impact in Maryland.

Contents

Preface

The recent advance in terahertz (THz) technology, which exploits the electromagnetic radiation spectrum between 0.1 and 10 THz, has generated intense interest in its application in a vast array of fields. Current applications of terahertz encompass nanotechnology, cancer therapy, quality control, pharmaceutical analysis, and material characterization, among many others.

This book, *Terahertz Spectroscopy - A Cutting Edge Technology*, focuses on the state-of-the-art research in terahertz spectroscopy and imaging and their application in different fields of study. Up-to-date innovative techniques for the generation, manipulation, and detection of terahertz radiation are also presented and discussed in this book.

This volume contains a wide variety of topics including the application of terahertz in medicine, spectroscopy, superconductors, metamaterial, remote sensing, and nanotechnology.

We (InTech publisher and editor) also owe our gratitude to the experts who gave much of their time and expertise in determining the scientific merit of the chapters submitted to this special book.

Dr. William Ghann, postdoctoral fellow, Center for Nanotechnology, Coppin State University, helped greatly with the review of chapters in this book, and I really appreciate his critical work.

Dr. Jamal Uddin
Professor of Chemistry
Founding Director, Center for Nanotechnology
Department of Natural Sciences
Coppin State University
Baltimore, Maryland, USA

Section 1

THz Spectroscopy

Chapter 1

Terahertz (THz) Spectroscopy: A Cutting-Edge Technology

William Ghann and Jamal Uddin

Additional information is available at the end of the chapter

http://dx.doi.org/10.5772/67031

1. Introduction

Terahertz (THz) represents the portion of the electromagnetic radiation between the microwave and the infrared region as displayed in **Figure 1**. It is within the frequency range of 0.1–10 THz, corresponding to wavelengths of radiation from 3000 to 30 μm. Terahertz radiation is also known as terahertz gap, terahertz waves, T-waves, terahertz light, T-light, or T-lux. This form of electromagnetic radiation is less known, due to the limited access to technology for generating and detecting radiation [1]. There are a number of reviews on the different technologies for generating and detecting terahertz [2, 3]. Terahertz waves are nonionizing, noninvasive, and penetrable to many materials with a depth of penetration lower than that of microwave radiation. Terahertz radiation also tends to be very sensitive to various kinds of resonances such as vibrational, translational, rotational, torsional, and conformational states, enabling it to provide information on molecules that are inaccessible with other analytical and imaging techniques. These unique characteristics make them suitable for identifying, analyzing, or imaging a variety of materials.

Unlike Raman and Infrared spectroscopy, the development of techniques for generating, manipulating, and detecting terahertz radiation is still in its early stages. Although terahertz radiation was deemed suitable for imaging and other applications, technologies to harness such capabilities were practically nonexistent [4]. On one hand, high terahertz frequencies do not lend themselves to be estimated with electronic counters employed in the detection of optical waves; instead various properties of wavelength and energy are harnessed to characterize terahertz radiation [5]. On the other hand, at such higher frequencies electronic devices utilized in generation and manipulation of radio waves and microwaves become less efficient. Thus with the exception of the terahertz gap, technology for the generation, manipulation, and everyday application of a vast majority of the electromagnetic spectrum has been developed. Significant progress made in terahertz technology has led to an increase in the different types of terahertz instrumentation available on the market with application for diverse fields. This review focuses on the generation, characterization, and application of terahertz radiation.

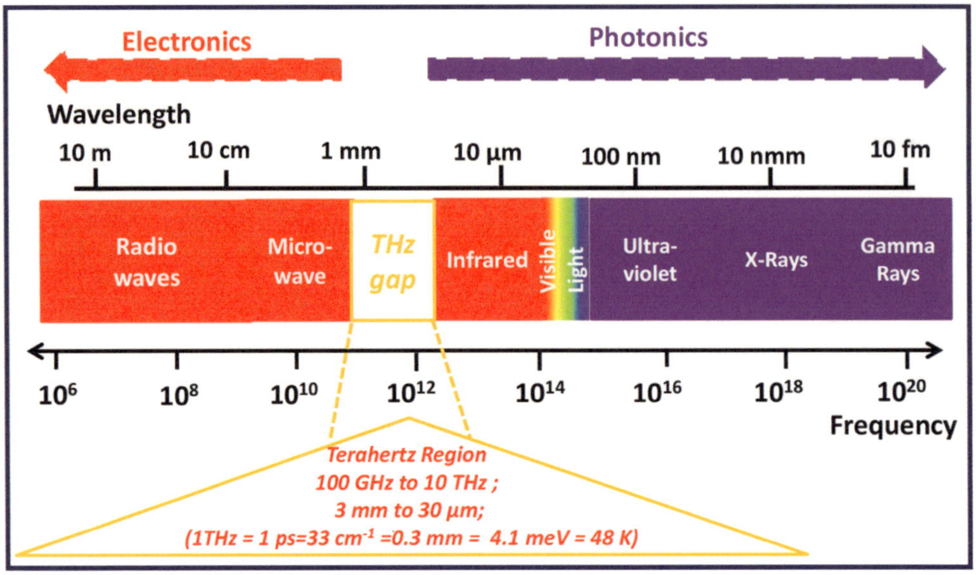

Figure 1. Frequency and wavelength regions of the electromagnetic spectrum.

2. Generation and detection of terahertz radiation

A typical terahertz system, in a majority of cases, consists of a source, components, and a detector. Some terahertz system may possess two or more of these components. A terahertz source generates a wide range of terahertz radiation; components such as lenses, mirrors, and polarizers manipulate the radiation; and a detector measures the radiation reaching it. R. A. Lewis list six main types of terahertz sources [6]. These include thermal [7, 8], vacuum electronic [9–11], solid-state electronic [12, 13], lasers [14, 15], sources pumped by lasers [16, 17], and mechanical-excitation types of terahertz radiation sources [18, 19]. Sources pumped by lasers employ either continuous or pulsed lasers. We will discuss the generation of terahertz radiation using a source pumped by continuous-wave (CW) diode lasers using an electro-optic dendrimer. In this method, an electro-optic dendrimer generates terahertz radiation through the difference-frequency technique (DFG). Rahman et al. pioneered work on the generation of terahertz radiation using this technique. The technique uses two continuous lasers to pump an electro-optic dendrimer emitter to give a combined pump power capable of generating a stable terahertz radiation from 0.1 THz to approximately 10 THz at room temperature.

Dendrimers are highly branched synthetic polymers composed of a central core, a treelike interior structure, and an external surface decorated with functional groups. Dendrimers have nanoscale dimensions and are radially symmetric molecules. The size of a dendrimer is dependent on its generation or the number of concentric shells around the core; the higher the generation, the larger the size. Their unique structure and size make them suitable for a variety of applications including their use as drug delivery agents, imaging agents, solubilizing agents, and radio-ligands. An electro-optic dendrimer has the capacity to generate terahertz radiation [8, 20]. When the structural features of dendrimers are carefully manipulated, the optical and electro-optic properties of the functional groups at the periphery of the dendrimers are enhanced in a very controlled manner. Dendrimers are able to form self-assembled multilayers. The thickness of such multilayer assemblies can be adjusted by controlling certain parameters such as the number of concentric layers or generation of the dendrimer. The surface and solution chemistry also affect the thickness of dendrimers.

Dendrimer dipole excitation (DDE) is the mechanism for the creation of the electro-optic dendrimer. Electro-optic properties are critical for applications that normally rely on nonlinear optical parameters such as electro-optic coefficient (EOC), and the second-order susceptibility, $\chi^{(2)}$. A third-generation dendrimer (**Figure 2a**) undergoes chromophore doping and poling to become an electro-optic dendrimer with high second-order susceptibility $\chi^{(2)}$ and a high electro-optic coefficient. A high $\chi^{(2)}$ is essential for both the up conversion and the bandwidth. The chromophore poling optimizes the dipole alignment of dendrimer film. There are a number of poling techniques including corona poling, optical poling, contact electrode poling, and photo-assisted poling. The electro-optic coefficient (r_{33}) is deduced from the difference in the refractive index (Δn) of poled and unpoled electro-optic dendrimer in accordance with the Pockel's law as shown in Eq. (1), where E_p is the poling field

$$|\Delta n| = \frac{1}{2} n^3 r_{33} \mathrm{E_p} \tag{1}$$

The electro-optic coefficient of dendrimers is significantly higher than inorganic crystalline materials since the second-order susceptibility is proportional to the relative electric constant (ε) and r_{33} as shown in Eq. (2):

$$\chi^{(2)} \propto \varepsilon^2 r_{33} \tag{2}$$

As displayed in **Figure 2(a)**, dendrimer doping creates a dipole moment (μ) distribution of charge (Q) carriers in the dendrimer such that the equation for dipole moment ($\mu = ql$) is adjusted to $\mu(r) = Ql(r)$. The separation (l) between the negative and positive charge centers is therefore a function of the coordinate, $l(r)$, of the charge centers. **Figure 2(c)** shows a schematic of the possible energy levels in an electro-optic dendrimer molecule that has undergone doping and poling.

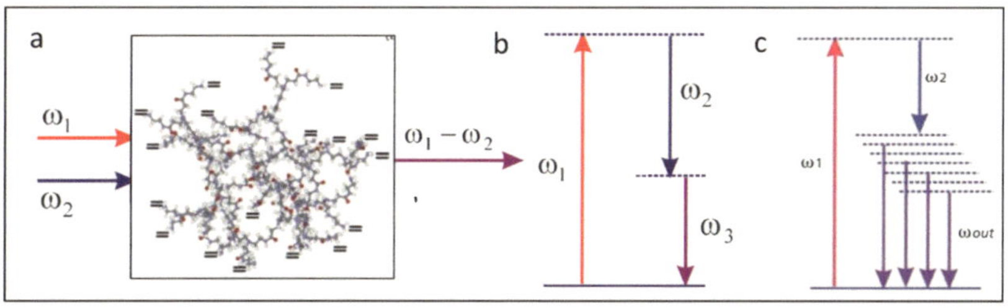

Figure 2. (a) Molecular structure of the third-generation dendrimer and interaction geometry for difference generation; (b) energy-level diagram of difference-frequency generation; (c) probable energy-level diagram in dendrimer molecule resulting from chromophore doping and poling. Adapted with permission from Ref. [20].

Difference-frequency generation (DFG) is used to generate terahertz radiation from the electro-optic dendrimers by a nonliner polarization as illustrated in Eq. (3).

$$P(\omega_1 - \omega_2) = 2\chi^{(2)} E_1 E_2^n \quad (3)$$

where E_1 and E_2 are the field strengths, ω_1 and ω_2 are the frequencies of the input lasers. **Figure 2** displays the energy-level diagram of the difference-frequency generation (panels a and b). Every photon created at the difference-frequency (Eq. (4)) requires the destruction of a photon at a higher input frequency (ω_1) and the creation of a photon at the lower input frequency (ω_2) to meet the requirements for energy conservation:

$$\omega_3 = \omega_1 - \omega_2 \quad (4)$$

A terahertz source generates the radiation, which goes through various components and reaches the detector. There are many methods used to detect terahertz radiation. The electro-optic sampling method of terahertz detection will be discussed in this section. Electro-optic sampling relies on the linear electro-optic effect, also known as the Pockel's effect. In the Pockel's effect, an externally applied electric field modulates the birefringence of an electro-optic material [17]. Typical components of this detection system include an electro-optic medium, a broadband quarter-wave plate, a Wollastron prism and differential detectors. When terahertz waves reach the detector, the incident THz pulse induces a birefringence in an electro-optic medium, which is proportional to the electric field of the pulse. The change in polarization state of a probe near-infrared probe pulse is measured with varying birefringence. The terahertz electric field can be mapped by measuring the degree of polarization rotation as a function of the delay between the terahertz pulse and the near-infrared probe pulse [17].

3. Applications of terahertz technology

The aforementioned distinctive capabilities of terahertz waves have been exploited in a plethora of applications in sensing and imaging. These applications of terahertz radiation are utilized in a variety of fields such as spectroscopy, photovoltaics, medicine, security screening, pharmacy, quality assurance, dentistry, communication, and astronomy (**Figure 3**).

3.1. Spectroscopic analysis

Terahertz has been used for the spectroscopic characterization of various materials. The sensitivity of terahertz radiation to resonances in different molecules has been exploited for the characterization of various compounds. Infrared spectroscopic techniques investigate intramolecular interactions but terahertz probes are unique in the sense that they probe the intramolecular and intermolecular interactions and provide information characteristic of a material. Huang et al. studied the terahertz time-domain spectroscopy of 1,3,5-trinitro-s-triazine. The results showed an absorption peak at 0.8 THz attributable to intermolecular action. The results were corrobo-

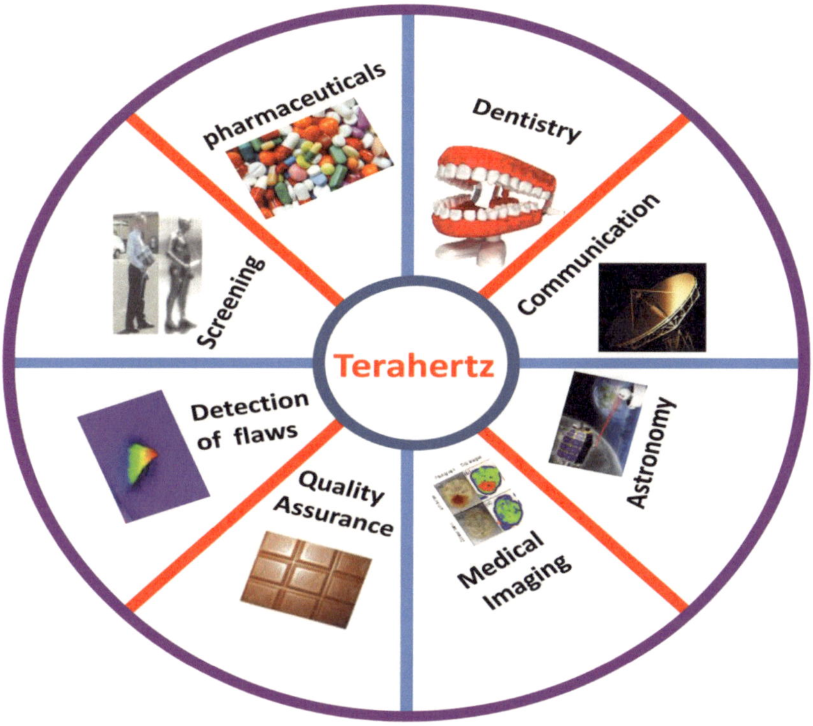

Figure 3. Examples of application of terahertz radiation (detection of flaws in photovoltaic materials, medical imaging, screening, pharmaceuticals, quality control, dentistry, communication, and astronomy).

rated with Fourier transform infrared spectroscopy (FTIR) study, which was carried out by the same researcher for comparison [21]. Upadhya et. al investigated the spectral features of glucose and uric acid using terahertz time-domain spectroscopy. It was found out that the unique features on the absorption spectra of the two biological molecules were as a result of the intermolecular vibrational modes existing in the molecules [22]. Terahertz spectroscopy has also been used to study the spectral absorption features of methamphetamine, a well-known illicit drug. Absorption spectra measured from 0.2 to 2.6 THz agreed with experimental data and were traced to the collective vibrational modes of the compound. Likewise, time-domain terahertz spectroscopic investigation of 2,4-dinitrotoluene carried out side by side with FTIR measurement of the same compound revealed a better correlation of results of two techniques at lower frequencies than at higher frequencies. The result obtained experimentally correlated well with theoretical study showing that terahertz spectroscopy is a powerful tool for studying the low-frequency modes of several compounds [23]. Other compounds that have been studied with terahertz spectroscopy include trialanine, short-chain polypeptides [24], and tryptophan [25].

3.2. Terahertz characterization of photovoltaics

We have used terahertz time-domain spectroscopy to obtain spectroscopic information on dye-sensitized titanium dioxide (TiO_2) films used in the fabrication of dye-sensitized solar cells [26]. **Figure 4** displays the probe setup used for this work. The dyes examined include two natural dyes, pomegranate and blackberry extract and a ruthenium bipyramidal complex. It was observed that the natural dyes, pomegranate and blackberry, had similar spectral features, which were different from that of the inorganic dye (rubpy).

In addition to the spectroscopic characterization of photovoltaic materials, terahertz reflectometry is also used to image photovoltaic material for detecting flaws and defects in the materials that could eventually affect the efficiency of solar cells. To confirm the unique capability of terahertz reflectometry to detect flaws present on the dye-sensitized TiO_2 films, dye-adsorbed TiO_2 films with size of 25 mm × 25 mm were scanned to examine any defects in an area of 14 mm × 14 mm. Pomegranate, blackberry, and rubpy dye-sensitized solar cell electrodes were scanned and with the measurements, 3D images, comparable to scanning electron microscopy images, were generated. As displayed in **Figure 5**, the cracks present in the TiO_2 films show up as large peaks on the surface plots due to the greater intensity of

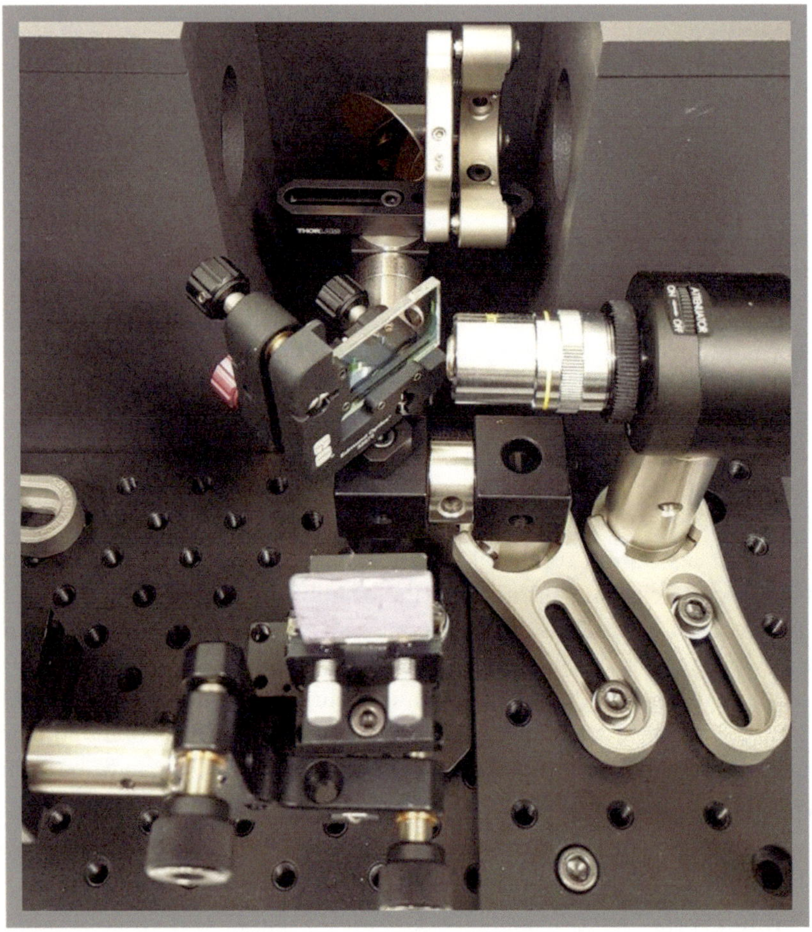

Figure 4. A typical terahertz probe setup used for reflectance measurement of photovoltaic materials.

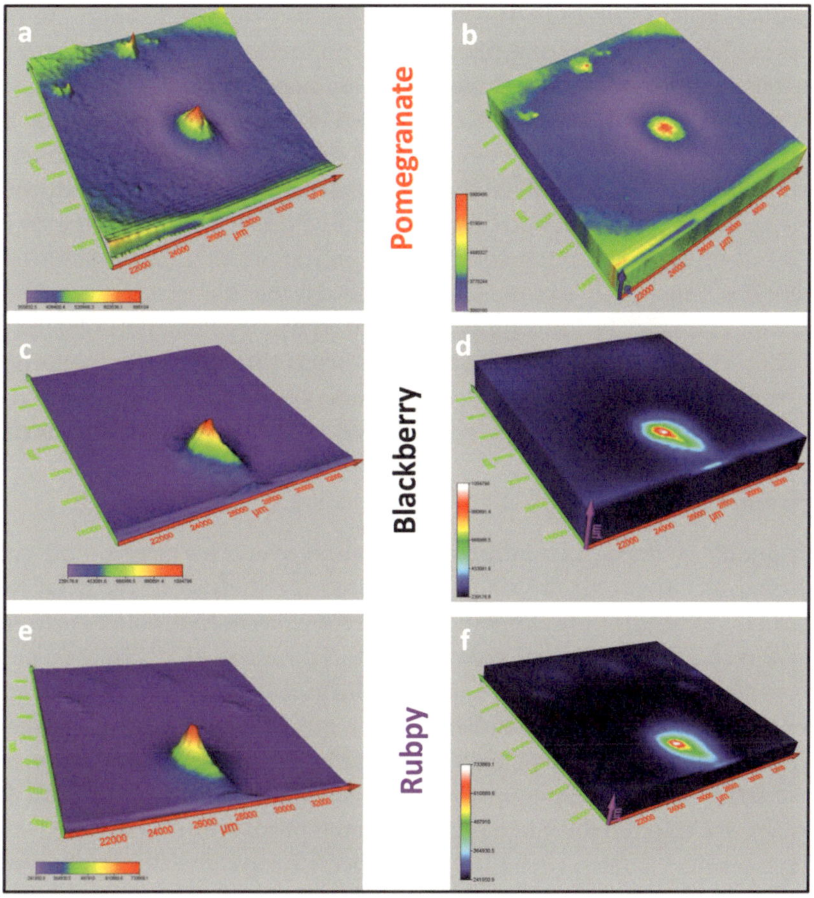

Figure 5. Comparison of terahertz surface (a, c, and e) plots and 3D (b, d, and f) images of Pomegranate (a and b), Blackberry (c and d), and Rubpy dye (e and f) sensitized TiO_2 film coated on FTO glass revealing defects present on these photovoltaic materials.

reflected light from the exposed glass surface. The magnitude of the peaks correlates with the size of the exposed surface area. Defects are easily noticeable as they reflect more light. As displayed in **Figure 5**, the protruding peaks in dye-sensitized TiO_2 film are the regions of higher light reflection occurring at the center of the 25 mm × 25 mm dye-sensitized TiO_2 film. These defects are partly attributed to the dye application and partly due to the lack of uniformity of the layer of TiO_2 on the glass substrate. The results demonstrate the unique defect detection capabilities of terahertz reflectometry. Terahertz reflectometry is therefore well suited to detect flaws and malfunctioning areas of solar cells.

Tiwana et al. used optical-pump terahertz to investigate the photo-induced conductivity dynamics of the semiconductor, titanium dioxide film, used in making the photoanode of dye-sensitized solar cells. In preparation of photoanode of dye-sensitized cell, a mesoporous titanium dioxide film is sensitized with the dye of interest. Upon absorption of radiant light, the dye molecules get excited to higher energy levels where they inject electrons into the conduction band of TiO_2. The injected electron travels through the mesoporous titanium dioxide layer to an external circuit via a conductive glass layer. The photoconductivity of the dye before and after dye adsorption was examined using the terahertz. The dye employed was a ruthenium bipyramidal complex. A biphasic charge injection with an initial sub-500 fs fast component and a slower 70–200 ps rise component was observed. Terahertz spectroscopy was also used to analyze the effect of $TiCl_4$ on the overall performance of the solar cell [27]. Similar work has been carried out by Nemec et al. where ultrafast studies were carried out on terahertz photoconductivity in nanocrystalline mesoporous titanium dioxide films. The transport of excited electron though the titanium dioxide film, prepared by the "brick and mortar" technology was enhanced upon calcination [28]. In addition to the studies involving the transport of excited electrons through the mesoporous titanium films, time-resolved terahertz spectroscopy has also been used to study the dynamics of interfacial charge transfer state and carrier separation in dye-sensitized solar cells [29].

3.3. Quality control

The role of quality inspection in industry cannot be overemphasized, and various techniques and methods are constantly being explored for the use in the quality control of various products. Terahertz time-domain spectroscopy has been used as a powerful tool for the analysis and characterization of various products. Rutz et al. utilized this technique in the evaluation of various polymeric compounds including a polyethylene compound with silver-coated titanium dioxide nanospheres and a glass-fiber-reinforced epoxy composite [30]. The water content in various materials could be estimated using terahertz radiation. Terahertz radiation is not transparent to water, and the attenuation of water to terahertz radiation has been exploited for quality control in the production of paper [31]. The humidity level of papers is evaluated using terahertz spectroscopy. In addition to the amount of water in papers, terahertz can also be used to assess the thickness and mass of paper. The ability of terahertz to measure the moisture or humidity in substances has been exploited in determination of the moisture content in leather as well as the thickness of the leather [32]. Various food processes have also been monitored using THz spectroscopic and imaging techniques [33].

3.4. Pharmaceutical applications of terahertz

Terahertz technology is exploited in the pharmaceutical industry owing to the uniqueness of its properties [34–36]. THz waves have the unique capability of penetrating different kinds of materials, which are usually not transparent to other forms of electromagnetic radiation. Some of the applications of terahertz in pharmaceutical industry include the identification and quantification of polymorphic forms or hydrates of new active pharmaceutical ingredients. This characterization makes it easy to optimize the stability, the bioavailability, and the production of drugs [3]. Russe et al. have used terahertz pulse imaging (TPI) for the quantification of the coating and thickness of coated pharmaceutical tablets. X-ray Microtomography, the traditional tomographic technique commonly used for the analysis of tablet coating, was used to validate the results of the terahertz pulsed imaging of the tablet coating. **Figure 6** depicts the results of the measurements of coating thickness of the coated pharmaceutical tablets comparing terahertz pulsed imaging to X-Ray Microtomography [37].

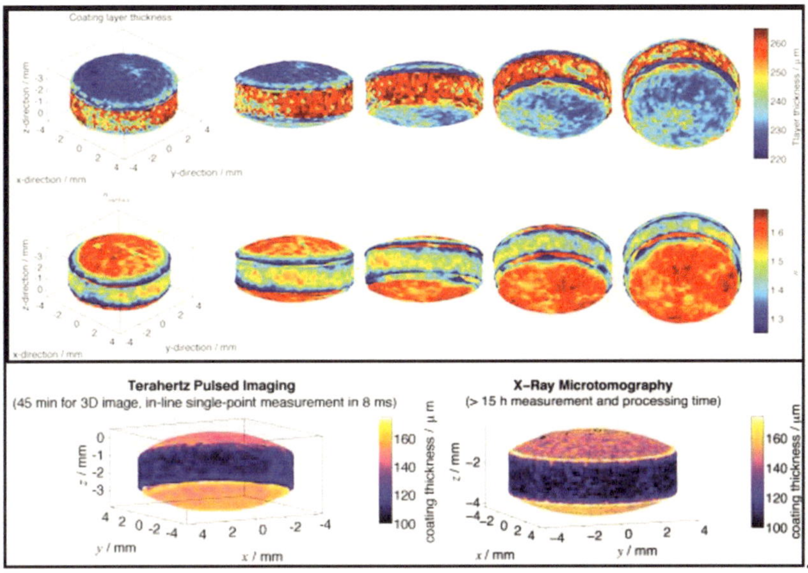

Figure 6. 3D maps of coating layer thickness (top) and surface refractive index (bottom) for a biconvex tablet. (Reprinted from http://thz.ceb.cam.ac.uk).

3.5. Medical applications of terahertz technology

The last couple of decades have seen the development of new terahertz techniques for imaging and sensing purposes. Like X-ray, terahertz radiation is capable of generating two-dimensional images of different kinds of objects owing to the fact that terahertz waves travel through certain materials such as semiconductors and plastics but are not transparent to others such as metallic and polarizing materials. Terahertz waves are preferable over X-rays for medical imaging since they are nonionizing and to a large degree harmless to the recipient. Terahertz waves are currently under investigation for use as an imaging modality to visualize cancerous tissues. According to the American Cancer Society, cancer is the second leading cause of death in the U.S., second only to heart disease, and accounts for nearly one in every four deaths. Early detection could mean the difference between life and death for many patients but there is a shortfall in the current imaging techniques. New terahertz spectroscopic studies on cancer could potentially provide novel techniques for the early detection of the disease. In fact, there has been a steady increase in the number of studies demonstrating the potential use of THz radiation for the imaging and detection of skin tissues and cancerous cells [20, 38–40]. Since terahertz radiation is readily absorbed by water and other polar liquids, the method lends itself well to imaging most organic tissue [38]. Water has intense absorption in the terahertz portion of the electromagnetic spectrum and with the concentration of water in cancerous cells significantly different from that in healthy cells, the variation in the water absorption of terahertz radiation becomes the basis of differentiation of normal cells from cancerous cells. Using THz pulse imaging, Woodward et al. were able to differentiate basal cell carcinoma, a type of skin cancer, from healthy tissues [39]. There have also been studies on the THz pulse imaging in breast cancer [41, 42].

A great number of biomolecules including DNA/RNA, carbohydrates, amino acids/peptides, proteins, cells, and tissues have been investigated using terahertz radiation [1]. The low-frequency internal helical vibration of DNA involving hydrogen bond of DNA base pairs is reflected in the terahertz frequency absorption spectra [43]. Such sensitivity of biomolecules to terahertz radiation has been utilized for the characterization of biological materials. Globus et al. used terahertz spectroscopy to probe DNA polymers in which they were able to prove the dependence of the mode strength on the polarization of the terahertz field with respect to the molecule alignment and thereby concluded that theoretical modeling, combined with measured data, may be used to directly assign vibrational modes to specific structural features of the macromolecule [43]. Using transmission time-domain terahertz spectroscopy and continuous wave terahertz imaging, Wahaia et al. were able to distinguish normal cells from cancerous cells and were, consequently, able to detect colon cancer [7].

3.6. Dental application of terahertz technology

X-ray imaging is the foremost imaging technique in dentistry. It is the principal tool employed to monitor caries and plan treatment for braces, dentures, extractions, and implants. However, X-rays are highly ionizing and frequent exposures are detrimental to health. Terahertz pulsed imaging (TPI) shows great potential as an alternative imaging technique in dental diagnosis [44–47]. It is low in energy, nonionizing, and less hazardous. Unlike other biological tissues with a huge percentage of water with the propensity for intensive absorption of terahertz radiation, teeth have a relatively low water content, which makes it convenient to image teeth samples using the terahertz pulsed imaging technique. Longbottom et al. employed a terahertz pulse imaging technique to perform *in vitro* experiments involving incisor teeth and concluded that terahertz has the ability to not only identify caries but also monitor erosion [48]. Using the transmission mode in a developed terahertz time-domain spectroscopy system, Kamburoglu et al. measured the properties of sliced teeth section. They analyzed the refractive and absorptive properties of primary and permanent teeth that were either healthy or had cavities in them and found the technique effective for characterizing dental structures [49].

Karagov et al. recently demonstrated how THz pulse imaging could be used to detect dental caries [50]. They employed a continuous wave (CW) and time-domain reflection mode raster scan THz imaging system to carry out 2D and 3D imaging of various teeth samples. The data were analyzed in both the spatial and frequency domains, and the results pointed to the sensitivity of terahertz pulse in detecting variations in the structure of teeth samples. Schirmer et al. conducted terahertz spectral imaging of teeth samples using a real-time terahertz color scanner [51]. To conduct the imaging studies, longitudinally sliced human molar specimen (**Figure 7a**) were heated to remove water molecules that could interfere with the measurements and then imaged to study differences in THz transmission in different part of the teeth samples. Spectral images of eight samples taken at different frequencies are displayed in **Figure 7(b)**.

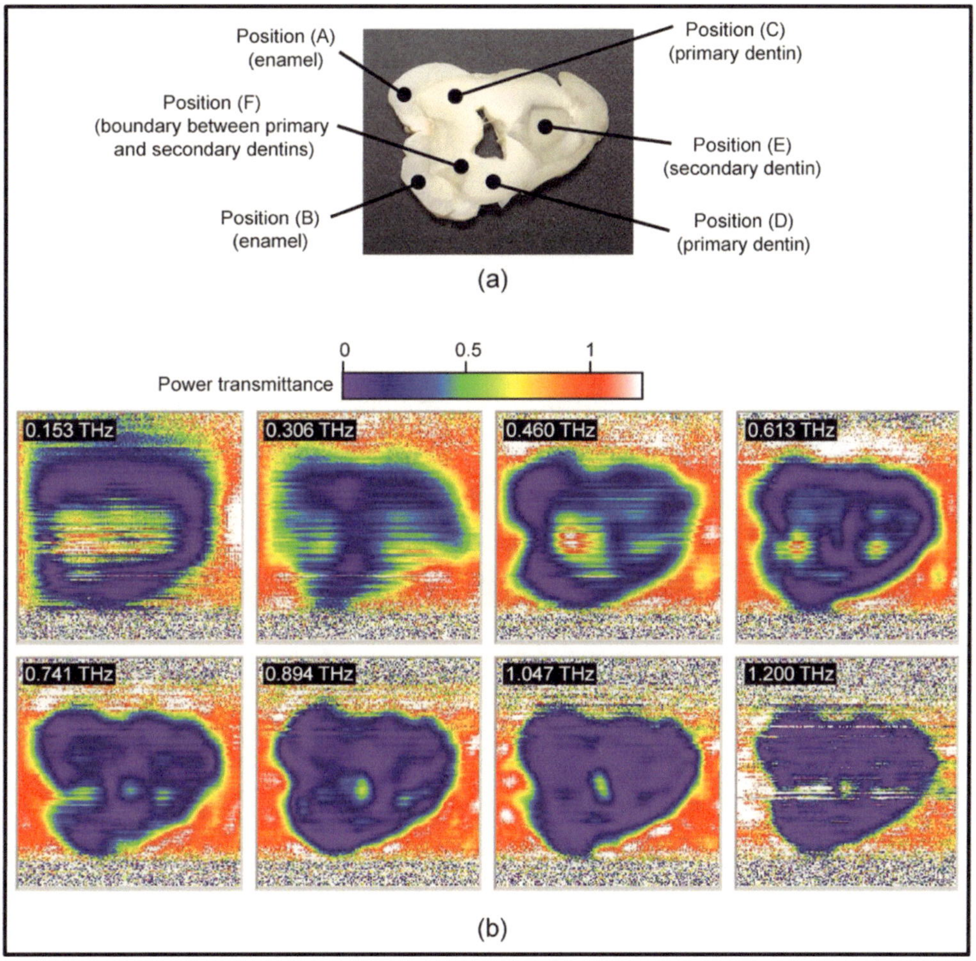

Figure 7. Visible and terahertz image of cavity in human tooth: (a) longitudinally sliced human molar specimen; (b) spectra images of eight samples taken at difference frequencies. (Reprinted from Markus Schirmer et al. "Biomedical applications of a real-time terahertz color scanner," Biomed. Opt. Express 1, 354–366, 2010) [51].

3.7. Security screening

Security screening is one of the most common applications of terahertz technology. Terahertz does not only give information about the presence of concealed items but also has the capacity to identify the composition of the materials in question. In recent years numerous investigations have been conducted on the capability of terahertz to detect explosives at airports and other sensitive places [52]. The advantage of terahertz radiation over X-ray, also used for security screening, is its low energy and nonionizing nature. The ionizing nature of X-rays makes the regular exposure of it harmful to people. The unique ability of Terahertz waves to penetrate a wide range of packaging materials makes it possible to detect weapons, explosives, and potentially explosive devices concealed within these materials [53]. **Figure 8** shows a terahertz image revealing a blade (c) and a metallic knife (d) concealed behind a cardboard and a thick dark cloth, respectively. Packaging materials such as cardboard, paper, plastics, and leather are all transparent to terahertz radiation.

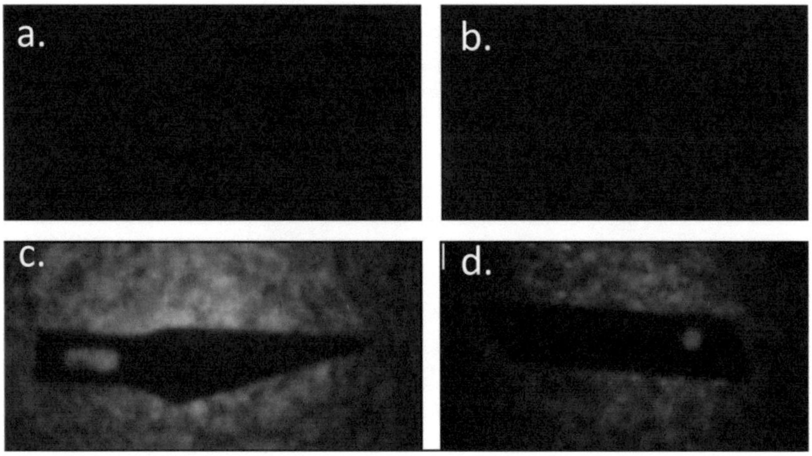

Figure 8. Security screening employing terahertz radiation: a hidden blade (c) behind a cardboard (a) is revealed with terahertz radiation; metallic knife (d) hidden behind a thick dark cloth (b) is revealed by the terahertz beam. Picture reprinted from Applied Research & Photonics, Inc.

3.8. Communication, remote sensing and astronomy

One other application of terahertz radiation is in the area of wireless data communication. The terahertz portion of the electromagnetic radiation has shown promise as a bandwidth suitable for data transmission. Ishagaki et al. have designed a terahertz oscillating resonate tunneling diode that has the ability to transmits signal at 542 GHz with a data transfer rate of 3 Gbit/s. This type of terahertz WIFI is limited to a region of 30 feet but has a potential to support data rates of up to 100 Gbit/s.

Remote sensing allows the acquisition of information about materials or locations from a distance, which could be from an aircraft or satellites. Terahertz spectroscopy is one of the promising tools for carrying out remote sensing. Lui et al. developed a technique based on broadband terahertz wave detection for remote sensing with which they were able to detect coherent terahertz wave at a distance of 10 m [54]. The technique has shown promise for the measurement of terahertz pulses at standoff distances with minimal water vapor absorption and unlimited directionality for optical signal collection. **Figure 9** shows a schematic of this terahertz wave remote sensing technique. The technique exploits the interaction between induced gas plasma and terahertz waves. Two lasers at different frequencies are focused at a target to induce gas plasma formation. Fluorescence emitted from the induced gas plasma is scattered in characteristic ways by the terahertz radiation of the material it hits. The terahertz wave sensing is then performed from remote distances [54]. The terahertz region of the electromagnetic radiation, furthermore, happens to be a very significant window for astronomical observation. Submillimeter astronomy is a branch of astronomy conducted with terahertz radiation providing a wide range of spectral lines useful for the investigation of many phenomena such as the formation of stars. It also holds spectral signatures of ions, atoms, and molecules that are necessary for understanding of the composition and origin of the Solar System, the evolution of matter in our Galaxy, and the star formation history of galaxies over cosmic timescales. The utility of these lines, their examples of the science they deliver, and the detail properties of successful low-resolution direct detection spectrometers for work in the THz regime have been reviewed by a number of researchers [55, 56].

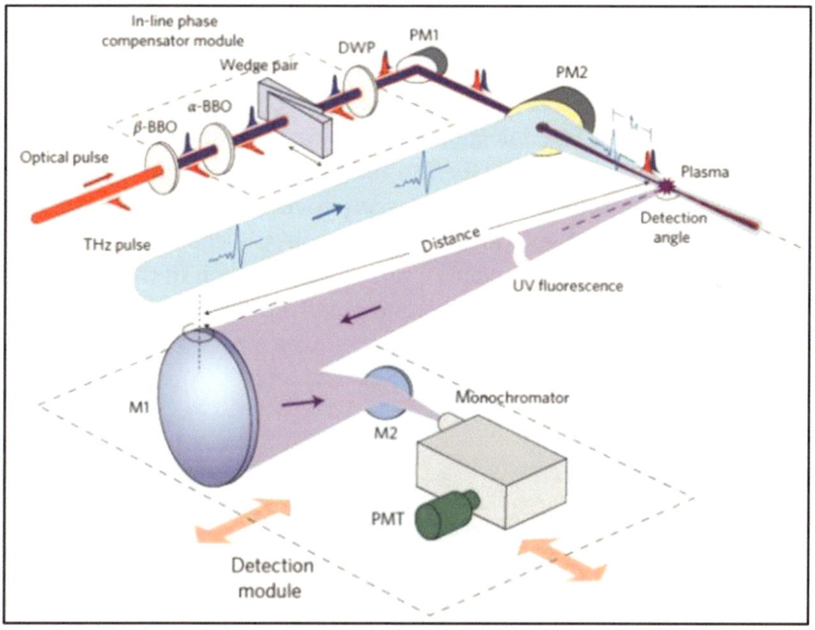

Figure 9. Schematic of the THz wave remote sensing technique (Obtained from Liu, J. et al, Nat. Photonic., 4, 627, 2010.) [54]. The 2ω pulse is generated by passing the fundamental beam through a type I β-BBO crystal. Both the fundamental and second-harmonic optical pulses are linearly polarized along a vertical direction. The relative phase change between the ω and 2ω pulses is tuned by the lateral translation of fused silica wedges in the optical beam path after the α-BBO. The two optical pulses are focused by a parabolic mirror (effective focal length, 150 mm) into air to generate plasma. The time delay t_d is defined as the delay between the optical pulse peak and terahertz pulse peak. The fluorescence detection system consists of a UV concave mirror (M1; diameter, 200 mm and focal length, 500 mm), a UV plane mirror (M2), a monochromator and a photomultiplier tube (PMT). The distance of remote sensing is varied by moving the fluorescence detection system with respect to the plasma. DWP, dual-band waveplate.

3.9. Other applications

Metamaterials are materials that are structurally adjusted to interact in a specific way to electromagnetic radiation. Terahertz spectroscopy is frequently used to characterize metamaterials. Metamaterials that are specifically fabricated to interact with terahertz frequencies are referred to terahertz metamaterials. They are artificial materials that are carefully engineered to provide a desired electromagnetic response. Metamaterial have a basic lattice structure composed of rudimentary elements and possess the characteristics of a crystal structure. These rudimentary elements are relatively bigger in size than atoms or individual molecules. Metamaterial can be used in making THz metamaterial sensors for the detection of an array of materials and systems. It has been used for the detection of microorganisms such as fungi and bacteria with high sensitivity [57].

Terahertz spectroscopy also has the potential to detect oil spills in water bodies [58, 59]. Using terahertz transmission spectroscopy, Gorenflo et al. determined the amount and structure of water in oil-water complexes by monitoring changes in absorption coefficient and refractive index [58]. Cunnell et al. carried out similar measurements using terahertz quantum cascade laser to quantify water in industrial water-oil emusions [59].

3.10. Conclusion

A remarkable progress has been made in the last couple of decades on the technology for the generation, manipulation, and detection of terahertz radiation leading to ever-increasing applications of this technology in industry and research. We have reviewed the applications of terahertz technology in spectroscopic analysis, photovoltaic characterization, pharmaceutical testing, security screening, biomedical, astronomy, oil spill characterization, pharmaceutical and quality control. With the exponential growth in research in the terahertz regime, we anticipate more advances in application of terahertz technology in the society.

Author details

William Ghann and Jamal Uddin*

*Address all correspondence to: juddin@coppin.edu

Center for Nanotechnology, Department of Natural Sciences, Coppin State University, Baltimore, MD, USA

References

[1] Plusquellic, D. F., Siegrist, K., Heilweil, E. J. & Esenturk, O. Applications of terahertz spectroscopy in biosystems. *Chemphyschem* **8**, 2412–2431 (2007).

[2] Ueno, Y. & Ajito, K. Analytical terahertz spectroscopy. *Analytical Science* **24**, 185–192 (2008).

[3] McIntosh, A. I., Yang, B., Goldup, S. M., Watkinson, M. & Donnan, R. S. Terahertz spectroscopy: a powerful new tool for the chemical sciences? *Chemical Society Reviews* **41**, 2072–2082 (2012).

[4] Wai, L. C., Jason, D. & Daniel, M. M. Imaging with terahertz radiation. *Reports on Progress in Physics* **70**, 1325 (2007).

[5] Pawar, A. Y., Sonawane, D. D., Erande, K. B. & Derle, D. V. Terahertz technology and its applications. *Drug Invention Today* **5**, 157–163 (2013).

[6] Lewis, R. A. A review of terahertz sources. *Journal of Physics D: Applied Physics* **47**, 374001 (2014).

[7] Thacker, C., Cooray, A., Smidt, J., De Bernardis, F., & Atlas, K. M. The cosmic abundance of dust from the far-infrared background power spectrum. *The Astrophysical Journal* **768**, 58 (2013).

[8] Zolotarev, V. M., Mamedov, R. K., Bekhterev, A. N. & Volchek, B. Z. Spectral emissivity of a Globar lamp in the 2-50-μm region. *Journal of Optical Technology* **74**, 378–384 (2007).

[9] He, W. *et al.* High power wideband gyrotron backward wave oscillator operating towards the terahertz region. *Physical Review Letters* **110**, 165101 (2013).

[10] Bratman, V. L., Kalynov, Y. K. & Manuilov, V. N. Large-orbit gyrotron operation in the terahertz frequency range. *Physical Review Letters* **102**, 245101 (2009).

[11] Byrd, J. M. *et al.* Observation of broadband self-amplified spontaneous coherent terahertz synchrotron radiation in a storage ring. *Physical Review Letters* **89**, 224801 (2002).

[12] Pérez, S., González, T., Pardo, D. & Mateos, J. Terahertz Gunn-like oscillations in GaAs/InAlAs planar diodes. *Applied Physics Letters* **103**, 094516 (2008).

[13] Maestrini, A. *et al.* A frequency-multiplied source with more than 1 mW of power across the 900-GHz band. *IEEE Transactions on Microwave Theory and Techniques* **58**, 1925–1932 (2010).

[14] Chassagneux, Y. *et al.* Electrically pumped photonic-crystal terahertz lasers controlled by boundary conditions. *Nature* **457**, 174–178 (2009).

[15] Williams, B. S. Terahertz quantum-cascade lasers. *Nature Photonics* **1**, 517–525 (2007).

[16] Wu, X. L. *et al.* Green light stimulates terahertz emission from mesocrystal microspheres. *Nature Nanotechnology* **6**, 103–106 (2011).

[17] Jotzu,G., Cooper, M., Parkinson, P. & Johnston, M.B. Virtual Terahertz Spectrometer. https://www-thz.physics.ox.ac.uk (2009).

[18] Horvat, J. & Lewis, R. A. Peeling adhesive tape emits electromagnetic radiation at terahertz frequencies. *Optics Letters* **34**, 2195–2197, doi:10.1364/OL.34.002195 (2009).

[19] Cortie, D. L. & Lewis, R. A. Terahertz surfoluminescence. *Surface Science* **606**, 1573-1576 (2012).

[20] Rahman, A. *et al.* Diffision kinetics & permeation concentration of human stratum corneum characterization by terahertz scanning reflectometry. *Drug Development and Delivery* **12**, 43–49 (2012).

[21] *Huang*, F. *et al.* Terahertz study of 1,3,5-trinitro-s-triazine by time-domain and Fourier transform infrared spectroscopy. *Applied Physics Letters* **85**, 5535–5537 (2004).

[22] Upadhya, P. C., Shen, Y. C., Davies, A. G. & Linfield, E. H. Terahertz time-domain spectroscopy of glucose and uric acid. *Journal of Biological Physics* **29**, 117–121 (2003).

[23] Chen, Y. *et al.* THz spectroscopic investigation of 2,4-dinitrotoluene. *Chemical Physics Letters* **400**, 357–361 (2004).

[24] Kutteruf, M. R. *et al.* Terahertz spectroscopy of short-chain polypeptides. *Chemical Physics Letters* **375**, 337–343 (2003).

[25] Yu, B. *et al.* Torsional vibrational modes of tryptophan studied by terahertz time-domain spectroscopy. *Biophysical Journal* **86**, 1649–1654 (2004).

[26] Ghann, W., Rahman, A., Rahman, A. & Uddin, J. Interaction of sensitizing dyes with nanostructured TiO_2 Film in dye-sensitized solar cells using terahertz spectroscopy. *Scientific Reports* **6**, 30140 (2016).

[27] Tiwana, P., Parkinson, P., Johnston, M. B., Snaith, H. J. & Herz, L. M. Ultrafast Terahertz Conductivity Dynamics in Mesoporous TiO_2: Influence of Dye Sensitization and Surface Treatment in Solid-State Dye-Sensitized Solar Cells. *The Journal of Physical Chemistry C* **114**, 1365-1371 (2010).

[28] Nemec, H. *et al.* Ultrafast terahert photoconductivity in nanocrystalline mesoporous TiO_2 films. *Applied Physics Letters* **96**, 062103 - 062103-3 (2010).

[29] Brauer, J. C., Marchioro, A., Paraecattil, A. A., Oskouei, A. A. & Moser, J.-E. Dynamics of interfacial charge transfer states and carriers separation in dye-sensitized solar cells: a time-resolved terahertz spectroscopy study. *The Journal of Physical Chemistry C* **119**, 26266–26274, doi:10.1021/acs.jpcc.5b06911 (2015).

[30] Rutz, F. *et al.* Terahertz quality control of polymeric products. *International Journal of Infrared and Millimeter Waves* **27**, 547–556 (2006).

[31] Banerjee, D., von Spiegel, W., Thomson, M. D., Schabel, S. & Roskos, H. G. Diagnosing water content in paper by terahertz radiation. *Optics Express* **16**, 9060–9066 (2008).

[32] Hernandez-Serrano, A. I., Corzo-Garcia, S. C., Garcia-Sanchez, E., Alfaro, M. & Castro-Camus, E. Quality control of leather by terahertz time-domain spectroscopy. *Applied Optics* **53**, 7872–7876 (2014).

[33] Gowen, A. A., O'Sullivan, C. & O'Donnell, C. P. Terahertz time domain spectroscopy and imaging: emerging techniques for food process monitoring and quality control. *Trends in Food Science & Technology* **25**, 40–46 (2012).

[34] Claybourn, M., Yang, H., Gradinarsky, L., Johansson, J. & Folestad, S. in *Handbook of Vibrational Spectroscopy* (John Wiley & Sons, Inc., Hoboken, NJ, USA., 2006).

[35] Taday, P. F. Applications of terahertz spectroscopy to pharmaceutical sciences. *Philosophical Transaction Series A Mathematical and Physical Engeneering Sciences* **362**, 351–363; discussion 363–354 (2004).

[36] Shen, Y.-C. Terahertz pulsed spectroscopy and imaging for pharmaceutical applications: a review. *International Journal of Pharmaceutics* **417**, 48–60 (2011).

[37] Russe, I.-S., Brock, D., Knop, K., Kleinebudde, P. & Zeitler, J. Validation of terahertz coating thickness measurements using X-ray microtomography. *Molecular Pharmaceutics* **9** 3551–3559 (2012).

[38] Pickwell, E., Cole, B. E., Fitzgerald, A. J., Pepper, M. & Wallace, V. P. In vivo study of human skin using pulsed terahertz radiation. *Physics in Medicine and Biology* **49**, 1595–1607 (2004).

[39] Woodward, R. M. *et al.* Terahertz pulse imaging of ex vivo basal cell carcinoma. *Journal of Investigative Dermatology* **120**, 72–78 (2003).

[40] Woodward, R. M. *et al.* Terahertz pulse imaging in reflection geometry of human skin cancer and skin tissue. *Physics in Medicine Biology* **47**, 3853–3863 (2002).

[41] Fitzgerald, A. J. *et al.* Terahertz pulsed imaging of human breast tumors. *Radiology* **239**, 533–540 (2006).

[42] Ashworth, P. C. *et al.* Terahertz pulsed spectroscopy of freshly excised human breast cancer. *Optics Express* **17**, 12444–12454 (2009).

[43] Globus, T. R. *et al.* THz-spectroscopy of biological molecules. *Journal of Biological Physics* **29**, 89–100 (2003).

[44] Crawley *et al.* Terahertz pulse imaging: a pilot study of potential applications in dentistry. *Caries Research* **37**, 352–359 (2003).

[45] Crawley *et al.* Three-dimensional terahertz pulse imaging of dental tissue. *Journal of Biomedical Optics* **8**, 303–307 (2003).

[46] Sim, Y. C., Park, J. Y., Ahn, K.-M., Park, C., Son, J.-H. Terahertz imaging of excised oral cancer at frozen temperature. *Biomedical Optics Express* **4** 1413–1421 (2013).

[47] Kamburoğlu K & Yetimoglu N. Applications of terahertz imaging in medicine. *OMICS Journal of Radiology* **3**, 1, http://dx.doi.org/10.4172/2167-7964.1000e127 (2014).

[48] Longbottom, C. *et al.* Potential uses of terahertz pulse imaging in dentistry: caries and erosion detection. Proc. SPIE 4610, Lasers in Dentistry VIII; doi: 10.1117/12.469310 (2002).

[49] Kamburoglu, K., Yetimoglu, N. O. & Altan, H. Characterization of primary and permanent teeth using terahertz spectroscopy. *Dentomaxillofacial Radiology* **43**, 20130404 (2014).

[50] Karagoz, H. Altan, & Kamburoglu, K. "Terahertz pulsed imaging study of dental caries," in Medical Laser Applications and Laser-Tissue Interactions VII, L. Lilge and R. Sroka, eds., Vol. 9542 of SPIE Proceedings (Optical Society of America, 2015), paper 95420N.

[51] Schirmer, M. *et al.* Biomedical applications of a real-time terahertz color scanner. *Biomedical Optics Express* **1**, 354–366 (2010).

[52] Shen, Y. C. *et al.* Detection and identification of explosives using terahertz pulsed spectroscopic imaging. *Applied Physics Letters* **86**, 241116 (2005).

[53] Davies, A. G., Burnett, A. D., Fan, W., Linfield, E. H. & Cunningham, J. E. Terahertz spectroscopy of explosives and drugs. *Materials Today* **11**, 18–26 (2008).

[54] Liu, J., Dai, J., Chin, S. L. & Zhang, X. C. Broadband terahertz wave remote sensing using coherent manipulation of fluorescence from asymmetrically ionized gases. *Nature Photonics* **4**, 627–631 (2010).

[55] Kulesa, C. Terahertz spectroscopy for astronomy: from comets to cosmology. *IEEE Transactions on Terahertz Science and Technology* **1**, 232–240 (2011).

[56] Graf, U. U., Honingh, C. E., Jacobs, K. & Stutzki, J. Terahertz heterodyne array receivers for astronomy. *Journal of Infrared, Millimeter, and Terahertz Waves* **36**, 896–921 (2015).

[57] Park, S. J. & Ahn, Y. H. "Terahertz metamaterials application in sensing bacteria and fungi," 40th International Conference on Infrared, Millimeter, and Terahertz waves (IRMMW-THz), Hong Kong, 1–1 (2015)

[58] Gorenflo, S. *et al.* Dielectric properties of oil–water complexes using terahertz transmission spectroscopy. *Chemical Physics Letters* **421**, 494–498 (2006).

[59] Cunnell, R. *et al.* Quantification of emulsified water content in oil using a terahertz quantum cascade laser. In *2009 34th International Conference on Infrared, Millimeter, and Terahertz Waves.* 1–2 (2009)

Chapter 2

Terahertz Nanoantennas for Enhanced Spectroscopy

Riccardo Piccoli, Andrea Rovere, Andrea Toma,

Roberto Morandotti and Luca Razzari

Additional information is available at the end of the chapter

http://dx.doi.org/10.5772/66349

Abstract

Terahertz spectroscopy has great potential for sensing a wide range of elementary excitations. However, terahertz studies are practically limited to macroscopic ensembles of compounds (e.g., thick pellets of crystallized molecules or highly concentrated solutions of nanomaterials) due to the long radiation wavelength (about 300 μm at 1 THz). In this chapter, we show how terahertz nanoantennas can overcome the current limitations of terahertz spectroscopy such as low sensitivity and low spatial resolution. We briefly discuss how to design the resonance characteristics of a dipole nanoantenna through a Fabry-Pérot model, and then we present the experimental characterization of the spectral response of ordered arrays of such devices. Furthermore, we demonstrate how nanoantenna arrays enable the possibility of retrieving the spectroscopic signature of a monolayer of quantum dots and, in principle, of many other organic or inorganic compounds. This technique, based on the idea of increasing the sensitivity through local field enhancement, is named *nanoantenna-enhanced terahertz spectroscopy* (NETS). A Fano-like interference between the fundamental mode of the nanoantennas and the phonon resonance of the quantum dots is observed, together with an enhancement of the absorption of the dots up to more than a million. Finally, we show how to extract the main spectroscopic information of the quantum dots through a simple coupled harmonic oscillator model. This novel technique can be widely applied in terahertz spectroscopic studies of nanocrystals and molecules, where extremely low concentrations are of concern.

Keywords: surface enhancement, terahertz spectroscopy, nanoantennas

1. Introduction

Terahertz (THz) spectroscopy is a promising technique for sensing low-frequency modes (e.g., collective vibrations, phonons, magnons, intra-excitonic transitions) of a variety of materials and compounds [1]. However, some of the main drawbacks of THz spectroscopy are its extremely low sensitivity and the difficulty to sense nano-objects as well as ultra-low amount of chemical compounds, due to the long radiation wavelength and the associated diffraction limitations. Recently, several research groups devoted their efforts to investigate metamaterials [2–4] and nanoslot-based [5–7] platforms tailored for the THz sensing of, for example, small molecules such as lactose, fructose, sucrose [6, 7] or microorganism such as fungi and bacteria [2]. However, THz studies employing these techniques were not able to provide any spectral information regarding the investigated specimen. In the last decades, *surface enhancement* [8] methods were widely explored, given the associated capability of increasing the sensitivity of traditional spectroscopies. More in detail, *surface-enhanced Raman spectroscopy* (SERS) was first developed [9, 10] and has been proven to be an effective tool for investigations down to the single molecule level [11–13]. Afterward, *surface-enhanced infrared absorption* (SEIRA) [14, 15] was proposed for direct absorption spectroscopy in the infrared region, enabling the study of molecules at the sub-attomolar level [16]. The central idea of these techniques stems from the fact that both Raman and absorption spectroscopies are sensitive to the local electric field, which can be significantly amplified on rough metal surfaces or in proximity of metallic nanostructures [17]. In this chapter, we show how the use of properly engineered metallic nanostructures can overcome the current limitations of THz spectroscopy, such as low spatial resolution and sensitivity. Indeed, the distinctive ability of metallic nanoantennas [18] to collect and concentrate incident light beyond the diffraction limit allows achieving high local fields, thus intensifying radiation-matter interactions by order of magnitudes. In Section 2.1, we provide the basic concepts for the design of the resonant properties of nanoantennas for THz light. In particular, we introduce the Drude model for the description of the complex dielectric function [19] of a conductor, and then, we present a simple quasi-analytical model to describe the antenna behavior as a Fabry-Pérot resonator [20]. Subsequently, in Section 2.2, we extend our investigation to the resonance properties of ordered arrays of THz dipole nanoantennas, both numerically and experimentally [21, 22]. In Section 3, after summarizing the recent discoveries and techniques regarding enhanced THz sensing, we demonstrate how arrays of dipolar nanoantennas can effectively be employed for THz spectroscopy, discussing what we have named: *nanoantenna-enhanced THz spectroscopy* (NETS) [23]. In particular, we present our test of NETS on a monolayer of cadmium selenide quantum dots, and by employing a simple model based on a mechanical analog (coupled harmonic oscillators), we show how to retrieve the main spectroscopic properties of the specimen under investigation. This technique enables unprecedented spatial localization of THz light and high sensitivity, which results in a new powerful tool for THz investigations of ultra-low amounts of compounds, potentially down to the single nano-object level.

2. Modeling, design and characterization of nanoantennas for terahertz light

2.1. Properties of metals and nanoantenna modeling at terahertz frequencies

In order to properly design and model THz nanoantennas, we first need to be able to describe the electromagnetic properties of a metal when it no longer behaves as a perfect conductor. In fact, already in the GHz domain, losses become an important constraint that microwave engineers have to deal with [18, 19]. In a first approximation, it is possible to retrieve the optical properties of a conductor by means of a simple model developed by Paul Drude in 1900, which considers the conducting material as an ideal gas of free electrons that move in a background of fixed positive ions. According to this picture, the valence electrons of the constituent atoms become conduction electrons and are able to freely move in the volume of the material. The complex dielectric function of the metal can therefore be obtained through a straightforward model describing the interaction of the electric field of the incoming radiation and this "sea" of electrons, which leads to the following expression [19]:

$$\varepsilon(\omega) = \varepsilon_1(\omega) + i\varepsilon_2(\omega) = \varepsilon_\infty - \frac{\omega_p^2}{\omega^2 + \gamma^2} + i\frac{\omega_p^2\gamma}{\left(\omega^2 + \gamma^2\right)\omega} \qquad {\omega_p}^2 = \frac{e^2 N}{m\varepsilon_0} \tag{1}$$

where ε_∞ is the dielectric constant at high frequencies, ω_p is the plasma frequency (N is the carrier concentration, e the electron charge, m the electron mass and ε_0 the vacuum permittivity) and $\gamma = 1/\tau$ is the Drude scattering rate (τ is the carrier lifetime). As one can see, in this model, the complex dielectric function and consequently all the others optical parameters (i.e., the refractive index and the conductivity) are fully characterized by the material plasma frequency and scattering rate [19]. Since the plasma frequency of noble metals lies in the visible UV region (e.g., for gold, $\omega_p/2\pi = 2067\ \text{THz}$; for silver, $\omega_p/2\pi\ = 2321\ \text{THz}$ [24]), at THz frequencies ($\omega \ll \omega_p$), the real part of the permittivity ε_1 results to be negative and significantly large in modulus. It is worth reminding that a finite and negative ε_1 is a fundamental requirement for the existence of a surface wave (named *surface plasmon polariton* [19] at optical frequencies) at the interface between a conductor and an insulator. The classical Drude model can be extended in order to consider the quantum nature of the carriers by introducing the *Drude-Sommerfeld model* [19]. The main consequence of this model in the derivation of the permittivity of metals is that, under an applied electric field, the carriers move in the material with an effective mass m_{eff}, which simply substitutes the free electron mass in the plasma frequency and conceptually incorporates the interaction between the lattice of positive ions and the sea of electrons. Finally, it is possible to take into account the contributions of bound electrons to the permittivity by using the *Drude-Lorentz model* [19], which assumes the presence of a restoring force between the carriers and the ions. In the most general scenario, considering a number k of real electronic transitions, we can therefore add to the complex dielectric function in Eq. (1) a collection of damped harmonic oscillators with resonance frequencies $\omega_{0,k}$ as follows: $\omega_p^2 \sum_k f_k/(\omega_{0,k}^2 - \omega^2 - i\omega\gamma_{0,k})$, where f_k, $\omega_{0,k}$ and $\gamma_{0,k}$ are the strength, the central frequency and the damping rate of the *kth* oscillator, respectively.

Taking advantage of the Drude description of the electromagnetic response of a metal and its extensions, we can develop a model able to predict the resonance characteristics of the basic element (i.e., a metallic nanoantenna) of the proposed NETS technique. This simplified model also allows a better understanding of the physical mechanism that gives rise to the optical response of such nanodevices. Let us consider the simplest nanoantenna geometry, a cylindrical wire of fixed radius R and length L (note that the cylindrical geometry is used to simplify the calculations, but the overall description is valid also for wires of arbitrary lateral section). Such wire can be described as a *Fabry-Pérot* resonator for a surface wave [18, 20], as illustrated in **Figure 1a**.

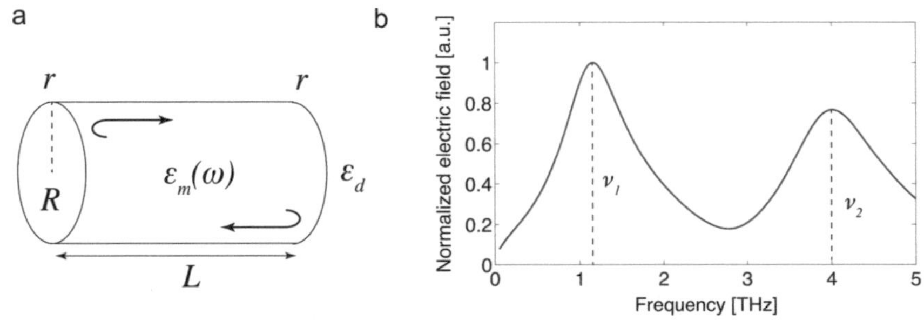

Figure 1. (a) Illustration for the modeling of a wire nanoantenna as a Fabry-Pérot resonator. (b) Normalized electric field amplitude at the tip of the nanoantenna obtained using Eq. (4) (permittivity of gold at THz frequencies taken from Ref. [32]).

For simplicity, we discuss only the propagation of the fundamental surface mode TM_0 (transverse magnetic, axially symmetric mode [18, 25]) along a thin metal wire with complex dielectric function $\varepsilon_m(\omega)$ (that can be extracted using the Drude description above). The wire is assumed to be surrounded by a dielectric medium of constant permittivity ε_d. The fundamental surface mode propagates along the antenna, and it is reflected at its ends, so that at specific frequencies standing waves (and the associated resonances) are formed, like in a traditional Fabry-Pérot resonator. These resonances occur when the antenna length is close to a multiple of half of the wavelength and, in particular, the lowest order resonance λ_{res} appears when the following relation is satisfied [26]:

$$\lambda_{res} / 2n_{eff} = L + 2\delta \qquad (2)$$

where n_{eff} is the effective index of the propagating surface mode and δ is introduced to take into account the apparent increase of the antenna length, due to the reactance of the antenna ends. The complex effective refractive index n_{eff} for a cylindrical wire can be derived from the Maxwell's equations by applying the proper boundary conditions at the dielectric—metallic cylinder interface [25], which leads to the following equation [27]:

$$\frac{\varepsilon_m}{k_m}\frac{I_1(\gamma_m R)}{I_0(\gamma_m R)}+\frac{\varepsilon_d}{k_d}\frac{K_1(\gamma_d R)}{K_0(\gamma_d R)}=0 \tag{3}$$

where I_j and K_j $(j = 0,\ 1)$ are the modified Bessel functions, $\gamma_{m,d} = k_0\sqrt{n_{eff}^2 - \varepsilon_{m,d}}$, $k_0 = \omega/c$, c is the speed of light in vacuum and $k_{m,d} = \sqrt{n_{eff}^2 - \varepsilon_{m,d}}$. The complex reflection coefficient r (thus including amplitude and phase) at the ends of the antenna can also be calculated in a closed-form for a subwavelength system, as detailed in Refs. [27, 28]. Basically, the electric and magnetic fields outside the antenna are expanded in terms of the free-space modes that are rotationally invariant (in order to preserve the symmetry of the problem). Then, the transverse electric and magnetic fields are matched at the interface (end of the antenna) to retrieve the reflection coefficient. Finally, we can calculate the electric field E_{tip} at the extremities of the wire antenna following the typical procedure used for a Fabry-Pérot cavity and considering that, for normal incidence, only the odd modes of the wire can be excited:

$$E_{tip} = E_0\frac{(1-re^{i\varphi})(1-e^{i\varphi})}{1-r^2e^{2i\varphi}} \qquad \varphi = k_0 n_{eff} L \tag{4}$$

where E_0 is the field that couples at the tips of the nanoantenna and φ is the phase accumulated along the wire by the propagating surface mode in half round-trip. **Figure 1b** shows the resonance characteristics of a 100-μm-long gold wire nanoantenna (80 nm diameter) in vacuum, as obtained using Eq. (4). In this example, one can clearly see the first two (odd) modes of the nanoantenna, corresponding to the conditions $\lambda_1 \approx 2n_{eff}L\left(\nu_1 = \frac{c}{\lambda_1}\right)$ and $\lambda_2 \approx 2n_{eff}L/3\left(\nu_2 = \frac{c}{\lambda_2}\right)$, respectively.

2.2. Design, fabrication and characterization of nanoantenna arrays

The Fabry-Pérot model presented above allows an initial and insightful investigation of the resonance properties of a single nanoantenna. However, the proposed NETS technique makes use of far-field measurements to retrieve the spectral properties of the specimen under investigation. In order to guarantee an easily detectable signal in the far field, arrays of nanoantennas are commonly used. Short- and long-range interactions between neighboring nanoantennas lead to changes in the resonance characteristics of an array when compared to a single isolated element [18]. In such scenario, a fine-tuning of the array resonances can be conducted by means of electromagnetic simulations, which allow taking into account the overall electromagnetic response of the system.

Once the proper architecture has been designed, it can be fabricated by employing electron-beam lithographic techniques. More specifically, for the arrays investigated in Refs. [21, 22], we used the following procedure: a 120-nm thick poly(methylmethacrylate) (PMMA) layer was spin-coated on a 500-μm thick, high-resistivity (>10 kΩcm) (100)-oriented silicon substrate. High-resistivity silicon was selected as a substrate since it is transparent and has a constant refractive index in the THz range. Charging effects that may occur during electron exposure over an insulating substrate have been prevented by means of a 10-nm-thick Al layer that was thermally evaporated on the PMMA surface. A high-resolution Raith150-Two e-beam writer at 15 keV beam energy and 520 $\mu C/cm^2$ exposure dose was used to prepare the nanoantenna patterns. After the Al removal in a KOH solution, the exposed resist was developed in MIBK/isopropanol (IPA) (1:3) for 30 s. Then, a 5-nm-thick adhesion layer of titanium was prepared using electron beam evaporation, with a 0.3-Å/s deposition rate in a high vacuum chamber (base pressure 10^{-7} mbar). In situ thermal evaporation (0.3-Å/s deposition rate) of a 60 nm gold film was obtained using a high temperature source mounted inside the vacuum chamber. After the film deposition, the unexposed resist was removed with acetone and rinsed out in IPA. The residual PMMA resist and organic contaminants were removed by means of O_2 plasma ashing for an improved lift-off. The two-dimensional arrays, composed of aligned nanoantennas with a fixed spacing $G = 20$ μm in both directions on the plane (see **Figure 2**), were fabricated on a 5 × 5 mm^2 area. In order to investigate the tunability of the resonance of such nanoantenna arrays in the THz range, we prepared a series of samples with different nanoantenna lengths ($L = 30$, 35, 40, 50 and 60 μm, respectively), fixed height (60 nm) and width ($D = 200$ nm).

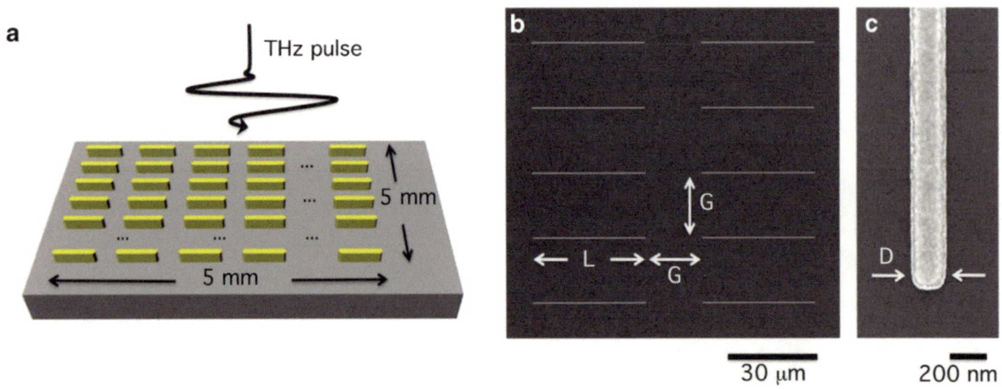

Figure 2. (a) Schematic illustration of the nanoantenna array. (b) SEM image of a detail of the array with L = 40 μm. (c) Magnification of the nanoantenna tip. Reproduced with permission from Ref. [22].

To characterize the spectral response of these samples, we employed a standard *time-domain spectroscopy* (TDS) setup [1]. In particular, THz pulses were generated via optical rectification of 130-fs-long pulses at 800 nm (1.6 mJ of energy at 1 kHz repetition rate) in a 500-μm-thick

(110)-ZnTe crystal. The arrays were placed in the collimated THz beam path (beam diameter of about 7 mm), and the transmitted radiation was then focused on a second ZnTe (500-μm thick) that was used to record the transmitted THz pulse trace in time, in a classical electro-optical sampling arrangement [29].

Because of the strong absorption of water molecules in the THz region, the measurements were performed in a nitrogen-purged environment. For the investigated arrays, the nanoantenna covering factor (defined as the ratio of the area covered by the nanoantennas to the total area of the array) is <1%, so that the transmitted spectrum when the THz radiation is polarized perpendicular to the nanoantenna main axis (nonresonant excitation) is found to be substantially equal to the one of a bare silicon substrate. For this reason, the array transmittance, named *relative transmittance* T_{rel}, can be simply calculated by dividing the spectrum collected exciting the nanoantennas with polarization aligned along their main axis (resonant excitation) by the one obtained under nonresonant excitation. The results of this procedure are shown in **Figure 3a**. As one can see, the transmission of the array is reduced up to 60% in correspondence of a resonance dip, which red-shifts by increasing the length L of the nanoantennas. From these data, we can estimate the nanoantenna *extinction efficiency* Q_{ext} as follows [21, 22]:

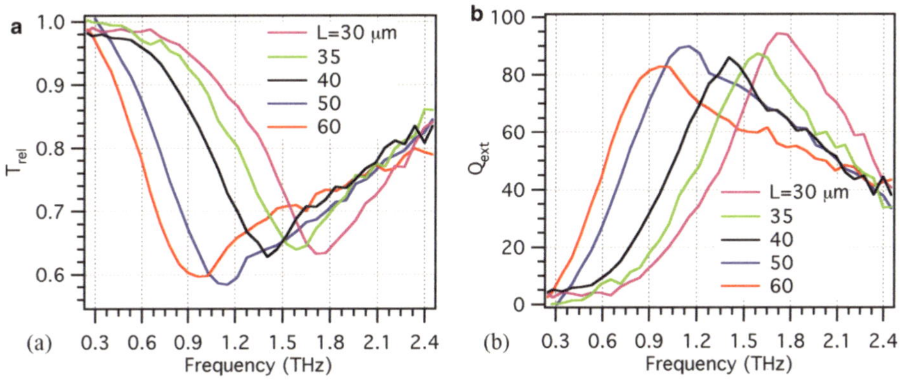

Figure 3. (a) Relative transmittance T_{rel} of the nanoantenna arrays as a function of frequency for the five samples with different nanoantenna lengths. (b) Extinction efficiency Q_{ext} as a function of frequency, calculated from the experimental data using Eq. (5). Reproduced with permission from Ref. [22].

$$Q_{ext} = \frac{\sigma_{ext}}{\sigma_{geo}} = \frac{A(1-T_{rel})}{NLD} = Q_{abs} + Q_{sca} = \frac{\sigma_{abs}}{\sigma_{geo}} + \frac{\sigma_{sca}}{\sigma_{geo}} \tag{5}$$

where σ_{ext} is the nanoantenna extinction cross section, σ_{abs} and σ_{sca} are the absorption and scattering cross sections respectively, $\sigma_{geo} = LD$ is the geometric cross section, A is the illuminated area, while N, L and D are the number, length and width of the illuminated nanoantennas, respectively. The extinction efficiency essentially quantifies the capability of nanoantennas of collecting free-space radiation. **Figure 3b** depicts Q_{ext} as a function of frequency. The peak value is about 90 for all the samples, meaning that, under resonance condition, the effective cross section of our nanostructures increases by about 90 times. The experimentally extracted resonance properties of the nanoantenna arrays were also compared with the results of numerical simulations, performed using a finite integration technique-based commercial software (CST Microwave Studio). We considered gold nanoantennas with a rectangular section of 200 × 60 nm^2, capped with hemi-cylindrical terminations of radius equal to 100 nm. To better resemble the fabricated structures, all the nanoantenna edges were blended with a radius of curvature of 20 nm.

Furthermore, to simplify the simulated geometry while taking into account the influence of the silicon substrate, we considered nanoantennas entirely embedded in a homogenous medium of effective dielectric constant: $\varepsilon_{bg} = (1 + n_{Si}^2)/2$ [30], where $n_{Si} = 3.42$ is the refractive index of silicon in the THz region [31]. The gold dielectric constant was extracted from Ref. [32]. In all simulations, the nanoantennas were excited by a plane wave set at normal incidence, with polarization along the long axis of the nanostructures. In order to retrieve the nanoantenna resonance properties, we numerically evaluated the extinction efficiency Q_{ext} by extracting, from the simulations, the total absorption and scattering cross sections as a function of frequency. **Figure 4** illustrates the peak resonance wavelength λ_{res} as a function of the nanoantenna length (L = 30, 35, 40, 50 and 60 μm) for three cases: individual nanoantenna (numerical simulation, magenta squares), array of nanoantennas (numerical simulation, red squares) and experimental data (black crosses).

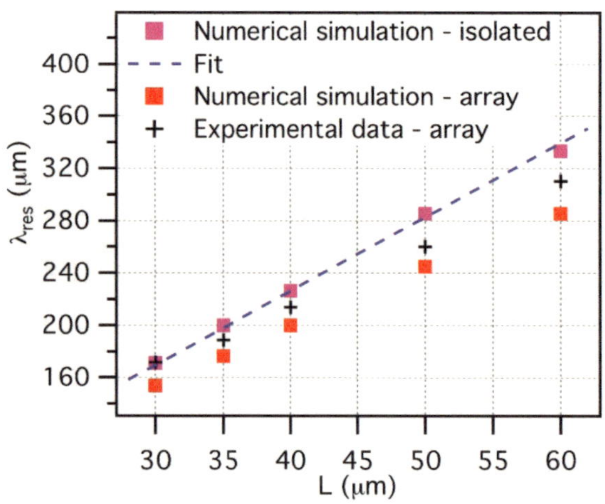

Figure 4. Resonance wavelength λ_{res} as a function of the nanoantenna length L. Numerical simulation for an isolated nanoantenna (magenta squares); best fit with the Fabry-Pérot model for an isolated THz antenna (dashed blue line); numerical simulations of the nanoantenna array (red squares); experimental data obtained from the resonance curves in **Figure 3** (black crosses). Adapted from with permission Ref. [22].

It is possible to observe that, as expected, the peak resonance wavelength increases linearly with the antenna length. This behavior can be further investigated using Eq. (2), and the simple Fabry-Pérot description discussed in the previous section (Section 2.1). Since δ is known to be of the order of the lateral dimension of the antenna [20, 26], it can be neglected in our case ($L \gg D$), and we can simply write: $\lambda_{res} = 2n_{eff}L$. The numerical results of the individual nanoantenna (magenta squares) can be therefore fitted with this equation to retrieve the effective refractive index. Notably, the obtained value, n_{eff} = 2.83, is higher than the background index employed in the simulations: $n_{bg} = \sqrt{\varepsilon_{bg}}$ = 2.52. This means that a gold nanowire with lateral section as in the studied case (200 × 60 nm^2) cannot be considered as an ideal conductor at THz frequencies (i.e., the electric field of the propagating surface mode penetrates inside the metal). Comparing the numerical values obtained for the resonance of an individual nanoantenna (magenta squares) with the experimental data (black crosses in **Figure 4**), we notice that the experimental values slightly shift toward shorter wavelengths. This shift is attributed to the long-range dipolar interaction between neighboring nanoantennas in the array, which is known to affect the resonance properties of the system [16, 18]. Indeed, when the nanoantennas are organized in an array configuration, each element of the array is excited by a superposition of the incident electromagnetic field and the field scattered by the other elements. In order to corroborate this observation, we performed numerical simulations with periodic boundary conditions, to accurately evaluate the response of an array with a 20 μm spacing in both directions on the plane. These results are reported in **Figure 4** (red squares) and show a blue-shift of the array resonance frequency, confirming the origin of the shift observed in the experimental data. The residual difference between simulated and experimental values may be due to the uncertainty in the effective dielectric constants of the materials involved [33]. For applications exploiting the *local field enhancement* in close proximity of the nanoantennas, as it is the case with NETS, it is also important to investigate the *near-field resonance* properties of these nanostructures and compare them with the above reported *far-field resonance* characteristics. For the sake of simplicity, we considered here only the case of the array with L = 40 μm.

Figure 5a (magenta curve) shows the field enhancement factor F at the nanoantenna tip as a function of frequency, which is defined as the ratio of the local to the free-space field. A broad resonance centered at around 1.3 THz can be observed, with a peak field enhancement of a few hundreds. **Figure 5a** also displays the values of Q_{abs} and Q_{sca}, as well as their sum Q_{ext} (right axis). It is worth noticing that both absorption and scattering significantly contribute to the far-field resonance properties of the nanoantennas, featuring a peak at the same frequency of 1.5 THz. The near-field resonance peak thus results to be red-shifted (of about 200 GHz) with respect to the far-field peak. This behavior was previously observed in the optical frequency region and was attributed to plasmon damping [34–36]. In fact, by modeling the nanoantenna as a damped harmonic oscillator driven by the external electric field, the oscillator energy dissipation can be associated to the far-field extinction cross section of the nanoantenna, while the oscillator amplitude corresponds to the nanostructure near-field response. When damping is present, the peak of the oscillator amplitude is known to appear at a frequency that is lower than the natural frequency of the oscillator, while the maximum of the energy dissipation remains un-shifted [34]. Therefore, the magnitude of the resonance shift between the near-field and far-field response is directly related to the total damping of the system, in terms of intrinsic ohmic loss in the metal and radiative damping [36]. In order to demonstrate the nature of this phenomenon at THz frequencies, we substituted gold and its realistic dielectric constant in the simulations with a perfect electric conductor (PEC, i.e., a material with infinite conductivity). **Figure 5b** shows the results of this procedure. As expected, in the case of a PEC, the contribution of absorption vanishes, and the far-field properties are ruled by scattering. In addition to a narrower resonance, the near-field peak was found to be almost coincident with the far-field peak. Thus, the red-shift of the near-field resonance observed in THz gold nanoantennas is mainly a consequence of ohmic damping. For an effective implementation of NETS, this effect has to be taken into account during the design of the nanostructures, in order to achieve the desired near-field response.

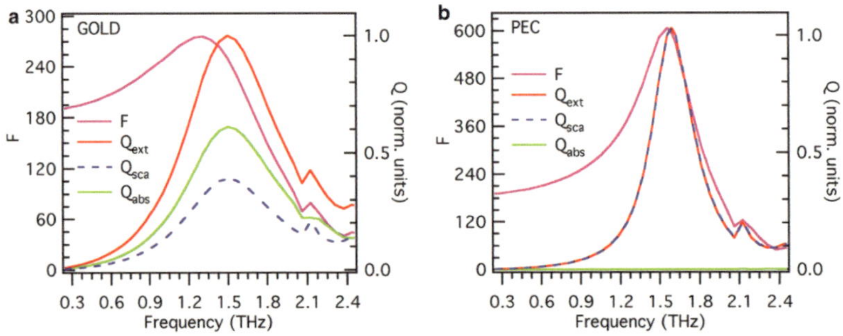

Figure 5. (a) Comparison between simulated near-field and far-field resonance spectra, for the array of gold nanoantennas with length L = 40 μm. Left vertical axis: field enhancement factor F (magenta line) as a function of frequency. Right vertical axis: normalized values of the extinction, scattering, and absorption efficiency Q_{ext}, Q_{sca} and Q_{abs} (red, dashed blue and green line, respectively) as a function of frequency. (b) Same as in (a), substituting gold with a perfect electric conductor (PEC). Reproduced with permission from Ref. [22].

3. Enhanced terahertz sensing and spectroscopy

As discussed in Section 2, metallic nanostructures can strongly increase the local THz electric field and can thus enhance the interaction of the incident radiation with a specimen placed in their proximity, this being the basic concept of NETS. Similar strategies, exploiting an augmented radiation-matter interaction, have been already successfully employed. For example, waveguide-assisted THz sensing [37] makes use of a guided geometry (e.g., in a parallel plate configuration), so that the effective interaction length can be increased up to several centimeters, enabling spectroscopic investigations of thin layers of biomolecules [38], explosives [39] and drugs [40]. The drawback of this approach is that the specimen has to be deposited throughout the whole length of the waveguide to effectively exploit the enhanced interaction. In a similar way, THz sensing using spoof plasmons (i.e., bound electromagnetic waves on corrugated metallic surfaces) also needs that the specimen under investigation is deposited along the entire propagation length of the THz surface wave [41]. On the other hand, THz sensors based on metamaterials [2–4] have shown to be effective in sensing thin films of various compounds, down to the sub-micrometer level. Metamaterials indeed possess a narrowband resonance, whose position in frequency is strongly sensitive to the surrounding environment. For instance, Park et al. [2] recently proposed a metamaterial-based THz sensor capable of detecting live microorganism such as molds, yeast cells and bacteria. In particular, they prepared metallic arrays of square rings with a micro-gap at the center and functionalized the sensor with an antibody specific to bacteria (*E. coli*) in aqueous environment, in order to enable selective detection. The presence of the microorganisms at low densities was detected by probing (via THz-TDS [1]) the shift of the inductive-capacitive resonance, due to a change of the dielectric constant of the environment. As this and other examples can show, such type of THz metasurfaces is excellent refractive-index sensors, but they are not suitable for the identification of the spectroscopic signatures of the specimen under investigation.

In recently years, a new approach based on nanoslots [5–7] resonating at THz frequencies has been proposed by Park et al. [6], which has shown the possibility of strongly enhancing the THz absorption coefficient of molecules. In particular, they fabricated and tested a set of single nanoslot antennas (see illustration and SEM images in **Figure 6**) of fixed length ($l = 90$ µm) and varying widths ($w = 50, 100, 200, 500, 1000$ and 5000 nm) realized on a thin gold film of thickness $h = 50$ nm, for the smallest width w, and $h = 100$ nm for all the other cases, on a 500-µm-thick quartz substrate. The slots were prepared to have a resonance at around 0.87 THz, which matched the frequency peak of an absorption band of RDX (1,3,5-trinitroperhydro-1,3,5-trizine) molecules. **Figure 6c** shows the normalized resonant transmission (measured via far-field THz-TDS [1]) of these nanoslots as a function of the RDX amount that was drop-casted over the sample surface. Remarkably, a significant decrease of the resonant transmission intensity is already observed for an amount of 40 ng spread over a sample area of 8 mm^2, which corresponds to only 22 fg inside the slot cavity. This was made possible thanks to the high field enhancement achieved inside the slots at resonance that, in turn, increased the absorption coefficient of RDX molecules up to values higher than 100,000 cm^{-1}.

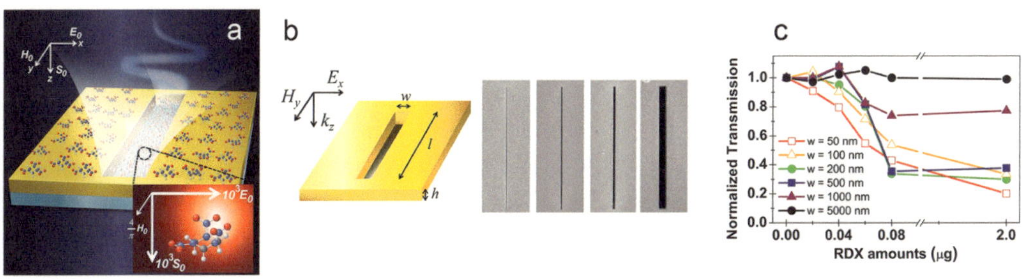

Figure 6. (a) Illustration of the RDX molecules spread over a THz nanoslot antenna. (b) Schematic of the nanoslot with definition of the relevant geometrical parameters, and SEM images of four samples: w = 50, 500, 1000 and 5000 nm. (c) Normalized resonant transmission measured at 0.87 THz for all the six samples, as function of the RDX amount. Adapted with permission from Ref. [6]. Copyright © 2013, American Chemical Society.

More recently, Lee et al. [7] demonstrated a nanoslot-array-based (see **Figure 7a**) THz sensing method which enables the selective detection of carbohydrate molecules (such as D-glucose, fructose, sucrose and cellulose). Here, two different nanoslot arrays with slot lengths l = 40 μm (f_{res} = 1.4 THz) and l = 35 μm (f_{res} = 1.7 THz) were designed, in order to match the absorption lines of D-glucose and fructose, respectively, following the sensing strategy reported in Ref. [6]. Even in this case, the strong field enhancement provided by the nanoslots (estimated to be about 50 for these slots featuring a width w = 500 nm) significantly increased the molecular absorption, enabling the detection of a specific sugar down to hundreds of micromoles. The selectivity of this system was verified by THz far-field imaging in transmission. In particular, the nanoslot array with its resonance matched to the absorption peak of fructose was used to discriminate this sugar from D-glucose. The upper-right corner of the array was covered with fructose and the lower-left corner with D-glucose (**Figure 7b**). **Figure 7c** shows the THz transmittance image of the sample under resonant conditions, highlighting the good selectivity of the sensor and its ability to discriminate the two types of sugars. The same system was also successfully employed to detect the sugars contained in popular sweetened beverages, including Coca-Cola, Pepsi-cola and Sprite. These works show the great potential of nanoslot-based sensors for the accurate detection of chemical compounds in the THz region. However, even in this case, no direct information regarding the main spectroscopic characteristics of the investigated sample (i.e., absorption peak position and bandwidth) was retrieved.

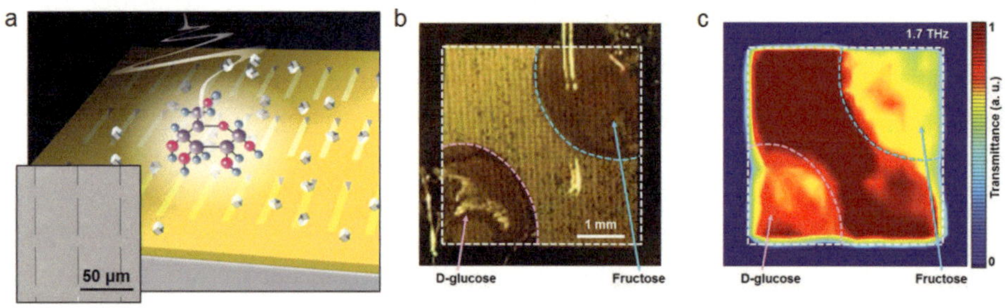

Figure 7. (a) Illustration and SEM detail of the nanoslot array-based THz sensing chip. (b) Photograph of the nanoslot array (l = 35 μm) after deposition of D-glucose and fructose (concentration: 250 mg/dL) in the two corners. (c) Normalized THz transmittance image of the array after deposition of the two sugar specimens. Adapted with permission from Ref. [7].

In the next section, we will discuss the use of engineered arrays of dipolar nanoantennas, reported in Section 2, to implement surface-enhanced spectroscopy in the THz spectral region. This technique basically translates the concept introduced by SEIRA for the infrared region into the THz domain and provides a valuable tool for THz spectroscopic investigations of ultra-low amounts of chemical compounds. In particular, we will summarize some of our recent results regarding the use of resonant dipole nanoantenna arrays to retrieve the spectroscopic response of a test sample: a monolayer of cadmium selenide quantum dots (CdSe QDs).

http://dx.doi.org/10.5772/66349

3.1. Resonant dipole nanoantenna arrays for enhanced THz spectroscopy

In Section 2, we discussed the properties of dipole nanoantenna arrays in the THz region. As mentioned, the simplest design for a resonant nanoantenna is represented by a metallic nanorod of length equal to about half of the effective wavelength of the exciting radiation (half-wavelength dipole nanoantenna) [20, 26]. In this configuration, the electric field concentrates into two sub-wavelength "hot-spots" at the antenna extremities. By moving from an isolated nanoantenna to nanostructures coupled end-to-end through a narrow gap, it is possible to increase and localize the electric field even further within such gap [18]. **Figure 8a** shows the nanoantenna arrangement employed in our investigation [23]. Several arrays of gold dipole nanoantennas (5 × 5 mm^2) were again fabricated on high-resistivity silicon substrates using e-beam lithography. We fixed the nanoantenna height and width at h = 60 nm and w = 200 nm, respectively. THz hot-spots featuring high field enhancement were obtained by coupling the nanoantennas along their long axis through nanogaps of nominal width of 20 nm (G_x in **Figure 8a**, see SEM image in **Figure 8b**). In order to demonstrate the capabilities of NETS, we selected CdSe QDs as test-bed nano-objects, since they are endowed with a clear phonon resonance in the THz range [42] (at 5.65 THz, as shown in **Figure 9b**). To tune the resonance of the nanoantenna arrays and match the QD phonon resonance, we performed extensive 3-D electromagnetic simulations, varying both the length L of the nanoantennas and their spacing G_y in the direction perpendicular to the antenna long axes (see **Figure 8a**). Indeed, as described in Section 2, its length [26] determines the resonance of a single nanoantenna and thus the main features of the spectral response of the whole array [22].

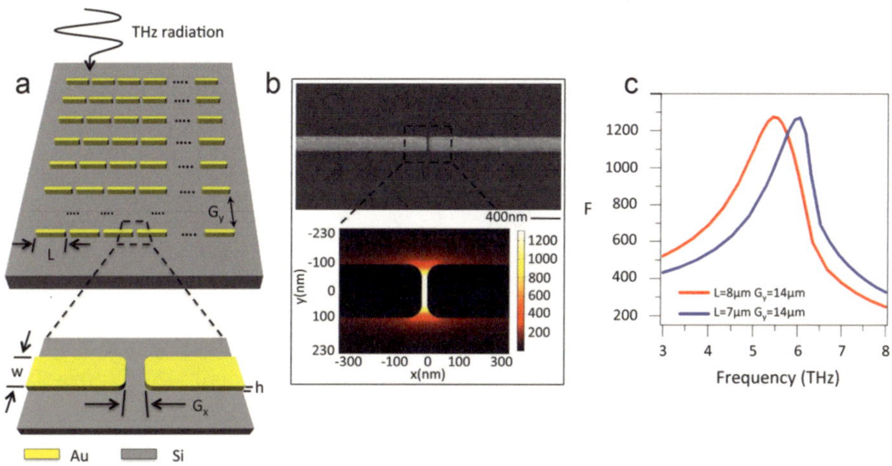

Figure 8. (a) Schematic illustration of the nanoantenna arrays used for the NETS demonstration. (b) SEM detail of a nanogap region (upper panel); 2D surface plot of the field enhancement factor F around the gap region at the QD resonance frequency, for L = 8 μm and G_y = 14 μm (lower panel). (c) F in the centre of the nanogap as a function of frequency, for two arrays with two different nanoantenna lengths L and same spacing G_y. Adapted with permission from Ref. [23]. Copyright © 2014, American Chemical Society.

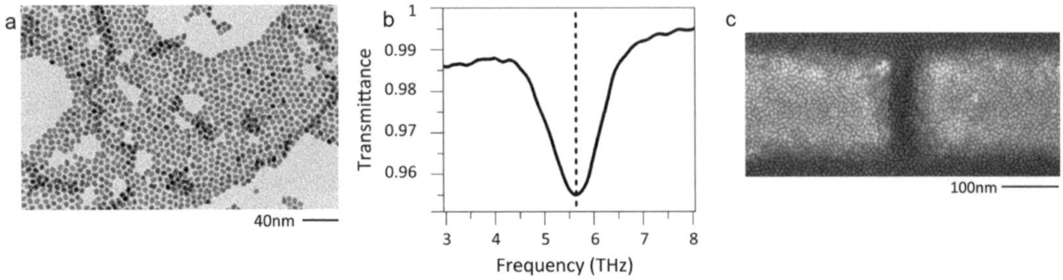

Figure 9. (a) TEM image of the synthesized CdSe QDs. (b) THz transmittance of a 100-nm-thick layer of CdSe QDs. (c) Magnification of a gap region covered with the dots, showing the uniformity of the QD layer. Adapted with permission from Ref. [23]. Copyright © 2014, American Chemical Society.

Additionally, the array spacing is also a critical parameter for engineering the frequency response of the nanostructures [22, 43]. Indeed, this spacing modifies the interaction between neighboring nanoantennas and has the ability to promote, by means of in-phase coupling, their collective excitation. This can result in resonance shift and narrowing, as well as lead to higher local fields. **Figure 8c** reports two examples of array geometries engineered to match the phonon resonance of the dots. In particular, it shows the field (amplitude) enhancement factor F in the center of the gap, being $F(x_0, y_0, z_0)$ defined here as the ratio of the local electric field at position (x_0, y_0, z_0) in the presence of the nanoantennas to the field in the same position considering a bare substrate with no structures. The effective localization of the THz electric field into sub-wavelength nano-volumes is illustrated in **Figure 8b** (lower part), which shows a 2D simulation of F around the gap region under resonant conditions. In the center of the nanogap, extremely high values of F (more than a thousand) are reached, which is fundamental for the successful realization of enhanced THz spectroscopy of nanomaterials. In fact, inside these gaps, the usually small effective absorption cross section of a nano-object at THz frequencies can be greatly amplified (up to more than a million times in the presented case), since it scales with $|F|^2$ [44].

3.1.1. Terahertz-enhanced spectroscopy of a monolayer of CdSe QDs

CdSe QDs were selected as a model system for our investigation since they can be prepared with high precision in shape and size, and they are known to form a compact and uniform layer, whose thickness can be accurately controlled. QDs with an average diameter of 5.2 nm were chemically synthesized using a previously developed protocol [45]. **Figure 9a** shows a transmission electron microscope (TEM) image of the dots, highlighting good size uniformity. The spectral response of the QDs was then retrieved through Fourier transform spectroscopy [46] (Bruker 70/v Fourier transform spectrometer) in a transmission configuration. **Figure 9b** presents the THz transmittance of a 100-nm-thick layer of QDs drop-casted on a bare silicon substrate. Their phonon resonance (Fröhlich mode) [42] is clearly visible as a transmission dip centered at ~5.65 THz. For the demonstration of NETS, a uniform monolayer of CdSe QDs was then spin-coated over the fabricated arrays (see detail of the gap region in **Figure 9c**).

Afterward, the THz transmittance of the four fabricated samples (with slightly shifted resonance frequencies) was again measured, and the results are shown in **Figure 10a**. For nonresonant excitation (not shown, polarization of the THz light set perpendicular to the long axis of the nanoantennas), the transmission of the samples was equivalent to the one of a bare silicon substrate, thus the presence of the QDs could not be detected. This result is in agreement with the QD response reported in **Figure 9b** for a 100-nm-thick layer. Indeed, for such thin layers, the transmittance change ΔT (i.e., the difference in transmittance between the reference silicon substrate and a substrate covered with the CdSe layer) results to be proportional to the layer thickness d ($\Delta T = 1 - e^{-\alpha d} \approx \alpha d$, where α is the layer attenuation coefficient). Considering that a transmission change of ~3.5% was measured for a 100-nm-thick layer (**Figure 9b**), a change of <0.2% should be expected for a monolayer, which is below the sensitivity of our experimental apparatus.

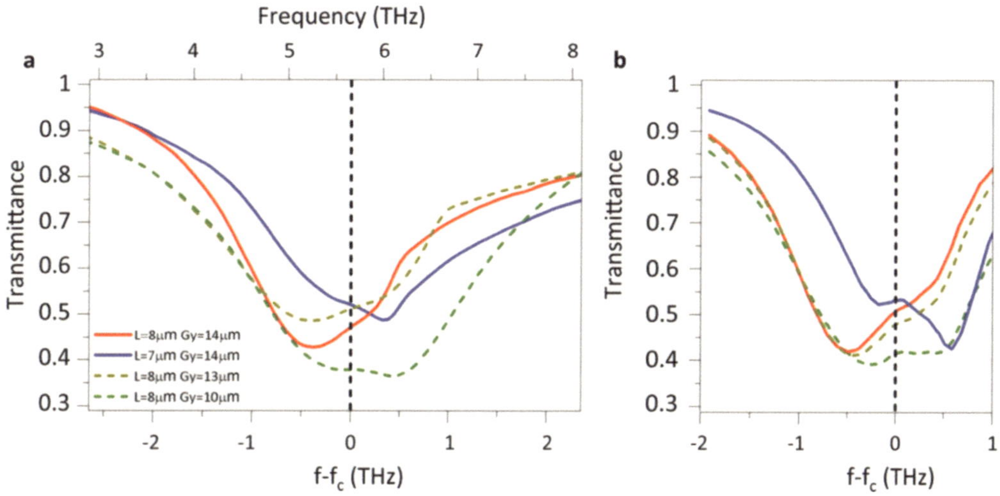

Figure 10. (a) Experimental transmittance of nanoantenna arrays covered with a monolayer of CdSe QDs, for different antenna lengths L and array spacings G_y (f_c is the central frequency of the phonon resonance). (b) Corresponding numerical simulations. Reproduced with permission from Ref. [23]. Copyright © 2014, American Chemical Society.

Conversely, when the nanoantennas were resonantly excited (polarization along their long axis), a clear anti-resonant peak, located in proximity of the QD phonon frequency f_c (black dotted line in **Figure 10**), was observed over the resonance response of the arrays (**Figure 10a**). This Fano-like behavior [47] arises from the interference between the nanoantenna mode and the QD phonon resonance. The result is similar to the one traditionally observed in SEIRA measurements [48–50]. It is worth underlining that the Fano-like anti-resonant feature is clearly visible for all the tested arrays, which possess distinct nanoantenna resonance frequencies. This demonstrates that only a coarse alignment between the phonon mode of the QDs and the nanoantenna resonance is necessary to observe the effect. Furthermore, the interference peak always appears in the spectrum in correspondence of the QD vibrational frequency (black dashed line), as expected for a Fano-like interference. The observed coupling between the nanoantenna resonant mode and the QD phonon line was also investigated through extensive numerical simulations. In order to reduce the required meshing elements and thus the computational time, we considered a uniform layer with thickness equal to the QD diameter (5.2 nm) and permittivity as in Ref. [51], which reports the THz response of CdSe QDs with a diameter (6.3 nm) similar to the ones investigated in our work. The results of the simulations are reported in **Figure 10b**. As can be seen, they well reproduce the overall behavior of the system and are in good agreement with the experimental measurements.

3.1.2. Absorption enhancement

The results presented in **Figure 10** show that NETS allows sensing ultra-low quantity of compounds (in our proof-of-concept experiment, a monolayer of CdSe QDs) through the formation of a Fano-like resonance. This is clearly visible and corresponds to a spectral feature of the specimen under investigation, which couples with the nanoantenna mode of the array. As already discussed in the beginning of this section, NETS relies on the fact that a specimen absorption is strongly sensitive to the local electric field, which can be greatly increased (see field enhancement values in **Figure 8c**) in proximity of nanoantennas. In this regard, we made use of numerical simulations to quantitatively estimate this effect in the case of our NETS arrays. The overall array absorption enhancement η_{tot} at the QD phonon resonance frequency ν_{ph} can be evaluated by calculating the surface integral of $|F|^2$at resonance, and dividing it by the total sensing area $A = 25$ mm^2:

$$\eta_{tot}\left(\upsilon_{ph}\right) = \frac{\iint_{tot}\left|F(\upsilon_{ph},x,y)\right|^2 dxdy}{A} \tag{6}$$

In particular, for the case of the array with $L = 8$ μm and $G_y = 14$ μm, whose characterization is reported in **Figure 10a** (red line), we find $\eta_{tot} \approx 70$, which is a significant result considering that the nanoantenna covering factor is only 1.4%. In a similar way, we can also estimate the contribution of the QDs situated in the nanogap regions to the overall absorption. To this end, we can compare the surface integral of $|F|^2$ calculated over the area covered by the nanogaps with the one calculated over the entire array:

$$\frac{\eta_{gap}\left(\upsilon_{ph}\right)}{\eta_{tot}\left(\upsilon_{ph}\right)} = \frac{\iint_{gap}\left|F(\upsilon_{ph},x,y)\right|^2 dxdy}{\iint_{tot}\left|F(\upsilon_{ph},x,y)\right|^2 dxdy} \tag{7}$$

Considering again the array with $L = 8$ μm and $G_y = 14$ μm, we find $\eta_{gap}/\eta_{tot} = 0.52$. This means that about half of the total absorption of the monolayer occurs in the nanogap regions. This result is of great relevance since the nanogaps cover an area ~30,000 times smaller than the total sensing surface, and each of them contains roughly 130 QDs (visual estimate from the SEM image in **Figure 9c**).

3.1.3. *Analytical description through a two coupled harmonic oscillator model*

A direct and simple analytical model can be used to describe the NETS measurements, in order to shed more light on the underlying physical mechanism and extract the main spectroscopic information of the investigated specimen. Indeed, the observed Fano-like interference can be modeled by considering a system composed of two coupled harmonic oscillators [52, 53]: one representing the nanoantenna resonance mode and the other the phonon mode of the QDs. The first oscillator is characterized by a resonance frequency $\omega_{pl} = 2\pi\nu_{pl}$ and damping $\gamma_{pl} = 2\pi\Delta\nu_{pl}$ ($\Delta\nu_{pl}$ being the full width at half maximum of the resonance), while the second one has a resonance frequency $\omega_{ph} = 2\pi\nu_{ph}$ and a damping factor $\gamma_{ph} = 2\pi\Delta\nu_{ph}$. The two systems can be connected together through the coupling constant g. Since the phonon mode of the QD monolayer is weakly excited by the far-field radiation, we consider the first (nano-antenna) oscillator to be the one excited by the external driving force: $fe^{i\omega t}$. Under this approximation, the equations of motion can be written as:

$$\begin{aligned} \ddot{x}_{pl} + \gamma_{pl}\dot{x}_{pl} + \omega_{pl}^2 x_{pl} + g x_{ph} &= f e^{i\omega t} \\ \ddot{x}_{ph} + \gamma_{ph}\dot{x}_{ph} + \omega_{ph}^2 x_{ph} + g x_{pl} &= 0 \end{aligned} \quad (8)$$

Assuming a harmonic displacement, $x_{pl,ph} = c_{pl,ph}e^{i\omega t}$, the amplitude of the nanoantenna oscillator can be written as:

$$c_{pl} = \frac{\omega_{ph}^2 + i\gamma_{ph}\omega - \omega^2}{\left(\omega_{pl}^2 + i\gamma_{pl}\omega - \omega^2\right)\left(\omega_{ph}^2 + i\gamma_{ph}\omega - \omega^2\right) - g^2} f \quad (9)$$

Figure 11 shows how the absolute value squared of the nanoantenna oscillator amplitude $\left|c_{pl}\right|^2$ (blue solid line) can properly reproduce the main characteristics of our experimental results (red circles, representing the normalized extinction efficiency, extracted from the experimental transmittance, for the array with L = 8 μm and G_y = 10 μm). In particular, the model well reconstructs the anti-resonant Fano-like feature in correspondence of the QD phonon resonance. A direct comparison between this model and NETS experimental results can be used to extract the main spectroscopic properties of the specimen under investigation, in terms of absorption peak position and bandwidth. In the studied case, by fitting the experimental data, we obtained values for the QDs phonon resonance frequency ($\nu_{ph,\text{model}}$ = 5.72 THz) and bandwidth ($\Delta\nu_{ph,\,\mathrm{model}} = 1.12$ THz) that were in good agreement with the experimental ones ($\nu_{ph,\,exp} = 5.64$ THz and $\Delta\nu_{ph,\,\mathrm{exp}} = 1.15$ THz, retrieved from the measurement shown in **Figure 9b**).

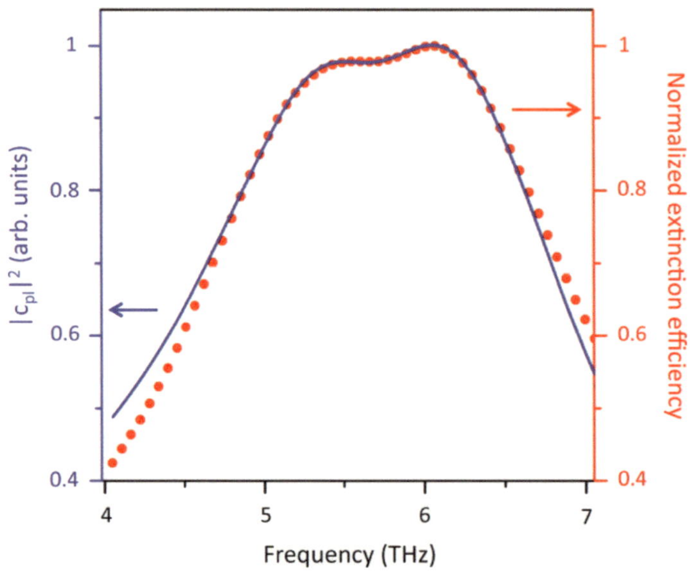

Figure 11. $\left|c_{pl}\right|^2$ as a function of frequency (blue) and normalized extinction efficiency σ_{ext} (from the experimental transmittance T as $\sigma_{ext} \propto 1 - T$ of the array (L = 8 μm and G_y = 10 μm) covered with QDs (red circles). Reproduced with permission from Ref. [23]. Copyright © 2014, American Chemical Society.

4. Conclusion

In this chapter, we have shown how to design arrays of THz nanoantennas to perform enhanced THz spectroscopy. In the beginning, we have discussed the properties of metals at THz frequencies, retrieving the complex dielectric function through the Drude model. In order to describe the resonance response of an individual dipole nanoantenna, the basic element of our investigation, we have introduced a Fabry-Pérot resonator model for a surface wave over a metallic wire. Through this simple quasi-analytical model, the resonance characteristics of a nanoantenna can be quickly evaluated, avoiding time-consuming numerical simulations. Afterward, we have presented both numerical and experimental results regarding the electromagnetic response of THz nanoantenna arrays. In particular, we have shown that, by varying the length of the nanoantennas, the resonance peak of the array can be tuned to cover the THz band offered by standard THz sources. In addition, we have discussed the resonance shift that arises between the near- and far-field responses of nanoantennas. We have shown that, in these devices, the near-field resonance peak can substantially red-shift in comparison to the far-field peak, due to the ohmic damping within the metal. This is an important information for the practical implementation of NETS, since the targeted absorption enhancement relies on the local field in proximity of the nanostructures. Subsequently, we have summarized some recent results reported in the literature on THz sensing via metamaterials and metallic nanostructures. Finally, we have presented the demonstration of NETS, reporting results obtained on a monolayer of CdSe QDs by means of engineered nanoantenna arrays coupled through nanogaps. As a result of the direct coupling between the nanoantenna mode and the phonon resonance of the QDs, the formation of an evident Fano-like interference (centered at the phonon resonance frequency) over the array response was observed. The high field enhancement (more than one thousand) obtained in the center of the nanogaps enabled an increase of the absorption cross section of the QDs up to more than a million times, which in turn allowed the ultrasensitive characterization of the QD spectroscopic signature. Moreover, we have shown that a simple model based on coupled harmonic oscillators can be employed to reproduce the Fano-like interference and extract the main spectroscopic characteristics (absorption peak frequency and bandwidth) of the investigated specimen. NETS has thus been proven to be a useful tool for the spectroscopic characterization of ultra-low quantities of nanomaterials. Very recently, using a similar strategy, Ueno et al. [54] performed surface-enhanced THz spectroscopy of amino acid molecules by means of arrays of gold dipole nanoantennas. This promising result shows that NETS can also be effectively extended to ensembles of molecules, and specifically to organic compounds of biological relevance.

Author details

Riccardo Piccoli[1*], Andrea Rovere[1], Andrea Toma[2], Roberto Morandotti[1] and Luca Razzari[1*]

*Address all correspondence to: riccardo.piccoli@emt.inrs.ca and luca.razzari@emt.inrs.ca

1 INRS-EMT, Varennes, Québec, Canada

2 Istituto Italiano di Tecnologia (IIT), Genova, Italy

References

[1] S.L. Dexheimer. Terahertz spectroscopy: principles and applications. Boca Raton, Florida, US: CRC Press; 2008.

[2] S. J. Park, J. T. Hong, S. J. Choi, H. S. Kim, W. K. Park, S. T. Han, J. Y. Park, S. Lee, D. S. Kim, Y. H. Ahn. Detection of microorganisms using terahertz metamaterials, Sci. Rep. 2014;4(4988).

[3] J. F. O'Hara, R. Singh, I. Brener, E. Smirnova, J. Han, A. J. Taylor, W. Zhang. Thin-film sensing with planar terahertz metamaterials: sensitivity and limitations. Opt. Express, 2008;16(3):1786–1795.

[4] F. D'Apuzzo, P. Candeloro, F. Domenici, M. Autore, P. Di Pietro, A. Perucchi, P. Roy, S. Sennato, F. Bordi, E. M. Di Fabrizio, S. Lupi. Resonating terahertz response of periodic arrays of subwavelength apertures. Plasmonics 2014;10(1):45–50.

[5] M. A. Seo, H. R. Park, S. M. Koo, D. J. Park, J. H. Kang, O. K. Suwal, S. S. Choi, P. C. M. Planken, G. S. Park, N. K. Park, Q. H. Park, D. S. Kim. Terahertz field enhancement by a metallic nano slit operating beyond the skin-depth limit. Nat. Photonics. 2009;3:152–156.

[6] H. R. Park, K. J. Ahn, S. Han, Y. M. Bahk, N. Park, D. S. Kim. Colossal absorption of molecules inside single terahertz nanoantennas. Nano Lett. 2013;13(4):1782–1786.

[7] D. K. Lee, J. H. Kang, J. S. Lee, H. S. Kim, C. Kim, J. H. Kim, T. Lee, J. H. Son, Q. H. Park M. Seo. Highly sensitive and selective sugar detection by terahertz nano-antennas. Sci. Rep. 2015;5:154–159.

[8] R. Aroca, editor. Surface-enhanced vibrational spectroscopy. Chichester, UK: John Wiley & Sons; 2006.

[9] M. Fleischmann, P. J. Hendra, A. J. McQuillan. Raman spectra of pyridine adsorbed at a silver electrode. Chem. Phys. Lett. 1974;26(2):163–166.

[10] K. Kneipp, M. Moskovits, H. Kneipp. Surface-enhanced Raman scattering: physics and applications. Berlin, Germany: Springer-Verlag; 2006.

[11] S. Nie, S. R. Emory. Probing single molecules and single nanoparticles by surface-enhanced Raman scattering. Science. 1997;275:1102–1106.

[12] K. Kneipp, Y. Wang, H. Kneipp, L. T. Perelman, I. Itzkan, R. R. Dasari, M. S. Feld. Single molecule detection using surface-enhanced Raman scattering (SERS). Phys. Rev. Lett. 1997;78:1667–1670.

[13] M. Chirumamilla, A. Toma, A. Gopalakrishnan, G. Das, R. Proietti Zaccaria, R. Krahne, E. Rondanina, M. Leoncini, C. Liberale, F. De Angelis, E. Di Fabrizio. 3D nanostar dimers with a sub-10-nm gap for single-/few-molecule surface-enhanced Raman scattering. Adv. Mater. 2014;26:2353–2358.

[14] A. Hartstein, J. R. Kirtley, J. C. Tsang. Enhancement of the infrared absorption from molecular monolayers with thin metal overlayers. Phys. Rev. Lett. 1980;45:201–204.

[15] M. Osawa, K.-I Ataka, K. Yoshii, Y. Nishikawa. Surface-enhanced infrared spectroscopy: the origin of the absorption enhancement and band selection rule in the infrared spectra of molecules adsorbed on fine metal particles. Appl. Spectrosc. 1993;47:1497–1502.

[16] R. Adato, A. A. Yanika, J. J. Amsdenc, D. L. Kaplanc, F. G. Omenetto, M. K. Honge, S. Erramillib, H. Altug. Ultra-sensitive vibrational spectroscopy of protein monolayers with plasmonic nanoantenna arrays. Proc. Natl. Acad. Sci. 2009;106:19227–19232.

[17] L. Novotny, B. Hecht. Principle of nano-optics. 2nd ed. Cambridge: University Press; Cambridge, UK 2012.

[18] P. Biagioni, J.-S. Huang, B. Hecht. Nanoantennas for visible and infrared radiation. Rep. Prog. Phys. 2012;75 (024402).

[19] M. Dressel, G. Grüner. Electrodynamics of solids: optical properties of electrons in matter. 1st ed. Cambridge: Cambridge University Press; 2002.

[20] E. Cubukcu, F. Capasso. Optical nanorod antennas as dispersive one-dimensional Fabry-Pérot resonators for surface plasmons. Appl. Phys. Lett. 2009;95 (201101).

[21] L. Razzari, A. Toma, M. Shalaby, M. Clerici, R. Proietti Zaccaria, C. Liberale, S. Marras, I. A. I. Al-Naib, G. Das, F. De Angelis, M. Peccianti, A. Falqui, T. Ozaki, R. Morandotti, E. Di Fabrizio. Extremely large extinction efficiency and field enhancement in terahertz resonant dipole nanoantennas. Opt. Express. 2011;19:26088–26094.

[22] L. Razzari, A. Toma, M. Clerici, M. Shalaby, G. Das, C. Liberale, M. Chirumamilla, R. Proietti Zaccaria, F. De Angelis, M. Peccianti, R. Morandotti, E. Di Fabrizio. Terahertz dipole nanoantenna arrays: resonance characteristics. Plasmonics. 2013;8(1):133–138.

[23] A. Toma, S. Tuccio, M. Prato, F. De Donato, A. Perucchi, P. Di Pietro, S. Marras, C. Liberale, R. Proietti Zaccaria, F. De Angelis, L. Manna, S. Lupi, E. Di Fabrizio, L. Razzari. Squeezing terahertz light into nanovolumes: nanoantenna enhanced terahertz spectroscopy (NETS) of semiconductor quantum dots. Nano Lett. 2015;15(1):386–391.

[24] M. G. Blaber, M. D. Arnold, M. J. Ford. Search for the ideal plasmonic nanoshell: the effects of surface scattering and alternatives to gold and silver. J. Phys. Chem. C. 1998;113(8):3041–3045.

[25] L. Novotny, C. Hafner. Light propagation in a cylindrical waveguide with a complex, metallic, dielectric function. Phys. Rev. E. 1994;50: 4094–4106.

[26] L. Novotny. Effective wavelength scaling for optical antennas, Phys. Rev. Lett. 2007;98(266802).

[27] R. Gordon. Reflection of cylindrical surface waves. Opt. Express. 2009;17(21):18621–18629.

[28] R. Gordon. Vectorial method for calculating the Fresnel reflection of surface plasmon polaritons Phys. Rev. B. 2006;74(153417).

[29] Q. Wu, X-C. Zhang. Free-space electro-optic sampling of terahertz beams. Appl. Phys. Lett. 1995;67:3523–3525.

[30] F. Neubrech, T. Kolb, R. Lovrincic, G. Fahsold, A. Pucci, J. Aizpurua, T. W. Cornelius, M. E. Toimil-Molares, R. Neumann and S. Karim. Resonances of individual metal nanowires in the infrared. Appl. Phys. Lett. 2006;89(253104).

[31] E. D. Palik. Handbook of optical constants of solids. San Diego: Academic; 1998.

[32] M. Walther, D. G. Cooke, C. Sherstan, M. Hajar, M. R. Freeman, F. A. Hegmann. Terahertz conductivity of thin gold films at the metal-insulator percolation transition. Phys. Rev. B. 2007;76(125408).

[33] K-P. Chen, V. P. Drachev, J. D. Borneman, A. V. Kildishev, V. M. Shalaev. Drude relaxation rate in grained gold nanoantennas. Nano Lett. 2010;10:916–922.

[34] J. Chen, P. Albella, Z. Pirzadeh, P. Alonso-González, F. Huth, S. Bonetti, V. Bonanni, J. Åkerman, J. Nogués, P. Vavassori, A. Dmitriev, J. Aizpurua, R. Hillenbrand. Plasmonic nickel nanoantennas. Small. 2011;7:2341–2347.

[35] B. M. Ross, L. P. Lee. Comparison of near- and far-field measures for plasmon resonance of metallic nanoparticles. Opt. Lett. 2009;34:896–898.

[36] J. Zuloaga, P. Nordlander. On the energy shift between near-field and far-field peak intensities in localized plasmon systems. Nano Lett. 2011;11:1280–1283.

[37] G. Gallot, S. P. Jamison, R. W. McGowan, D. J. Grischkowsky. Terahertz waveguides. Opt. Soc. Am. B. 2000;17:851–863.

[38] N. Laman, S. S. Harsha, D. Grischkowsky, J. S. Melinger. High-resolution waveguide THz spectroscopy of biological molecules. Biophys. J. 2008;94:1010–1020.

[39] J. S. Melinger, N. Laman, D. Grischkowsky. The underlying terahertz vibrational spectrum of explosives solids. Appl. Phys. Lett. 2008;93(011102)

[40] N. Laman; S. S. Harsha, D. Grischkowsky. Narrow-line waveguide terahertz time-domain spectroscopy of aspirin and aspirin precursors. Appl. Spectrosc. 2008;62:319–326.

[41] B. Ng, S. M. Hanham, J. Wu, A. I. Fernández-Domínguez, N. Klein, Y. Fook Liew M. B. H. Breese, M. Hong, S. A. Maier. Broadband terahertz sensing on spoof plasmon surfaces. ACS Photonics. 2014;1:1059–1067.

[42] M. I. Vasilevskiy, A. G. Rolo, M. V. Artemyev, S. A. Filonovich, M. J. M. Gomes, Yu. P. Rakovich. FIR absorption in CdSe quantum dot ensembles. Phys. Status Sol. B. 2001;224:599–604.

[43] W. Zhou, T. W. Odom. Tunable subradiant lattice plasmons by out-of-plane dipolar interactions. Nat. Nanotechnol. 2011;6:423–427.

[44] L. Novotny, N. van Hulst. Antennas for light. Nat. Photonics. 2011;5:83–90.

[45] L. Carbone, C. Nobile, M. De Giorgi, F. Della Sala, G. Morello, P. Pompa, M. Hytch, E. Snoeck,A. Fiore, I. R. Franchini, M. Nadasan, A. F. Silvestre, L. Chiodo, S. Kudera, R.

Cingolani, R. Krahne, L. Manna. Synthesis and micrometer-scale assembly of colloidal CdSe/CdS nanorods prepared by a seeded growth approach. Nano Lett. 2007;7:2942–2950.

[46] P. R. Griffiths, J. A. de Haseth. Fourier transform infrared spectrometry. Hoboken, NJ: John Wiley & Sons; 2007.

[47] U. Fano. Effects of configuration interaction on intensities and phase shifts. Phys. Rev. 1961;124:1866–1878.

[48] C. Huck, F. Neubrech, J. Vogt, A. Toma, D. Gerbert, J. Katzmann, T. Härtling, A. Pucci. Surface-enhanced infrared spectroscopy using nanometer-sized gaps. ACS Nano. 2014;8:4908–4914.

[49] F. Neubrech A. Pucci, T. W. Cornelius, S. Karim, A. García-Etxarri, J. Aizpurua. Resonant Plasmonic and Vibrational Coupling in a Tailored Nanoantenna for Infrared Detection. Phys. Rev. Lett. 2008;101(157403).

[50] C. D'Andrea, J. Bochterle, A. Toma, C. Huck, F. Neubrech, E. Messina, B. Fazio, O. M. Maragò, E. Di Fabrizio, M. Lamy de La Chapelle, P. G. Gucciardi, A. Pucci. Optical nanoantennas for multiband surface-enhanced infrared and Raman spectroscopy. ACS Nano. 2013;7:3522–3531.

[51] P. K. Mandal, V. Chikan. Plasmon–phonon coupling in charged n-Type CdSe quantum dots: a THz time-domain spectroscopic study. Nano Lett. 2007;7:2521–2528.

[52] Y. S. Joe, A. M. Satanin, C. S. Kim. Classical analogy of Fano resonances. Phys. Scr. 2006;74:259–266.

[53] B. Gallinet, O. J. F. Martin. Ab initio theory of Fano resonances in plasmonic nanostructures and metamaterials. Phys. Rev. B. 2011;83(235427).

[54] K. Ueno, S. Nozawa, H. Misawa. Surface-enhanced terahertz spectroscopy using gold rod structures resonant with terahertz waves. Opt. Express. 2015;23:28584–28592.

Chapter 3

Research on Hydrogen-Bonded Materials Using Terahertz Technology

Kei Takeya and Kodo Kawase

Additional information is available at the end of the chapter

http://dx.doi.org/10.5772/67640

Abstract

We measured terahertz (THz) characterization of hydrogen-bonded materials using THz time domain spectroscopy (TDS) with a gas-cooling cryostat. The temperature and frequency dependencies of the complex dielectric constants of icy materials were measured over a wide temperature range. We checked the dielectric parameters of ices and gas hydrates using a mathematical model. Ice exhibits increasing absorption with frequency in the THz range because of the low-frequency tail of the infrared-absorption band. This behavior is also observed in gas hydrates. The parameters describing the frequency dependence of ε'' are treated as functions of temperature. From the THz spectroscopy on gas hydrates, we showed that the dielectric constants of the gas hydrates in the THz range can be analyzed using methods for ice. The complex dielectric constants in the THz range contribute to the infrared polarization and phonon absorption of the water molecules on the hydrogen-bonding matrices, so we suggest that THz-TDS is useful for physical and chemical studies of gas hydrates.

Keywords: terahertz time domain spectroscopy, ice, gas hydrate

1. Introduction

A hydrogen bond is an interaction between two electronegative atoms through one hydrogen atom. The famous material as a crystal which consists of a hydrogen bond is an ice. Each oxygen atom in the structure of the ice is bonded to four oxygen atoms which are arranged as a tetrahedral around the oxygen atom through the hydrogen atom. Whereas materials which consist of the hydrogen bond contain the bond partially, ice is composed of only the hydrogen bond. The hydrogen bond is the controlling factor of the character of the ice [1]. Gas hydrates also consist of hydrogen bonds.

Gas hydrates have a curious icy crystalline structure which is stabilized under conditions of relatively high pressure and low temperature. Gas hydrates have attracted much attention as new materials for their use in the transport and storage of natural gases (**Figure 1**), since their structure contains high volume of gases. In particular, methane hydrate is abundant in many locations; it is found in sediment or permafrost regions and is expected to be a future energy resource. The structure of gas hydrates consists of many cages, which include guest gas molecules. Several structures of gas hydrates have been reported. The typical structure types are structures I, II, and H, and the type depends on the guest molecule size and temperature-pressure region. Structures I and II have a cubic unit cell, while structure H has a hexagonal cell. In structure-II hydrate, sixteen 12-Hedra cages (S-cages) and eight 16-Hedra cages (L-cages) are present in the unit cell. Structure I is generated by molecules with sizes 0.4–0.6 nm, while structure II is formed by somewhat larger (0.6–0.7 nm) molecules that occupy only the L-cages, such as tetrahydrofuran (THF) and propane. At high pressures, some of the smallest guest molecules (0.38–0.42 nm) such as argon or nitrogen form a structure-II hydrate and occupy both the S- and the L-cages. If the L-cages with structure II are fully occupied by compound A, the composition of the gas hydrate will be $A \bullet 17H_2O$ [2, 3].

Although ice and gas hydrates are important material for our life, several characters and phenomena are still unknown. The examples are as follows: the highest density of water at 4°C and the self-preservation effect of methane hydrate at around 250 K.

The stability, cage occupancy, structure, and other properties of gas hydrates have been investigated using Raman spectroscopy, X-ray diffraction, NMR, and so on [4–9]. The dielectric constants, which are fundamental parameters of a material, provide information on the water reorientation and proton disorder of the crystalline lattice of a hydrogen-bonded material. Although the dielectric parameters of ice have been reported in detail for a wide frequency and temperature range [1, 10–21], there are few such studies for gas hydrates. Davidson et al. provided a systematic discussion of the relaxation and reorientation of water molecules of gas hydrates using dielectric parameters [22–24]. Williams et al. reported measurements of the dielectric parameters to determine the dipole dispersion of guest molecules [25, 26]. Rick and Freeman reported a computational study of the proton disorder of structure-II hydrates using

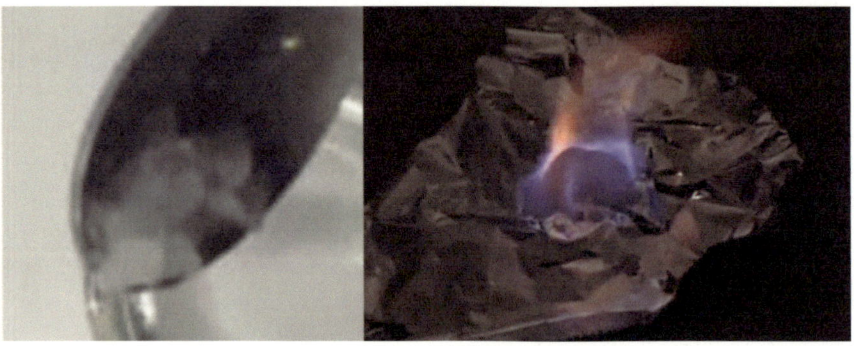

Figure 1. Burning methane hydrates.

dielectric parameters [27]. These studies were conducted for frequencies below GHz frequencies, and there are few studies of gas hydrates at higher frequencies in the terahertz (THz) region [28, 29]. The frequency-dependent dielectric constants of the hydrogen-bonded materials provide information on the hydrogen-bonded structures.

In terms of the dielectric constant, the real part of the dielectric constant below 10^4 Hz of ice is dominated by orientation polarization, and the value beyond around 10^5 Hz is dominated by ionic polarization. The value drops from 3.2 to 1.7 at around a few THz, that is, the region of infrared absorption. In the region between a few THz and deep ultraviolet, electronic polarization dominates the value of the dielectric [30]. Hence, a study of the dielectric parameter at THz frequency is able to investigate the nature of the hydrogen-bonded structure. The above studies have mainly reported in the case of ice, whereas the report for the gas hydrate is not sufficient. Further, the report of the dielectric parameter at THz region is not sufficient.

THz region lies between the optical region and microwave region. The energy of a THz wave corresponds to the motion of a relatively large molecule; thus, THz waves offer the ability to observe absorption of a highly polymerized compound [31]. The energy is also related to the rotation of a hydrogen bond of water molecules and the rotation of free water molecules [32], so THz waves have the potential to observe water dynamics in a solution. Although it has been difficult to use THz waves due to the lack of a good emitter to date, recent developments in the technology of THz emitters and detectors have allowed many applications of THz technology, such as security checks, communications, nondestructive inspection, and spectroscopy [33, 34].

Famous for one of the THz applications is the THz time domain spectroscopy (TDS). THz-TDS is a convenient analytical technique that allows one to determine the optical constants and absorption coefficients of samples without using the Kramers-Kronig relationship [35]. Several studies on water and ice using THz spectroscopy have been reported so far [14, 36]. However, the temperature range used in these studies was very limited. The optical and dielectric properties of ice over a wide temperature range are expected to provide meaningful information for the fields of astronomy, remote-sensing, and low-temperature science. Ice in the proximity of a celestial body or in space exists at very low temperatures [1, 37–40]; moreover, ice in the low-temperature range exhibits interesting phenomena, such as an amorphous structure associated with the glass transition temperature [41, 42]. This information is necessary for THz remote-sensing applications. The frequency and temperature-dependent dielectric constants of ice provide information on hydrogen-bonded structures. Therefore, a survey of the optical parameters of hydrogen-bonded materials over a wide temperature range is necessary.

2. Research methods

For the preparation of water ice samples, we used ultrapure water and pure heavy water (WAKO Chemical). The water was placed in a Teflon plate with 10-mm-diameter holes to form tablet samples. The water and heavy water were frozen using an environmental testing machine. The freezing of the water was slow and occurred at a temperature of 272 K to avoid mixing air bubbles into the ice. The tablet thickness (~1.0 ± 0.005 mm) was measured using a micrometer (Anritsu). The volume density of the tablet sample was over 99%, according to the volume and weight.

Propane and tetrahydrofuran (THF) hydrates were synthesized by stirring THF + water solution or pressurized gas and distilled water at 275.2 K in a stainless steel cell [28]. After stirring, the hydrates were extracted from the stainless cell at ca. 253 K in a low-temperature chamber. The obtained hydrate particles were crushed by the mortar and the pestle. The hydrate crystals were made into tablets with a diameter of 10 mm and varying thicknesses (1–2 mm) by using the tablet-making apparatus (Ichihashi-Seiki, HANDTAB-Jr) in the low-temperature chamber. The thickness of each sample was accurately calculated from the weight and the volume of water after melting. The error in the thickness is within 0.5%.

The tablet samples were measured using a THz-TDS system equipped with a dipole-type low-temperature-grown GaAs (LT-GaAs) photoconductive switch as an emitter and a detector and a Ti:sapphire laser (Mai-Tai, Spectra-Physics; $\lambda = 780$ nm; repetition rate: 82 MHz; pulse width: 100 fs) as a femtosecond pulsed laser source. The femtosecond laser pulse is first separated into a "pump pulse" and a "trigger pulse." The THz pulse is generated at the photoconductive switch by the pump-pulse illumination and collimated onto the sample by using a set of off-axis parabolic mirrors and then collimated again onto the photoconductive switch antenna detector using another set of off-axis parabolic mirrors. Since the temporal photocurrent is generated only when the triggered laser pulse and the radiated THz pulse simultaneously arrive at the photoconductive switch, we can directly obtain both the amplitude and the phase of the radiated THz pulse by changing the optical delay and measuring the corresponding change in the photocurrent.

For the low-temperature measurement, a gas-cooled cryostat made by PASCAL was used. The temperature of the sample enclosure in the cryostat was maintained in the temperature range of 20–240 K using compressed He. The sample enclosure, equipped with quartz windows for transmitting THz waves, was positioned in the intermediate focal plane between the two sets of parabolic mirrors. The ice tablets were held at the top of a sample rod placed in the center between optical windows. During the initial sample setting, the sample rod was cooled by liquid nitrogen. The temperature distribution of the sample enclosure was kept uniform by circulating the helium cooling gas. Three thermistors surrounded the sample rod to monitor and maintain the set sample temperature within ±0.2 K. All measurements were done multiple times, and each data is an averaged value.

The real and imaginary parts of the complex refractive indices, $n(\omega)$ and $k(\omega)$, of the samples can be directly calculated from the following equations:

$$\frac{\mathrm{Esam}(\omega)}{\mathrm{Eref}(\omega)} = \rho(\omega) \exp\left[-i\phi(\omega)\right] \tag{1}$$

$$n(\omega) = \frac{c\phi(\omega)}{\omega L} + 1 \tag{2}$$

http://dx.doi.org/10.5772/67640

$$k(\omega) = \frac{c}{\omega L} \ln \left\{ \frac{4n(\omega)}{\rho(\omega)[n(\omega) + 1]^2} \right\} \tag{3}$$

where ω is the frequency, $E_{\mathrm{sam}}(\omega)$ is the fast Fourier transform (FFT) spectrum of the THz pulse propagating through the samples, $E_{\mathrm{ref}}(\omega)$ is the reference FFT spectrum, $\rho(\omega)$ is the transmission, $\varphi(\omega)$ is the phase shift, c is the light speed, and L is the thickness of the sample [35]. Using these estimated complex refractive indices, we can obtain the real $\varepsilon'$and imaginary ε'' dielectric constants of the samples, as follows:

$$\varepsilon'(\omega) = [n(\omega)]^2 - [k(\omega)]^2 \tag{4}$$

$$\varepsilon''(\omega) = 2n(\omega)k(\omega) \tag{5}$$

3. Dielectric parameters of water ice and heavy water ice

The frequency dependence of the real part of the dielectric constants ε' of ice and heavy water ice is shown in **Figure 2**. Both ices were measured over the frequency range of 0.2–1.4 THz at 20, 100, 200, and 240 K. As the frequency increased from 0.2 to 1.4 THz, ε' increased from 3.12 to 3.24 for water ice and from 2.96 to 3.07 for heavy water ice at 200 K. The data expressed by open squares in **Figure 2(a)** are calculated, and values obtained from the reported complex index of refraction are shown in Ref. [14]. Our experimental data shows good agreement with the reported data at the same temperature. Lattice vibrations of the hydrogen-bonded water molecules contributed significantly to the amplitude of ε' in the THz frequency range [14]. The dielectric constant of heavy water ice is lower than that of water from comparison of both results. This is caused by a difference of polarization in water or heavy water ice. The polarization in heavy water ice is smaller than that in water ice, since a deuterium atom is heavier than hydrogen atom.

According to previous studies, the ε' values for ice are constant from 0.1 MHz to 1 THz and show complicated behavior beyond 1 THz, with the maximum occurring at a few THz [10, 19, 37]. The frequency range of our measurements corresponded to the range used in these

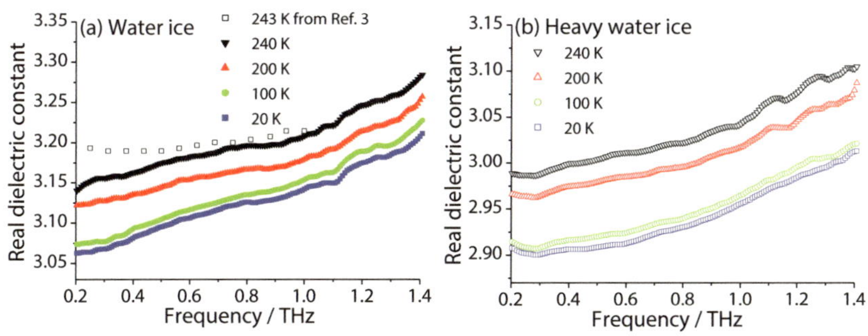

Figure 2. Frequency dependence of the real dielectric constant ε' of (a) water and (b) heavy water ice at 20, 100, 200, and 240 K. The reported data of water ice [14] are plotted in graph (a).

previous studies; thus, in our measurements, it was expected we observe a local maximum value for ε' of ice at several THz [10].

Figure 3 shows the temperature dependence of the real dielectric constant ε' for both ices at 0.26, 0.50, 0.75, and 1 THz. As the temperature increased from 10 to 240 K, ε' increased from 3.14 to 3.21 for water ice and from 2.95 to 3.04 for heavy water ice at 1 THz.

In previous reports of ε' of ice in the GHz regime, the temperature dependence of ε' has been discussed with regard to the contribution of inharmonic effects to the molecular polarizability α of lattice water [10, 12]. We can obtain α for water and heavy water ice in the THz frequency range using the Clausius–Mossotti equation:

$$\frac{N_i}{3}\alpha = \left(\frac{\varepsilon' - 1}{\varepsilon' + 2}\right) \tag{6}$$

where N_i is the number of molecules per unit volume. According to previous reports, the density for ice is 0.917 Mgm^{-3}; for heavy water ice, it can be estimated as 1.019 Mgm^{-3}. The estimated values of α are 41.51 $Å^3$ for water ice and 39.67 $Å^3$ for heavy water ice at 1 THz at a temperature of 240 K (**Figure 4**). The values of heavy water ice are lower than those for water ice under the same conditions. This resulted from the mass of deuterium atoms. The molecular polarizability α is the sum of the electronic polarizability α_{el} (displacement of the molecules within the lattice) and the infrared polarizability α_{IR} (the vibration of the molecules within the lattice).

For ice, α_{el} is independent of temperature; thus, the temperature dependence of α appears in the α_{IR} term. Additionally, the α_{IR} of ice increases linearly with the square of the temperature (T^2) [10, 12]. Applying this discussion of the polarizability of ice to the present results, we derived the following equation for the molecular polarizability α:

$$\alpha = \alpha_{el} + \alpha_{IR} = \alpha_{el} + \alpha_{0,IR} + aT^2 \tag{7}$$

where $\alpha_{0,IR}$ is the infrared polarizability at 0 K and a is an empirical constant. Using Eq. (**7**), we determined that the fitted curves were in good agreement with the experimental results

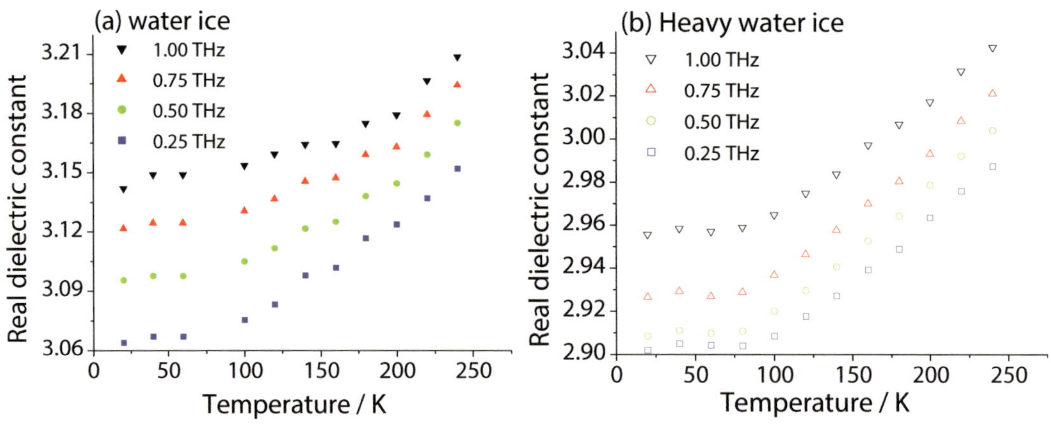

Figure 3. Temperature dependence of the real dielectric constant ε' at 0.25, 0.50, 0.75, and 1 THz for (a) water ice and (b) heavy water ice.

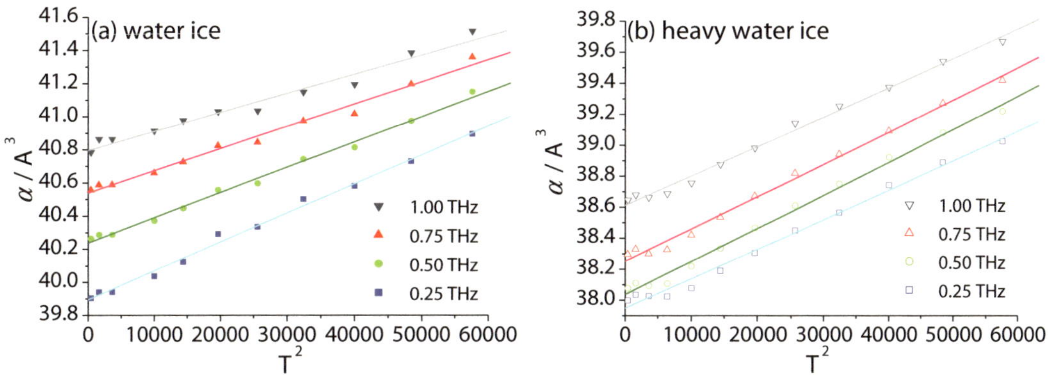

Figure 4. Molecular polarizability α of (a) water and (b) heavy water ice plotted versus temperature squared. The solid lines are the fitted curves obtained from Eq. (7) at 0.25, 0.50, 0.75, and 1.00 THz.

(**Figure 4**). The values of $\alpha_{el} + \alpha_{0,IR}$ and a derived from the fitted curves (**Figure 4**) are listed in **Table 1**. The value of a in the range of 0.26–0.75 THz was ~ 8.6 × 10^{-6}, which was nearly constant, but it increased beyond 0.75 THz. The real dielectric constant ε' of ice exhibited a local maximum value at a few THz and showed complex behavior beyond 1 THz, corresponding to the skirt region of the local maximum value [11, 19]. In the case of ice, the real dielectric constant ε' was almost constant in the range from GHz to sub-terahertz, and it showed complicated behavior beyond 1 THz [11, 14, 19]. This behavior was consistent with the results for a; thus, we concluded that the local maximum of ε'' contributed to the orientation polarization and affected the frequency dependence of a.

Figure 5 shows the frequency dependence of the imaginary dielectric constants ε'' of water and heavy water ice over the range of 0.4–1.4 THz at 40, 100, 200, and 240 K. A block of data with error is discarded in **Figure 5**. As the frequency and temperature increased, ε'' increased over the range considered in this chapter. Our ε'' values approximately agreed with those reported by Zhang et al. [14] at the same temperature. The frequency range below 1 THz corresponds to the low-frequency tail of the infrared-absorption band of ice; thus, ε'' is dominated by the contribution from molecular reorientation. Furthermore, the effect of Debye relaxation may extend to this frequency range. However, the THz range is far from the region dominated by the Debye relaxation. We therefore suggest that in the THz region, the contribution from Debye relaxation which appears in a decrease with increasing frequency is small for ε''. Thus, a simple model for the frequency dependence of ε'' is the single term that increases with increasing frequency because it corresponds to the low-frequency tail of the infrared-absorption band at high frequency [11, 14]. Considering previous studies, the equation for ε'' in this frequency range is denoted by:

$$\varepsilon'' = A\omega^B \tag{8}$$

where the coefficients A and B are temperature-dependent empirical constants. Applying this formula to the results, we obtained values for A and B that gave fitted curves at each

	Water ice		Heavy water ice	
	0.50 THz	1.00 THz	0.50 THz	1.00 THz
$\alpha_{el} + \alpha_{0,IR}$ / Å^3	40.2	40.8	38.0	38.6
a/Å^3 K^{-2}	1.52×10^{-5}	1.15×10^{-5}	2.13×10^{-5}	1.89×10^{-5}

Table 1. Values of $\alpha_{el} + \alpha_{0,IR,}$ and a derived from fitted curves in Figure 4.

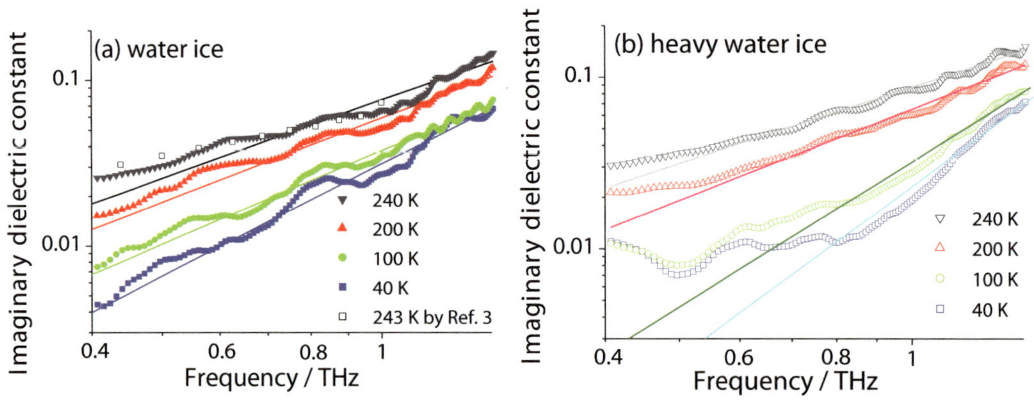

Figure 5. Imaginary dielectric constant ε'' of water ice (a) and heavy water ice (b) at four temperatures plotted double logarithmically versus frequency. The solid lines are fitted curves obtained from Eq. (8). The reported data of water ice [14] are plotted in graph (a).

temperature, as shown in **Figure 5**. As the temperature increased, parameter A increased; however, parameter B decreased. This behavior is consistent with that of ice in the GHz range. The fitted curves of heavy water ice are different from the experimental data at low frequency and low temperature (**Figure 5(b)**). This may be caused by experimental uncertainty or events that appear on heavy water ice at low temperatures.

Figure 6 shows the temperature dependence of ε'' of both ices from 20 to 240 K at two frequencies. For all four parameters, ε'' increased with temperature. According to a previous optical study of ice in the GHz regime, the temperature dependence of ε'' in the THz range is determined by the A term in Eq. (**8**). However, in this chapter, parameters A and B of Eq. (**8**) showed temperature dependence. We can approximate the temperature dependence of A and B using the least squares method, as follows:

For water ice

$$A = 1.9 \times 10^{-2} + 2.2 \times 10^{-4} \times T \quad \text{in THz}^{-1} (R^2 = 0.954) \tag{9}$$

$$B = 2.3 - 3.1 \times 10^{-3} \times T \quad \text{in THz}^{-1} (R^2 = 0.926) \tag{10}$$

For heavy water ice

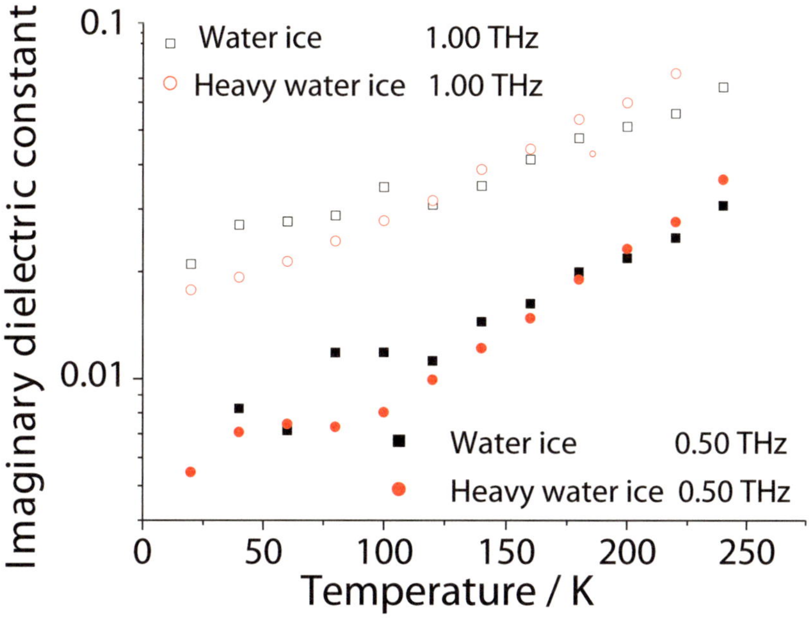

Figure 6. Temperature dependence of the imaginary dielectric constant ε'' of water and heavy water ice at 0.50 and 1 THz.

$$A = 5.7 \times 10^{-3} + 3.2 \times 10^{-4} \times T \qquad \text{in THz}^{-1} (R^2 = 0.961) \tag{11}$$

$$B = 3.7 - 9.3 \times 10^{-3} \times T \qquad \text{in THz}^{-1} (R^2 = 0.997) \tag{12}$$

These formulae fit the experimental results with good agreement, as noted by the correlation coefficients R^2.

In order to analyze the frequency and temperature dependencies of ε'', approximation of these equations should be considered from the viewpoint of the following factors. For ice, phenomena such as Debye relaxation, acoustic phonon excitation, and infrared phonon absorption are reported in detail in various ways [1, 19, 20]. The dielectric and optical constants of ice material in the THz region may contribute to previous findings. For example, the glass transition of ice at low temperatures is well known. This transition strongly dominates the hydrogen-bond network; thus, information on the dielectric constants of ice over the given temperature range may facilitate astronomy and remote-sensing applications.

4. Dielectric parameters of sulfur hexafluoride hydrate

Figure 7 shows our results for the frequency dependence of the real dielectric constant ε' of sulfur hexafluoride hydrate in the range 0.3–1.0 THz at 100.6, 199.9, and 239.9 K. As the frequency increases from 0.3 THz to 1.0 THz, ε' increases from 3.02 to 3.09 at 199.9 K. These

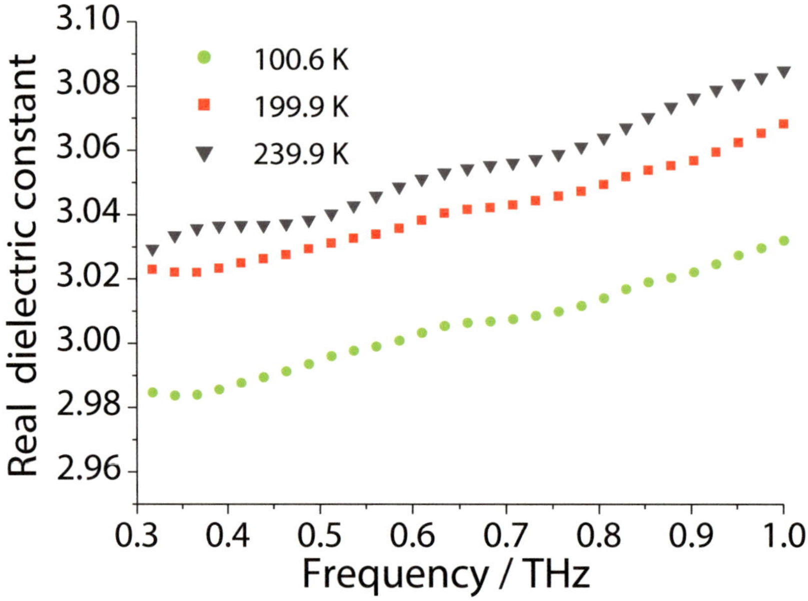

Figure 7. Frequency dependence of real dielectric constant ε' of sulfur hexafluoride hydrate at 100.6, 199.9, and 239.9 K.

values are similar to but slightly lower than those for ice under the same conditions [14]. As for the hydrogen-bonded crystal of the water molecules, a lattice vibration significantly contributes to the amplitude of ε' in the THz frequency range [14]. Hence, on the basis of a comparison of the ε' values, the states of the hydrogen-bonding matrices of ice and gas hydrates, including the nondipolar molecules, are observed to be similar. The difference in the polarizability and the molecular density between ice and the gas hydrate is likely to contribute to a slight difference in these ε' values.

According to previous studies, the ε' values for ice are constant from 0.1 MHz to 1 THz and show complex behavior beyond 1 THz. The maximum occurs at a few THz [1, 18, 19]. The frequency range of our measurements corresponds to the above range, so we expect that a local maximum for theε' of the gas hydrate exists at a few THz, similar to the local maximum for ice [19].

Figure 8 shows the temperature dependence of the real dielectric constant ε' at 0.50, 0.75, and 1 THz. As the temperature increases from 50 to 240 K, ε' increases from 3.00 to 3.03 at 500 GHz and from 3.02 to 3.08 at 1 THz. The rate of increase of ε' with temperature ($d\varepsilon'/dT$) clearly increases beyond 0.75 THz.

Figure 9 shows the frequency dependence of the imaginary dielectric constants ε'' of sulfur hexafluoride hydrate in the range 0.3–1.2 THz at 100.6, 199.9, and 239.9 K, together with the reported values for ice [14]. As frequency and temperature increase, ε'' increases over the range considered in this chapter. Our ε'' values approximately agree with those reported by Zhang et al. for ice at the same temperature [14]. For ice, the frequency range below 1 THz is the low-frequency tail of the infrared-absorption band so ε'' is dominated by the contributions from

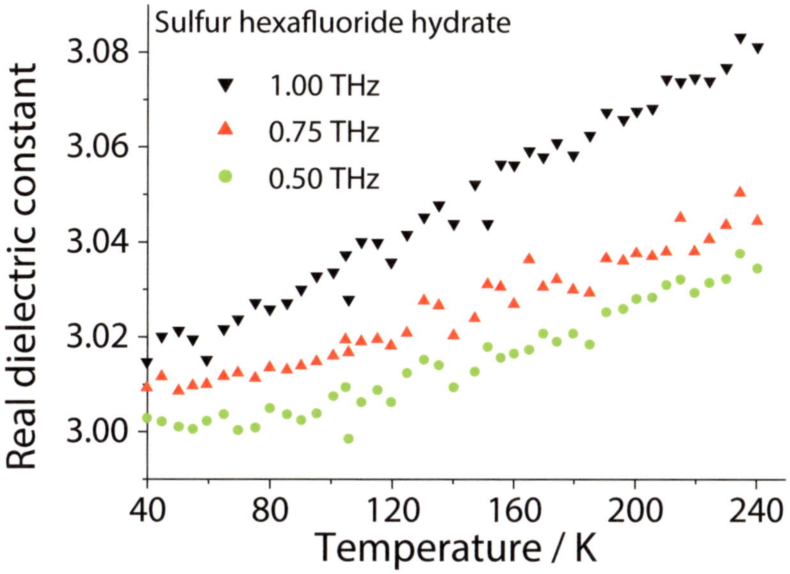

Figure 8. Temperature dependence of the real dielectric constant ε' of the sulfur hexafluoride hydrate at 0.50, 0.75, and 1 THz.

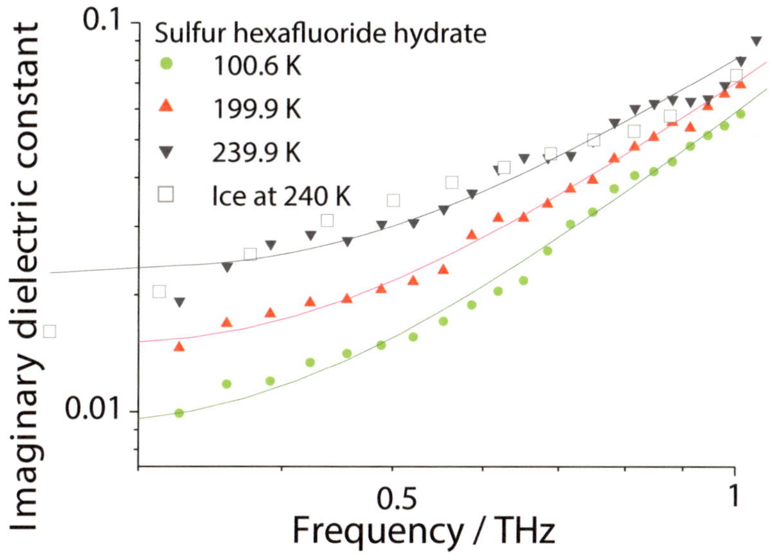

Figure 9. Imaginary dielectric constant ε'' of sulfur hexafluoride hydrate at three temperatures plotted double logarithmically versus frequency. The reported values for ice are shown as white squares. The solid lines are fitted curves obtained from Eq. (**13**).

molecular reorientation. Moreover, the influence of Debye relaxation extends to this frequency range, and thus, a simple model for the frequency dependence of ε'' is the sum of two components. One term is contributed by the Debye relaxation and decreases with increasing frequency. The other term increases with increasing frequency since it is the low-frequency tail of the infrared-absorption band at a high frequency [12, 14, 28]. From previous studies, the equation for ε'' in this frequency range is

$$\varepsilon'' = (C/\omega) + D\omega^E \tag{13}$$

where the coefficients *C*, *D*, and *E* are temperature-dependent empirical constants. Applying this formula to our results, we obtain the constants *C*, *D*, and *E*, which give the fitted curves at each temperature, as shown in **Figure 9**.

As the temperature increases, parameters *C* and *D* increase but parameter *E* decreases (**Table 2**). This behavior is consistent with that of ice in the GHz range.

Figure 10 shows the temperature dependence of ε'' from 50 to 240 K at four frequencies. At all four frequencies, ε'' increases with temperature. According to a previous optical study of ice, the temperature dependence of ε'' in the THz range is determined by term *B* in Eq. (**13**). However, in this chapter, parameters *C*, *D*, and *E* of Eq. (**13**) all show temperature dependencies. We can approximate the temperature dependencies of *C*, *D*, and *E* using the least squares method, which gives

Temperature (K)	*C* (THz)	*D* (THz^{-1})	*E*
10.3	0.00137	0.0474	2.364
100.6	0.00172	0.0574	2.246
200	0.00289	0.0675	2.081
240	0.00477	0.0765	1.894

Table 2. Empirical coefficients *C*, *D*, and *E* obtained via least squares.

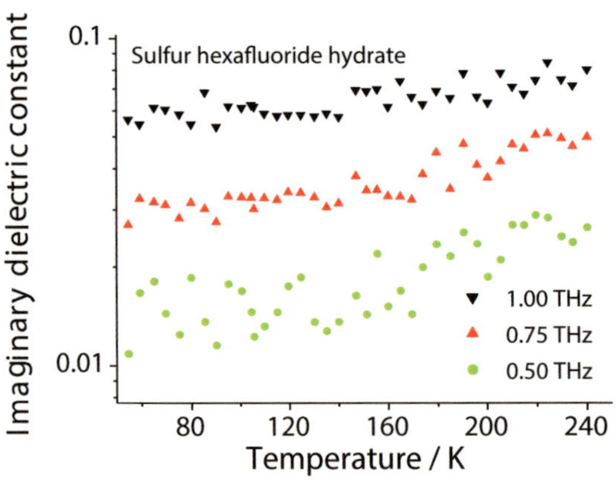

Figure 10. Temperature dependence of the imaginary dielectric constant ε'' of sulfur hexafluoride hydrate at 0.50, 075, and 1 THz.

http://dx.doi.org/10.5772/67640

$$C = 1.4 \times 10^{-4} + 3.0 \times 10^{-5} \ \exp(T \times 2.0 \times 10^{-2}) \tag{14}$$

$$D = 4.6 \times 10^{-2} + 1.2 \times 10^{-4} \times T \tag{15}$$

$$E = 2.4 - 5.8 \times 10^{-2} \ \exp(T \times 9.1 \times 10^{-3}) \tag{16}$$

The curves fitted using these formulas fit the experimental results with good correlation coefficients R^2.

To analyze the frequency and temperature dependencies of ε'', we have to consider the approximation of these formulas in terms of the following factors. For ice, phenomena such as Debye relaxation, the excitations of acoustic phonons, and infrared phonon absorption have been reported in detail using various methods [11, 14, 19, 28]. For gas hydrates, there have been few reports for this frequency range. In addition, the structures of gas hydrates are different from those of ice. To investigate the dielectric parameters physically and chemically, we need more information on gas hydrates in this frequency range.

Further, we have to consider the contribution of guest molecules to the polarizability and the properties of the hydrogen bond of the gas hydrate. The hydration number of sulfur hexafluoride hydrate is 17, and the sulfur hexafluoride molecule is nonpolar. We therefore ignored the contribution of guest molecules. In the future, we will carry out a more detailed dielectric study of gas hydrates, including the influence of guest molecules. Nonetheless, since our results in this chapter show a good agreement between the experimental data and the theoretical description, our analysis is reasonably accurate for the investigation of the dielectric behavior of the gas hydrate in the THz frequency range.

5. Temperature dependence of the absorption of tetrahydrofuran

Conversely, the tetrahydrofuran hydrate of the same structure-II of gas hydrate has a broad absorption peak around 0.5 THz (**Figure 11**). This characteristic absorption is not due to the cage structure of the gas hydrate but due to the kinetics of the tetrahydrofuran molecule in the cage or

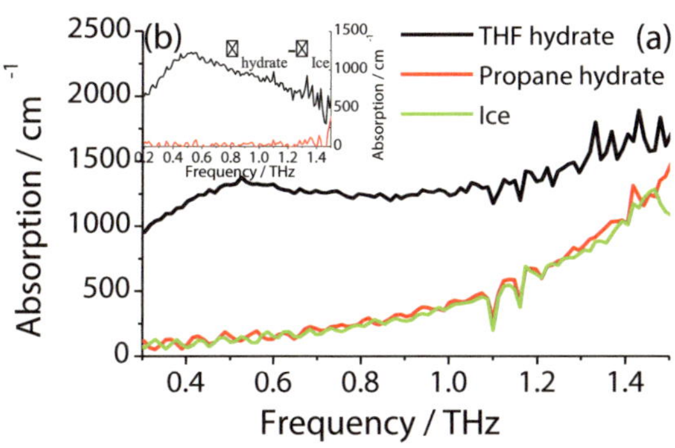

Figure 11. (a) Absorption coefficients of tetrahydrofuran and propane hydrates and ice at 243 K. Inset (b): Subtract spectra of gas hydrates (α_{hydrate} - α_{ice}).

the interaction between the molecule and the cage. This is because propane hydrate does not exhibit such an absorption peak and the tetrahydrofuran molecule has a larger dipole moment than the propane molecule. A similar absorption peak was found by Klug and Whalley during IR spectroscopy [6]; in their study, tetrahydrofuran hydrate showed two broad absorption peaks at 25 cm^{-1} (~0.75 THz) and 38 cm^{-1} (~1.14 THz) because of rotational oscillations of the tetrahydrofuran molecule at 17 K. Since these absorption peaks are integrated and broadened as the temperature increases due to the reorientation of the tetrahydrofuran molecules and the occupancy of higher-potential minima increases [6], the absorption peak observed in the tetrahydrofuran hydrate probably originates from the rotational oscillations.

6. Summary

We reviewed optical properties of hydrogen-bonded materials that are ice, heavy ice, and gas hydrates within a terahertz frequency and at a wide temperature region of 20–240 K. Frequency and temperature dependencies of the complex dielectric parameter obtained from THz-TDS measurements are described by a mathematical model. An increase in the real part of the dielectric constants with frequency and temperature was observed on all materials.

In terms of the imaginary part of the dielectric parameter of the hydrogen-bonded materials, the THz frequency corresponds to the skirt region of the infrared-absorption band. The imaginary parts of the dielectric parameter of the materials increased with frequency and temperature. The parameters are described by a function of a simple form.

Meanwhile, a gas hydrate which includes polar molecules showed specific absorption at THz frequency. The absorption is based on the rotation of guest molecules inside the structure. Hence THz-TDS can observe the dynamics of molecules in gas hydrates.

Although the optical behavior of the ice and gas hydrates including nonpolar gas molecules was described by a similar model, small but certain differences between ice and gas hydrates were observed on the value of the parameter. The difference corresponds to the difference of the structures; hence, the optical properties in THz frequency provide meaningful information for the hydrogen-bonded materials.

Acknowledgements

This chapter contains many quotes from Refs. [21] and [29]. We thank both journals.

Author details

Kei Takeya* and Kodo Kawase

*Address all correspondence to: takeya@nuee.nagoya-u.ac.jp

Nagoya University, Nagoya city, Aichi, Japan

References

[1] Petrenko VF, Whitworth RW (2006). Physics of Ice. Oxford University Press, Oxford.

[2] Sloan ED, Koh CA (2007). Clathrate Hydrates of Natural Gases, 3rd ed. CRC Press, Boca Raton.

[3] Ripmeester JA, Tse JS, Ratcliffe CI, Powell BM (1987). A new clathrate hydrate structure. Nature 325, 135–136.

[4] Ohgaki K, Makihara Y, Takano K (1993). Formation of CO_2 hydrate in pure and sea waters. J. Chem. Eng. Japan 26, 558–564.

[5] Takeya S, Honda K, Yoneyama A, Hirai Y, Okuyama J, Hondoh T, Hyodo K, Takeda T (2006). Observation of low-temperature object by phase-contrast x-ray imaging: non-destructive imaging of air clathrate hydrates at 233 K. Rev. Sci. Instr. 77, 053705.

[6] Klug DD, Whalley E (1973). The rotational oscillations of tetrahydrofuran in tetrahydrofuran clathrate hydrate. Can. J. Chem. 51, 4062–4071.

[7] Mak TCW, McMullan RK (1965). Polyhedral clathrate hydrates. X. Structure of the double hydrate of tetrahydrofuran and hydrogen sulfide. J. Chem. Phys. 42, 2732–2737.

[8] Stern LA, Circone S, Kirby SH, Durham WB (2001). Anomalous preservation of pure methane hydrate at 1 atm. J. Phys. Chem. B 105, 1756–1762.

[9] Subramanian S, Kini RA, Dec SF, Sloan ED (2000). Evidence of structure II hydrate formation from methane + ethane mixtures. Chem. Eng. Sci. 55, 1981–1999.

[10] Johari GP, Jones SJ (1976). Infrared polarisability of hexagonal ice. Nature 263, 672–673.

[11] Bertie JE, Labbé HJ, Whalley E (1969). Absorptivity of ice I in the range 4000–30 cm^{-1}. J. Chem. Phys. 50, 4501–4520.

[12] Matsuoka T, Fujita S, Mae S (1996). Effect of temperature on dielectric properties of ice in the range 5–39 GHz. J. Appl. Phys. 80, 5884–5890.

[13] Koh G (1997). Dielectric properties of ice at millimeter wavelengths. Geophys. Res. Lett. 24, 2311–2313.

[14] Zhang C, Lee K-S, Zhang X-C, Wei X, Shen YR (2001). Optical constants of ice Ih crystal at terahertz frequencies. Appl. Phys. Lett. 79, 491–493.

[15] Whalley E, Labbé HJ (1969). Optical spectra of orientationally disordered crystals. III. Infrared spectra of the sound waves. J. Chem. Phys. 51, 3120–3127.

[16] Matzler C, Wegmuller U (1987). Dielectric properties of freshwater ice at microwave frequencies. J. Phys. D. 20, 1623–1630.

[17] Jiang JH, Wu DL (2004). Ice and water permittivities for millimeter and sub-millimeter remote sensing applications. Atmos. Sci. Lett. 5, 146–151.

[18] Takei I, Maeno N (1997). Dielectric low-frequency dispersion and crossover phenomena of HCl-doped ice. J. Phys. Chem. B. 101, 6234–6236.

[19] Warren SG (1984). Optical constants of ice from the ultraviolet to the microwave. Appl. Opt. 23, 1206–1225.

[20] Mishima O, Klug DD, Whalley E (1983). The far infrared spectrum of ice Ih in the range 8–25 cm^{-1}. Sound waves and difference band, with application to Saturn's rings. J. Chem. Phys. 78, 6399–6404.

[21] Takeya K, Fukui T, Takahashi R, Kawase K (2014). Dielectric constants of H_2O and D_2O ice in the terahertz frequency regime over a wide temperature range. J. Opt. 16, 094005.

[22] Gough SR, Hawkins RE, Morris B, Davidson DW (1973). Dielectric properties of some clathrate hydrates of structure II. J. Phys. Chem. 77, 2969–2976.

[23] Davidson DW, Wilson GJ (1963). The low-frequency dielectric properties of ethylene oxide and ethylene oxide hydrate. Can. J. Chem. 41, 1424–1434.

[24] Majid YA, Garg SK, Davidson DW (1968). Dielectric and nuclear magnetic resonance study of the hydrate of sulfur hexafluoride. Can. J. Chem. 46, 1683–1690.

[25] Davies M, Williams K (1968). Dielectric relaxation in clathrates. Trans. Faraday Soc. 64, 529–548.

[26] Davidson DW, Davies MM, Williams K (1964). Dielectric absorption and molecular motion in gas hydrates. J. Chem. Phys. 40, 3449–3450.

[27] Rick SW, Freeman DL (2010). Proton disorder and the dielectric constant of type II clathrate hydrates. J. Chem. Phys. 132, 054509.

[28] Takeya K, Zhang C, Kawayama I, Murakami H, Jepsen PU, Chen J, Wu P, Ohgaki K, Tonouchi M (2009). Terahertz time domain spectroscopy on structure-II gas hydrates. Appl. Phys. Exp. 2, 122303.

[29] Takeya K, Kawayama I, Murakami H, Tonouchi M, Ohgaki K (2011) Dielectric study on gas hydrates using terahertz time domain spectroscopy. In: Proceedings of the 7th International Conference on Gas Hydrates (ICGH 2011) 736–742.

[30] Fletcher NH (1970). The Chemical Physics of Ice. Cambridge University Press, 198–246.

[31] Tonouchi M (2007). Cutting-edge terahertz technology. Nat. Photonics. 1, 97–105.

[32] Yada H, Nagai M, Tanaka K (2008). Origin of the fast relaxation component of water and heavy water revealed by terahertz time-domain attenuated total reflection spectroscopy. Chem. Phys. Lett. 464, 166–170.

[33] Kawase K, Ogawa Y, Watanabe Y, Inoue H (2003). Non-destructive terahertz imaging of illicit drugs using spectral fingerprints. Opt. Exp. 11, 2549–2554.

[34] Song HJ, Nagatsuama T (2011). Present and future of terahertz communications. IEEE Trans. Terahertz Sci. Technol. 1, 256–263.

[35] Duvillaret L, Garet F, Coutaz J-L (1996). A reliable method for extraction of material parameters in terahertz time-domain spectroscopy. IEEE J. Sel. Top. Quantum Electron. 2, 739–746.

[36] Rønne C, Åstrand PO, Keiding SR (1999). THz spectroscopy of liquid H_2O and D_2O. Phys. Rev. Lett. 82, 2888–2891.

[37] Quirico E, Douté S, Schmitt B, Bergh C, Cruikshank DP, Owen TC, Geballe TR, Roush TL (1999). Composition, physical state, and distribution of ices at the surface of triton. Icarus 139, 159–178.

[38] Cruikshank DP, Schmitt B, Roush TL, Owen TC, Quirico E, Geballe TR, Bergh C, Bartholomew MJ, Ore CMD, Douté S, Meier E (2000). Water ice on triton. Icarus. 147, 309–316.

[39] Klinger J (1983). Extraterrestial ice. A review. J. Phys. Chem. 87, 4209–4214.

[40] Rothery DA (1999). Satellites of the Outer Planets, 2nd ed. Oxford University Press, 49–62.

[41] Mishima O, Calvert LD, Whalley E (1984). 'Melting ice' I at 77 K and 10 kbar: a new method of making amorphous solids. Nature. 310, 393–395.

[42] Handa YP, Klug DD (1988). Heat capacity and glass transition behavior of amorphous ice. J. Phys. Chem. 92, 3323–3325.

Chapter 4

Terahertz Fiber Sensing

Borwen You and Ja-Yu Lu

Additional information is available at the end of the chapter

http://dx.doi.org/10.5772/66345

Abstract

Terahertz fibers used for optical-sensing applications are introduced in this chapter, including the dielectric wires, ribbons and pipes. Different analyte conformations of the liquid, solid particle, thin film and vapor gas are successfully integrated with suitable fibers to perform high sensitivities. Based on the optimal sensitivities, analyte recognitions limited in traditional terahertz spectroscopy are experimentally demonstrated by the terahertz fiber sensors. Using the cladding index-dependent waveguide dispersion and high fractional cladding power of terahertz wire fiber, 20 ppm concentration between polyethylene and melamine particles can be distinguished. When the evanescent mode field of a terahertz ribbon fiber is controlled by a diffraction metal grating, subwavelength-confined surface terahertz waves potentially enable the near-field recognition for nano-thin films. Resonance waveguide field surrounding the terahertz pipe fiber is able to identify the macromolecule deposition in subwavelength-scaled thickness, approximately $\lambda/225$. For inner core-confined resonance waveguide field inside the terahertz pipe fiber, low physical density of the vaporized molecules around 1.6 nano-mole/mm^3 can also be discriminated.

Keywords: optical sensing, terahertz spectroscopy, terahertz optics

1. Introduction

Various dielectric fibers are introduced in this chapter for terahertz (THz) wave transmission and sensing. Most of THz-electromagnetic power guided along the proposed fiber is distributed in the air region, realizing long distance delivery of THz waves with low propagation loss and dispersion. The detection sensitivity consequently increases due to the sufficient field-analyte interaction length. Large portion of THz field evanescent in the air also makes the waveguide mode quite sensitive to the slight refractive index variation of the analyte surrounding the fiber. The sensitive feature of THz refractive index is advantageous to minute material detection based on analyzing the propagation property change of waveguide mode. In this chapter, we address not only the waveguide characteristics and sensing mechanisms of the proposed THz fibers, but also the fiber-sensing schemes with the highest sensitivities, experimentally achieved via optimizing the waveguide geometric parameters. The introduced low-loss THz fibers are categorized based on their geometries including the wire-, ribbon- and pipe-fibers. The former two fibers have the subwavelength-scaled waveguide core and air cladding, allowing single mode transmission with loosely field confinement. The analytes distributed in the air cladding significantly influences the fiber dispersion of the waveguide modes. The pipe-fiber delivers and confines terahertz waves in the hollow air core based on the anti-resonant reflecting waveguide principle, leading to multiple resonant dips and exhibiting in the transmission spectrum. The analyte inducing slight variation of the optical path in the pipe wall or hollow core is able to significantly shift the resonant spectral dips. Such variation of waveguide dispersion or spectral shift for the proposed terahertz fibers are successfully measured by the fiber-based terahertz time domain spectroscopy to identify various minute analytes, such as subwavelength films, particles and volatile vapors.

2. Subwavelength plastic wire sensor

2.1. THz waveguide field along a plastic wire

The THz wire fiber has a circular cross section as the fiber core and an infinite air cladding, where the wire core size is smaller than the THz wavelengths and called as the subwavelength fiber [1]. The subwavelength fiber used to guide electromagnetic waves was first demonstrated in infrared frequency, performing a large portion of the evanescent power outside the fiber core [2]. On the basis of the evanescent waveguide field, the wire fiber is demonstrated to deliver THz wave in a long distance with very low waveguide loss. In this section, the polystyrene (PS) wire with a step refractive-index profile is taken as one example of a THz subwavelength fiber and its evanescent power in the air cladding is expressed in **Figure 1(a)**, which is calculated from the solutions of a general step-profile fiber [3]. The core diameter and THz refractive index of the PS core material are, respectively, 300 μm and 1.59 at 0.300 THz. **Figure 1(a)** shows the fractional THz power in the air cladding which is increased with wavelength and the guiding power at 3 mm-wavelength almost entirely expands outside the fiber core. Obviously, the power percentage of a waveguide mode in the air cladding exceeds 60% when the wavelength is longer than 1 mm. The large power ratio in the air cladding greatly reduces the THz-wave propagation loss along the wire. Three THz wavelengths, 0.75, 1.00 and 1.25 mm are observed for the relating across power distributions in radial (inset of **Figure 1(a)**) and their modal sizes based on the full width at half maximum (FWHM) are, respectively, 0.118, 0.438 and >6 mm. The waveguiding power obviously enhances near the core-cladding interface for the long wavelength waves. Such power enhancement is thus highly sensitive to detect analytes surrounding the wire core [3].

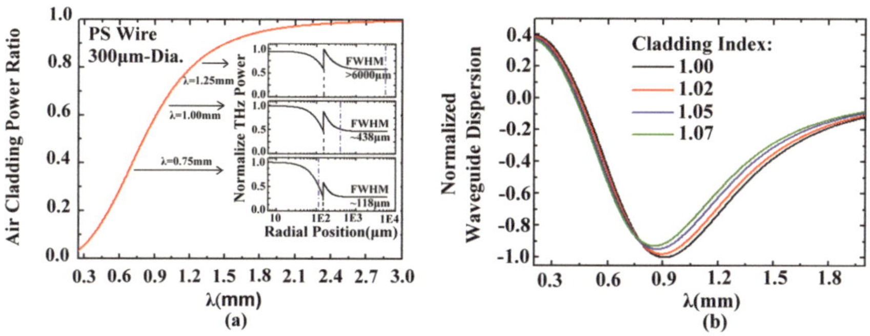

Figure 1. Calculation of the fiber-guiding THz-wave power ratio in the air cladding around a PS wire with a 300 μm core diameter. (Inset) THz-wave power distributes in the cross section of a PS wire, where the black dash line and blue dash dot line, respectively, indicate the wire radius, 150 μm and the HE_{11} modal size range at the full with of half maximum (FWHM) power. (b) Calculation results of PS wire waveguide dispersion, which are normalized with the air cladding index of 1.00 to observe the negative waveguide dispersion variation for cladding index increment up to 1.07 (Reprinted from Ref. [7] © 2009 OSA).

To assess the refractive index-sensing ability of the THz plastic wire fiber, the waveguide dispersion of a PS wire (300 μm core size) under various cladding indices are simulated (**Figure 1(b)**). The simulated results are normalized with respect to the lowest minimum value as "−1"for the negative waveguide dispersion in the waveguide dispersion spectrum. When the cladding refractive index is modified from 1.00 to 1.07, the waveguide dispersion dip becomes less negative and shifts toward a short wavelength [4]. The variation percentage of the waveguide dispersion dip, denoted as ΔD_{WG} is approximately proportional to the cladding index. For the case of a PS wire, ΔD_{WG} varies around 1% when 0.01 index increment at the fiber cladding is performed.

According to the theoretical results in **Figure 1(b)**, it reveals that the variation of waveguide dispersion dip is caused by the evanescent THz waves transmitted along different specimens with various refractive indices, where the air-cladding index is partially replaced by the analytes with higher refractive indices, i.e., >1.00. The plastic wire-sensing scheme is advantageous to guide long-wavelength THz waves within a narrow bandwidth because of the loose confinement and low transmission loss properties [5]. The fiber-sensing scheme, therefore, provides one possibility to use a simple THz spectral system avoiding strict spectrum criteria in the system, such as the broadband, high power or high frequency radiation. Advanced applications in remote sensing or the integration with lab-on-a chip probably becomes workable in THz frequency region by implementing this concept into practice.

2.2. Liquid-sensing results

Figure 2 schematically shows the optical configuration of a THz wire fiber-sensing scheme. THz radiation is emitted and detected by a pair of photoconductive switches, excited by a mode-locked pulse laser with a 100 fs pulse width, 80 MHz repetition rate and central wavelength of 800 nm. Two off-axis parabolic mirrors with effective focal length of 75 mm couple THz radiation from free space into the fiber. A 15-cm-long PS wire with a 300-μm-core diameter is applied as a THz subwavelength fiber. After the 15-cm-long fiber propagation, the output THz waves are collimated by a lens and a pair of parabolic mirrors to couple the guided THz waves into a THz detector. The signal process of the measured THz waveform from the plastic wire waveguide follows the operation principle of THz time-domain spectroscopy, involving the amplitude and phase of an electric field oscillation [6].

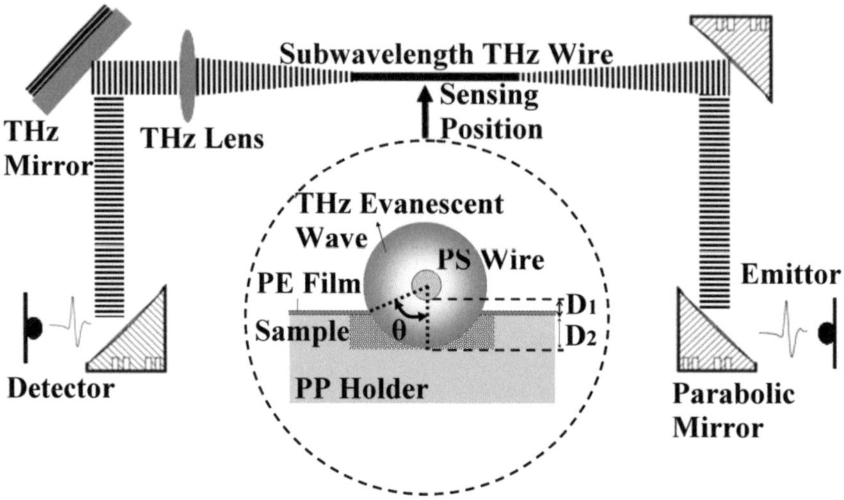

Figure 2. Optical configuration of a THz fiber system. (Inset) Geometric arrangement between the PS-wire-guided evanescent wave and the liquid sample holder (Reprinted from Ref. [7] © 2009 OSA).

We conduct the liquid-sensing experiment by integrating a sample holder, made of polypropylene (PP), beneath the PS wire for loading the liquid analyte. The PP holder contains two liquid channels with different lengths of 1 and 3 mm and the same width (6 mm) and depth (0.5 mm). A 10-μm-thick polyethylene (PE) film is covered on the top of liquid channels to prevent liquid evaporation under the ambient temperature and pressure. The waveguide dispersion can be derived from the phase difference of the transmitted THz waveform acquired from PS wire integrated with PP holder with and without the analytes loading.

The phase difference for THz wave transmitted along the wire with and without liquid samples is described as follows,

$$\Delta\phi \equiv \left(\phi_{sL1} - \phi_{airL1}\right) - \left(\phi_{sL2} - \phi_{airL2}\right) \tag{1}$$

where ϕ_{sL1} and ϕ_{airL1} represent the phases of THz waves pass through a L_1-long PP channel with and without a liquid analyte loading, respectively. Similarly, ϕ_{sL2} and ϕ_{airL2} refer to the phases of THz waves propagating through L_2-long PP channel with and without liquid analyte, respectively. The phase difference according to Eq. (1) is only contributed from THz wave along the $|L_1 - L_2|$ -long PS wire with the liquid analyte in the cladding region and the 2 mm-length (i.e., difference between 1- and 3-mm-long liquid channels) is specifically designed to sense liquids with a sufficient signal-to-noise ratio to get the reliable phase difference. For other analytes with lower THz refractive index comparing to the liquid analyte, increasing the length difference of the two liquid channels is suggested. Based on the definition of a propagation constant, $\beta = |L_1 - L_2| \cdot \Delta\phi = 2\pi n_{eff}/\lambda$, the effective refractive index (n_{eff}) of the THz wavelength (λ) can be derived by substituting the measured $\Delta\phi$ in Eq. (2).

$$n_{eff} = \frac{\lambda \cdot \Delta\phi}{2\pi\left(L_2 - L_1\right)} + 1 \quad (2)$$

The measured waveguide dispersion (D_{WG}) along a PS wire with the sample cladding is then obtained from Eq. (3), relating to the group velocity (V_g) as formulated in Eq. (4). For different liquid samples loaded in the sensing scheme of **Figure 2**, the performed propagation constants (β) along the PS wire are certainly distinct to shift the feature of the negative waveguide dispersion curves, approximate to the calculation results of **Figure 1(b)**.

$$D_{wg} = \frac{d\left(V_g^{-1}\right)}{d\lambda} \quad (3)$$

$$V_g = -\frac{2\pi C}{\lambda^2} \cdot \frac{1}{d\beta / d\lambda} \quad (4)$$

On the basis of the geometric parameters of the wire radius (r), channel depth (D_2) and the separation between the wire and the liquid surface (D_1) as illustrated in the inset of **Figure 2**, the percentages of the waveguide mode spreading in the liquid and air space can approximately be estimated. Based on the spreading ratio of waveguide mode and effective medium concept, the theoretical effective refractive indices of waveguide mode relating the liquid samples can be estimated and the theoretical waveguide dispersion can consequently be calculated based on Eqs. (3) and (4). In experiment, the pure water and ethanol are employed as standard analytes to verify the fiber-sensing modality. The variation of waveguide dispersion at the spectral dip is successfully observed while the water and ethanol individually interact with the evanescent waves of the PS wire. The measured variation percentage of the waveguide dispersion dip well agrees to the theoretical calculation [7]. The result further reveals only the variation in the negative waveguide dispersion without the significant spectral shift as shown in **Figure 1(b)**.

Figure 3 shows the sensing result to recognize minute polymer powder grains dissolved in the ethanol, where the polymer powder grains are PE (434,272 ultra-high molecular weight, surface-modified polyethylene powder, Sigma–Aldrich Inc.) and melamine (Melamine, Nippon Bacterial Test Co., Ltd.) materials. A small amount of the both grains are mixed in ethanol solutions separately with various concentrations, 20–100 ppm (part per million, i.e., one milligram powder per one liter of ethanol in this case). The pure PE grains usually cannot be dissolved in ethanol; however, the used PE grains are modified as polar surfaces for a higher solubility in ethanol. In the ethanol solution, melamine powders are slightly solute and its solubility is larger than that of PE grains due to the polar molecular structure [8].

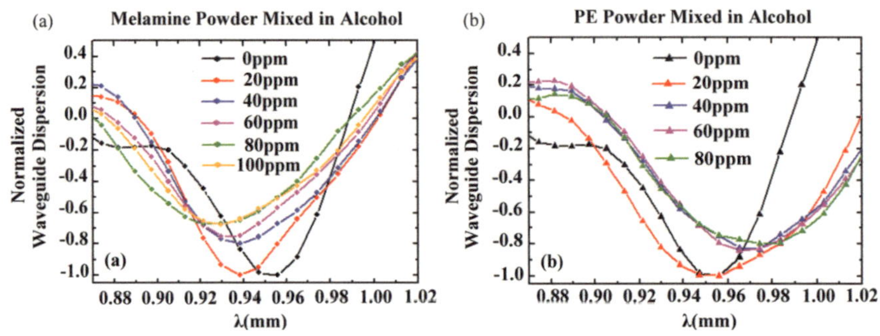

Figure 3. PS wire waveguide dispersions for liquid samples in the cladding, which are pure ethanol mixed with different concentrations of (a) melamine and (b) PE powders. It is normalized by the pure ethanol condition at the negative dispersion dip to observe their dispersion variation performance (Reprinted from Ref. [7] © 2009 OSA).

As illustrated in **Figure 3(a)**, apparent variation at waveguide dispersion dips is indeed observed while increasing the melamine concentration from 20 to 80 ppm in ethanol solutions. However, the variation of waveguide dispersion dips in the PE-ethanol solution occurred only when the PE concentration is altered from 20 to 40 ppm as shown in **Figure 3(b)**. Because the effective-cladding index of PS wire is changed with the solute concentration of analyte, the waveguide dispersion of THz waveguide mode is also modified and agrees with the theoretical prediction as shown in **Figure 1(b)**. Comparison between **Figure 3(a)** and **Figure 3(b)**, the waveguide dispersion dip variations (ΔD_{WG}) at the powder concentration of 40 ppm are 21 and 18%, respectively, for melamine- and PE-ethanol solutions. It implies more melamine grains can be dissolved in an ethanol solution, comparing to the PE grains. Such different dissolving abilities is reasonable because the polarity of melamine is higher than that of PE, which is easier to be dissolved by a polar liquid, ethanol.

3. Plastic ribbon fiber sensor

3.1. Subwavelength confinement using a metal grating integrated THz ribbon fiber

A subwavelength plastic ribbon has been demonstrated as a low-loss THz waveguide to deliver loosely confined THz waves with bandwidth of about 1 THz [9] for a long distance. Similar to the aforementioned plastic wire, the waveguide mode power of the ribbon fiber also mostly spreads in the air cladding. However, the ribbon fiber core is nonsymmetrical different from the circular core of a plastic wire and able to guide a broader bandwidth than that of a plastic wire. To excite the waveguide mode of the subwavelength ribbon fiber, the polarization direction of the input THz waves should be parallel to the ribbon thickness, i.e., perpendicular to the thin-film surface.

Although the evanescent wave of a subwavelength fiber is highly sensitive to the refractive index variation in the air cladding, the much extensive modal field is not suitable to sense micro-particles or over layers in subwavelength scales. Subwavelength optics provides the schemes to visualize or sense those subwavelength-scaled analytes in the near-field region and the plasmonic waveguide [10] is one of the notable near-field sensors. To confine the evanescent waves along a THz ribbon fiber, the feasibility to increase the waveguide refractive index via integrating the ribbon fiber with a period metal structure is presented in this section. **Figure 4(a)** schematically depicts the hybrid THz plasmonic waveguide, constructed by one plastic ribbon and a 1D periodic metal grating. The diffraction metal grating has a length of 50 mm, a width of 15 mm and a thickness of *h*, machined on a brass sheet with a lattice constant of 1.5 mm and a slit width of 1 mm. It is attached on one side near the output end of the ribbon fiber. The ribbon fiber is made of a 20-μm-thick, 15-mm-wide and 22-cm-long PE plastic thin film.

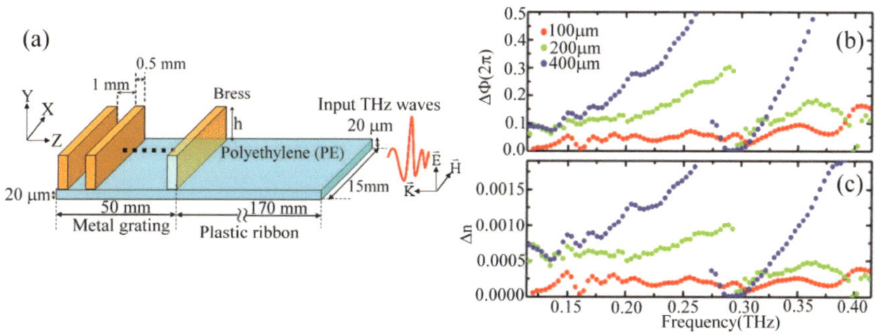

Figure 4. (a) Configuration drawing of the planar hybrid plasmonic waveguide. (b) Phase-variation spectrum contributed by different metal grating thicknesses. (c) Variation spectra about the effective waveguide refractive indices in different metal grating thicknesses (Reprinted from Ref. [13] © 2013 OSA).

Using a one-dimensional metal grating to generate and deliver THz spoof plasmons has been presented for the waveguide loss and dispersion properties [11]. The slit width and depth of a diffraction grating have a significant impact on the field confinement and propagation distance of the transverse magnetic (TM) polarized THz waves [12]. The ribbon fiber delivers TM waveguide mode, matching the polarization criterion of the THz spoof plasmons. As presented in **Figure 4(a)**, TM-polarized THz waves are edge-coupled to one end of the ribbon fiber from the free space, forming the weakly confined waveguide mode to propagate through a 170-mm-long fiber length. The stabilized waveguide mode would then directly enter a 50-mm-long integrated metal grating waveguide. When the loosely confined THz waves illuminate on the metal grating, partial transmission along grating periods occurs due to the mismatch in the cross-modal field and effective waveguide index. With the same slit width of 1 mm and grating period of 1.5 mm, we discuss the impact of various metal grating thicknesses, including 100, 200 and 400 μm, on the modal confinement of the propagated TM-wave.

The propagation vectors of the reflected and transmitted THz waves on the diffraction grating follows the momentum conservation relation as Eq. (5),

$$\vec{K}_{in} + \vec{K}_{\Lambda} = \vec{K}_R \tag{5}$$

where the vectors of K_{in}, K_R and K_Λ are, respectively, the propagation constants of the input and reflected THz waves along the ribbon fiber as well as a grating wave vector. The grating wave vector, K_Λ, equals $2\pi m/\Lambda$ where m and Λ are individually the Bragg diffraction order and the grating period. The directions of propagation constants, K_{in} and K_R, are opposite but have the same magnitude of $2\pi v n_{eff}/C$ for THz waves propagated along the integrated ribbon fiber, where v, C and n_{eff} are, respectively, the THz wave frequency, speed of light in a vacuum and an effective waveguide refractive index. The diffraction grating works to reflect THz waves exactly at the Bragg frequencies in different orders, which can be derived as $mC/2n_{eff}\Lambda$ from Eq. (5). In the case of 1.5-mm-period metal grating, there are two narrow low-transmission bands found in experiment, caused by the Bragg reflections, respectively, at near 0.300 and 0.400 THz. Such spectral dip exactly at Bragg frequency is the result of phase-matching between the ribbon- and grating-waveguide modes. In experiment, the low order Bragg reflection dips, near the 0.100 and 0.200 THz, cannot be found because the large cross section of input THz wave leads to the low efficiency of coupling with the grating [13].

Figure 4(b) shows the measured results of the phase retardation ($\Delta\Phi$) of THz waves after the 50-mm-long propagation of the metal gratings with different thicknesses. The retarded THz wave phase is only considered from the metallic grating contribution because the PE ribbon fiber is normalized with the defined phase difference, $\Delta\Phi = \Phi_G - \Phi_{Blank}$, where Φ_G and Φ_{Blank} are, respectively, THz-wave phases of the electric-field oscillation passing the 220-mm-long ribbon fiber with and without the 50-mm-long metal grating. Equation (6) expresses refractive-index variations (Δn) relating to the phase difference ($\Delta\Phi$).

$$\Delta \mathrm{n} = {}^{C\Delta\phi}\!\big/\!_{(2\pi \upsilon L)} \tag{6}$$

The light speed in vacuum, THz wave frequency and the length of a metal grating are, respectively, denoted as C, v and L in Eq. (6). The refractive index increment resulted from the metal gratings with different thicknesses is therefore illustrated in **Figure 4(c)** based on **Figure 4(b)** and Eq. (6). The results show 100-μm-thick grating slightly performs the phase retardation of the guided THz waves, which is much smaller than those of the 200- and 400-μm-thick gratings. When the grating thickness increases to 200-μm thickness, zero phase difference occurs at 0.300 and 0.400 THz, representing the phase-matching condition, i.e., $\Phi_{Blank} = \Phi_G$, between the ribbon guided THz waveguide mode (K_{in}) and the reflected surface plasmonic (SP) modes of the hybrid waveguide (K_R) via the metal grating (K_Λ) structure. For those ribbon-guided THz waves with the zero phase difference, their power can be completely transferred to the spoof surface plasmon polaritions (SPP)modes and reflected through the Bragg reflection, corresponding to the zero-refractive index variation for index matching between the two modes (**Figure 4(c)**).

The effective waveguide modal indices along the metal grating are obviously larger than those of a blank ribbon waveguide and proportional with the frequency increment within the deliverable frequency range. The waveguide theory expresses the high waveguide refractive index is able to confine the extended EM fields close to the waveguide core as a small spatial mode [14]. The metal grating delivering THz waves approximate to the modal field of a dielectric waveguide. It is found the 400-μm-thick grating performs the largest phase difference and effective waveguide index within 0.300–0.360 THz and <0.260 THz because of the largest across section to interact input THz waves. More closely confined THz SPPs can therefore be guided along the 400-μm-thick grating, instead of the 200- and 100-μm-thick gratings.

Figure 5(a) shows the measured and simulated decay lengths on one side of the bare ribbon and hybrid grating waveguides with different grating thicknesses. The waveguide decay lengths with and without metal grating decrease as the frequency increment of THz waves. The minimum decay length of the blank ribbon is around 0.56 mm at frequencies beyond 0.300 THz. However, the decay length is apparently extended for THz frequency below 0.350 THz while the ribbon-guided waves input the 100-μm-thick metal grating. Contrarily, the decay lengths shrink above the 200- and 400-μm-thick metal gratings for THz frequencies, respectively, approaching the 0.300 and 0.250 THz. The trend about these decay lengths can also be obtained in FDTD calculation.

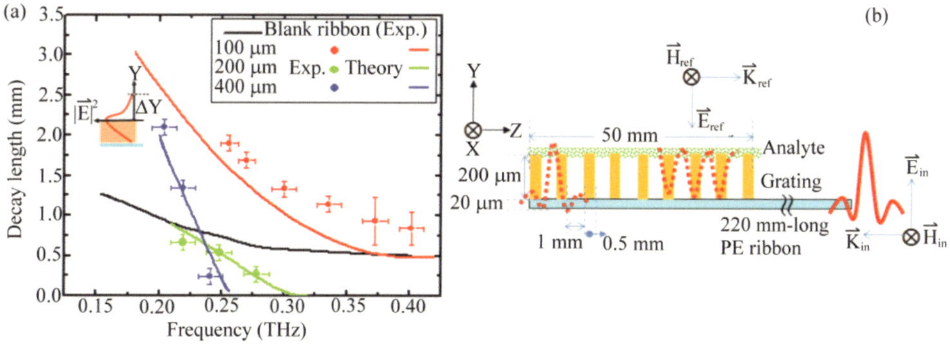

Figure 5. (a) Measured and calculated decay lengths of the guiding THz waves along a blank ribbon and different integrated gratings with 100, 200 and 400 μm thicknesses (Reprinted from Ref. [13] © 2013 OSA). (b) Sensing configuration of a plastic ribbon fiber integrated with a metal grating (Reprinted from Opt. Express 21, 21087–21096 (2013). © 2013 OSA).

3.2. Particle sensing using a hybrid THz waveguide

In a traditional THz spectroscopy, to recognize solid particles with size much smaller than THz wavelength, the required analyte amount should be sufficient to form tablets and the thickness should be precise in order to obtain the correct THz information. For minute material detection, a versatile and flexible THz-sensing platform is necessary. This section introduces the particle-sensing methodology using the metal-grating-integrated hybrid THz waveguide and assesses its detection ability. The sensing scheme is schematically plotted in **Figure 5(b)**, where the particle analyte is located on a dielectric substrate which is attached on the metal grating of the hybrid waveguide. The particle loaded membrane substrate is entirely covered above the metal grating, becoming one part of the hybrid plasmonic waveguide and to facilitate passing through the subwavelength confined THz surface plasmonic waves (SPWs). The evanescent fields of the THz-SPWs interact with the particle analyte in a sufficiently long distance, leading to the enhanced THz response in spectrum. Due to the multiple reflections from the metal grating at Bragg frequencies, the optical paths of THz waves in the hybrid waveguide are obviously longer than those on the bare ribbon. The long optical path makes the guided THz-SPWs quite sensitive to any slight variation of dielectric analytes located within the evanescent field range of waveguide mode. The propagation constants of the resonant waves are equal to $2\pi n_{eff}\nu/C$, where n_{eff}, ν and C denote the effective refractive index, resonant frequency of the THz-SPWs and light speed in vacuum, respectively. For a specific grating period Λ and an incident angle, the resonant frequency of a THz-SPW is inversely proportional to the effective refractive index of the hybrid plasmonic waveguide (n_{eff}).The change of the effective refractive index related to surrounded analyte would thereby change the resonant dip in THz transmission spectrum, providing a straightforward spectral feature for the sensing application.

To conceptually demonstrate the minute particle-sensing capability using the hybrid waveguide sensor, a chemical compound of Bi_2CuISe_3 with irregular grain shapes and nonuniform particle sizes, <300 μm, is considered as a sample under test in the experiment. Different amounts of compound analyte are prepared to produce distinct effective indices of the THz-SPW propagated on the hybrid waveguide for sensitivity test. A 60-μm thick, 50-mm-long and 15-mm-wide PE film was used as a sample substrate and attached to the grating surface to adhere the analyte (Bi_2CuISe_3) in different quantities (e.g., 16.5, 29.3 and 55.3 μmole). The analytes with different amounts are individually loaded on the PE substrate to be sensed by the SPWs and the fourth-order resonance dip can be found performing evident spectral shifts (**Figure 6(a)**). For the blank PE-film substrate and sample loading amounts of 16.5, 29.3 and 55.3 μmole, the transmission dips are, respectively, at 0.276, 0.266, 0.259 and 0.237 THz. Obvious spectral shift can be found because of the granular particles contributing the high material dispersion in THz frequency as illustrated in the inset of **Figure 6(a)**.

Figure 6(b) summarizes the resonant dip-shifts of THz-SPWs that relate to waveguide-index variations obtained from different thicknesses of blank PE substrates and different quantities of chemical compounds on a 60 μm-thick substrate. For different thicknesses of the blank PE substrates, their spectral responses are resulted from the distinct waveguide index variation, ranging from 0.2 to 0.467, because of their different substrate thicknesses. A fitting curve (black line in **Figure 6(b)**) is presented as a quadratic function. The detection sensitivity of the hybrid plasmonic waveguide can be considered as the tangential slope of each data point, defined as the frequency-shift range per refractive index unit, i.e., $\Delta\nu/\Delta n$. Using different PE substrate thicknesses in the sensing scheme, the estimated sensitivities based on 20, 50 and 90 μm thicknesses are, respectively, 261, 197 and 159 GHz/RIU. The particle-sensing ability using a 60 μm thick PE film as substrate shows the spectral shift is almost linearly proportional to the waveguide index variation (Δn) within Δn range of 0.395–0.626, as indicated by the red fitting line in **Figure 6(b)**. Based on the slope of the linear fitting curve (red line), the sensitivity is estimated as 163 GHz/RIU which is less than the sensitivity of the 60 μm thick thin film (~187 GHz/RIU acquired from the slope of the quadratic fitting curve in **Figure 6(b)**). Therefore, a uniform overlayer can be more easily detected than a powder analyte based on the waveguide-sensing scheme even though the granular material has considerably high THz-wave dispersion.

The sensing performance of the hybrid THz plasmonic waveguide dependent on the polynomial fits in **Figure 6(b)** is then discussed on the sample substrate of a 20 μm thick PE film, whose sensitivity is about 261 GHz/RIU. Its thin-film detection ability approaches to 1.8 μm when the thin film index is 1.50 and the spectral resolution is down to 0.004 THz. The optical path difference (OPD) is approximately 2.7 μm and the thickness resolution can approach $\lambda/289$ of the 4th-order resonance wave at 0.380 THz. For sensing the powder on the substrate of a 60 μm thick PE film (163 GHz/RIU), the minimal detectable molecular quantity is approximately 13 μmole on the 15 mm-wide and 50 mm-long substrate area, corresponding to the molecular density of 17.3 nano-mole/mm^2. When the spectral resolution is further decreased to 1 GHz, 454 nm-thick thin film (681 nm-OPD) could be resolved in the resonance spectral shift. Therefore, the hybrid waveguide-sensing scheme is certainly compatible with the planar terahertz metamaterial [15] to act as a sensitive thin film sensor in THz frequency for nano-thin film detection. Metal hole array and planar terahertz metamaterials are difficult to sense nonuniform and random distributed particles, even though the highest sensitivity of metal hole array approaches 2 THz/RIU [16]. In the hybrid waveguide-sensing scheme, the particles layer can be considered as an effective waveguide cladding of the hybrid plasmonic waveguide sensor to induce the waveguide index change from various particle quantities.

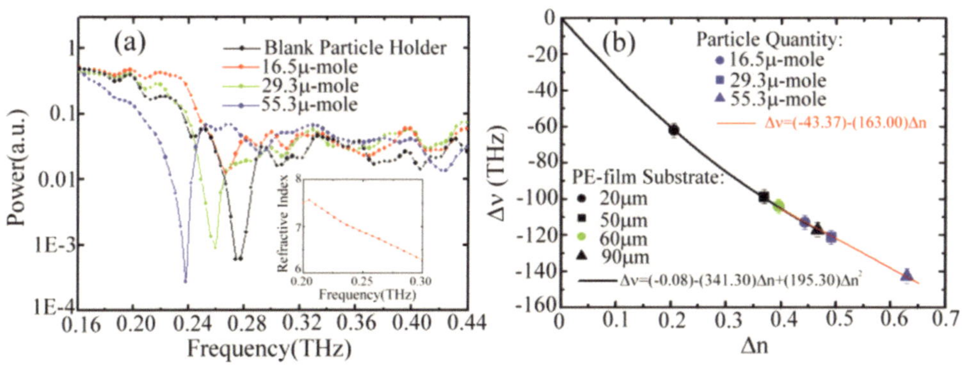

Figure 6. (a) Waveguide transmission spectra for different amounts of powder analytes loaded on a PE film substrate; (inset) THz refractive index spectrum of a tablet-formed compound Bi_2CuISe_3; (b) relation between effective waveguide refractive index variation (Δn)and the spectral shift range ($\Delta\nu$) of the fourth-order SPW resonance, which is compared with the blank grating condition (Reprinted from Opt. Express 21, 21087–21096 (2013). © 2013 OSA).

4. Dielectric pipe fiber sensor

4.1. Overlayer-sensing principle

A THz pipe fiber is a hollow tube which is composed of a ring-shaped dielectric cladding and an air-core; it has been demonstrated as a broadband and low-loss THz waveguide [17]. The dielectric ring cladding, inherently served as a Fabry-Pérot resonator, can transmit THz waves in the hollow core based on anti-resonant reflecting optical waveguide principle (ARROW) [18] to form multiple transmission peaks in a transmission spectrum, called as antiresonant modes. The resonant modes of the ring-shaped Fabry-Pérot resonator leak out of the ring cladding and form dips in a transmission spectrum [19]. The spectral dip position of a resonant mode is closely related to the thickness and refractive index of the ring-shaped cladding. A thin film attaching to the ring waveguide cladding is able to change the resonant condition of the THz pipe fiber and results in resonant-frequency-dip shift [20]. Sensing the tiny variation in refractive index or thickness of an adsorbed molecular film near the ring cladding is therefore applicable.

A double-layered Fabry-Pérot model is proposed to calculate the resonant mode of the THz pipe fiber with a composite cladding [21], where the effective thickness of the sample or the cladding is defined as Eq. (7),

$$\tau_{cld\ or\ fim} = d_{cld\ or\ fim}\sqrt{n_{cld\ or\ fim}^2 - n_0^2}, \tag{7}$$

where the $d_{cld\ or\ fim}$ and $n_{cld\ or\ fim}$ are the physical thickness and the refractive index of pipe wall or thin-film analyte, respectively. A calibration experiment is conducted by attaching different thicknesses of PE films on the ring-shaped pipe wall to estimate the sensitivity for thin-film detection. It is found that the sensitivity of thin-film detection is proportional to THz resonance wavelength, where long THz resonance wavelengths are equivalent to the low-order resonant modes of the pipe fiber based on the fundamental of ARROW waveguide. For thin-film-sensing application by the THz-pipe fiber, the low order resonant mode with long wavelength is thus able to perform high sensitivities for those analyte thicknesses down to subwavelength scales. The calibration results also show the pipe with a thin pipe wall has a higher sensitivity, comparing to the thick pipe wall made of the same material. We take into account the effects of pipe-wall thickness, index and THz resonance wavelength on sensing capability and the thin-film detection sensitivity (S_λ) can be deduced in Eq. (8), where τ_{cld} and τ_{fim} are defined in Eq. (7).

$$S_\lambda = \frac{\lambda_m}{\tau_{cld} + \tau_{fim}}. \tag{8}$$

From Eq. (8), approaching highly sensitive THz pipe-fiber sensor can be described as following two conditions. One way is that the effective thickness of a pipe cladding (τ_{cld}) should be small when the dip shift is observed at a specific resonant wavelength (λ_m) for a certain thickness of the thin-film analyte (τ_{fim}). The other way is the low-order resonant modes, i.e., long resonance wavelengths, λ_m, are suggested to be employed under the geometry of a pipe sensor for sensing analyte. More evanescent power is surrounded the ring cladding for the lower-order resonant modes to perform the higher sensitivity, comparing to the shorter resonant wavelengths. For a low index or a small physical thickness of pipe-wall, a resonant THz wave will be loosely confined in the waveguide cladding and results in more THz evanescent power surrounding the pipe-wall for efficient interaction with the analyte thin films near the waveguide cladding.

Based on the result of the calibration experiment, a PP tube with the highest sensitivity is used to probe the macro-molecular overlayer adhered on the pipe inner-wall. The PP pipe fiber used to sense the macro-molecular layer has a length of 15 cm, a cladding (i.e., pipe wall) thickness of 0.29 mm and an inner-core diameter of 12 mm. Different thicknesses of the macro-molecular overlayers are prepared by dissolving different amounts of carboxypolymethylene powders, named as carbopol 940 (Boai Nky Pharmaceuticals Ltd.), in water with different weight-concentrations, including 1, 2, 3 and 4%. The Carbopol 940 powder belongs to macromolecular material with molecular weight about 104,000 g per mole, corresponding to 1450 monomer units [22]. The carbopol liquids have high viscosity, adhering to the inner pipe wall and forming macro-molecular layers via gravity-driven flowing and natural evaporation.

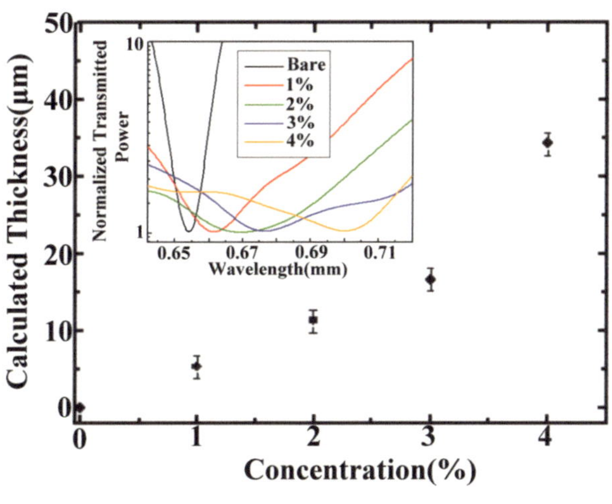

Figure 7. Estimated thicknesses of carbopol micro-molecular layers for different concentrations of the carbopol solutions. (Inset) The transmission spectra of the first resonant mode along a 0.29-mm-thick pipe-wall PP tube, loaded with various thicknesses of the macro-molecular layers (Reprinted from Ref. [21] © 2010 OSA).

The waveguide spectral features in **Figure 7** show different macro-molecular layers with different physical thicknesses can be detected from the spectral shift of resonance waveguide modes. The 1st resonance mode wavelength is at 0.654 mm and shifted toward the long wavelength range when increasing the concentrations of carbopol aqueous solutions. The spectral shift of the resonant dip is owing to the different thickness of macro-molecular layer adhered on the PP cladding and the distinct thickness of molecular over layer is formed by different viscosities of carbopol solutions [18]. The wavelength spectral positions (λ_m) of the resonant dips are 0.661, 0.669, 0.676 and 0.700 mm, respectively, for 1, 2, 3 and 4%- carbopol aqueous solutions. The thicknesses of the carbopol molecular layer can thus be estimated through the double-layered Fabry-Pérot models illustrated in **Figure 7**, where the THz refractive index of the carbopol-layer is around 1.2. In Eq. (9), the resonance wavelength, physical thicknesses and refractive indices of PP-pipe wall, macromolecular layer and the inner core refractive index are, respectively, denoted as λ_m, d_1, n_1, d_2, n_2 and n_0.

$$\lambda_m = \frac{2\left(d_1\sqrt{n_1^2 - n_0^2} + d_2\sqrt{n_2^2 - n_0^2} \right)}{m} \tag{9}$$

The estimation shows the thicknesses of carbopol-molecular layers are around 5.3, 11.3, 16.7 and 34.3 μm, respectively, for the 1-, 2-, 3- and 4%- solutions. The minimum detectable thin film thickness of the PP-tube-sensing scheme can therefore be estimated with the sensitivity of 0.0030 mm/μm and the system spectral resolution, ~0.0057 mm. Taking a carbopol-layer as an example, the detectable thickness increment can be down to 2.9 μm (~λ/225), corresponding to the resolution of carbopol molecule quantities as low as 1.22 picomole/mm^2. However, molecular resolution limited in the spectral dip shift would be resulted from the broadened bandwidth at the resonance wavelength, which is observed in the inset of **Figure 7**. The interference theory explains the same intensity of interfered waves performs high visibility [23], but the case of carbopol-layer has obvious THz waves absorption to decay the interference visibility with broadening resonance bandwidth.

4.2. Vapor-sensing principle

The THz pipe fiber can provide a sharp spectral line-width of resonant mode with a large portion of evanescent field spreads outside the pipe cladding, facilitating to integrate with various biochips for highly sensitive molecular sensing. Such evanescent wave characteristic is difficult to achieve in the metallic cavity type bio-sensors despite they possess the ultrahigh quality factor (or Q-factor) for sensing applications [20]. In addition, the pipe fibers are easy to commercially acquired without any complex fabrication. Based on above advantages, the dielectric-pipe-fiber-based refractive index sensor could potentially be applied in micro-fluidic systems for bio-chemical detection and inspecting industrial pollutants. Many industrial and environmental pollutants are toxic gases and harmful to animals. This section further introduces the capability to identify various vapors using the resonant modes of pipe fiber. To demonstrate vapor sensing, a pipe made of glass material is taken as the sensing fiber to prevent the volatile analytes from eroding/damaging the pipe-wall. In the vapor-sensing experiment, a small amount of different volatile liquid droplets is individually injected into the glass pipe and completely evaporate into vaporized molecules in the sealed hollow core under normal ambient condition. The transmitted THz waves and the vaporized analytes are simultaneously confined in the same channel to provide a sufficient interaction length and result in sensitive detection of minute vapors.

According to Eq. (10), the resonant dip wavelength (λ_m) changes with inner core refractive index (n_{cor}), and various vapors in the hollow core can thus be identified by detecting the spectral shift of a certain resonant dip (m).

$$\lambda_m = \frac{2d\sqrt{n_{cld}^2 - n_{cor}^2}}{m}, \; m = 1, 2, 3\ldots \tag{10}$$

To realize a highly sensitive detection of vaporized analyte in the hollow core, the resonant waves in the pipe wall must be sufficiently evanescent toward the hollow core because the sufficient interaction is required. Based on the transverse power distributions of resonant waves in a pipe waveguide [24], the higher THz transmission power in the hollow core is observed in the higher-order resonant-dip, i.e., the dips in the high frequency range. **Figure 8(a)** shows the vapor-sensing example based on a 30 cm-long glass pipe, possessing the inner core diameter and the pipe-wall thickness are, respectively, 5.57 and 1.17 mm. The vapor inducing dip-frequency-shift only occurs in a resonance dip of 0.452 THz and other resonance dips in the low-frequency range do not exhibit any spectral shift. It does not follow the simulated results in **Figure 8(b)** because the leaky power of the low-order resonance-wave in the hollow core is insufficient to sense the vapor molecules. It also practically presents the evanescent field strength of a pipe resonant modes within the inner core is critical for the vapor sensing.

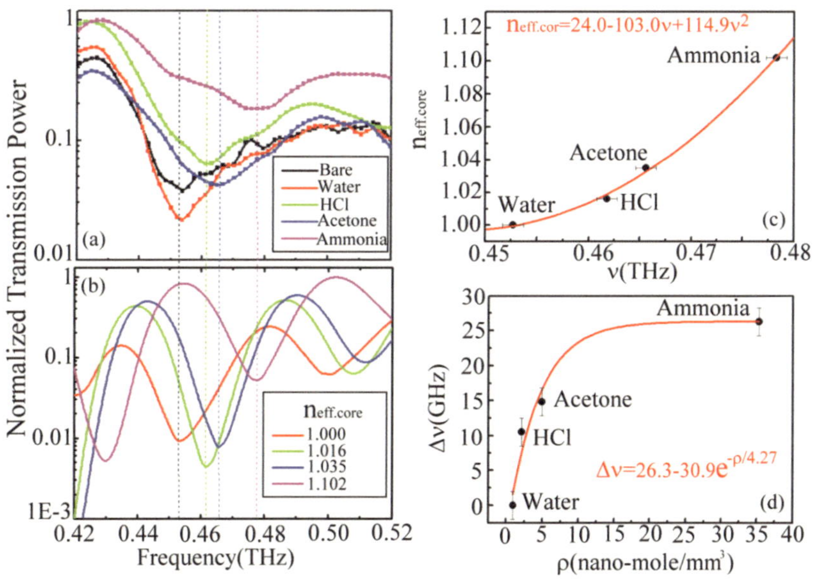

Figure 8. (a) Spectral dip shift of the glass-pipe resonance wave at 0.452 THz for sensing different volatile liquids. (b) Theoretical estimation of the inner core refractive indices to match the related spectral dip positions for the vapor-sensing results. (c) The spectral relation of the inner core refractive indices for different vapors. (d) Relation between the spectral dip shift and the related molecular densities (Reprinted from Ref. [24] © 2012 OSA).

The vapors are prepared by dropping 0.05 cm^3 liquids individually of water, hydrochloric acid (HCl), acetone and ammonia inside the hollow core of glass pipe with vapor pressure saturation. The sensing result shows the spectral dip at 0.452 THz can be shifted with different levels (**Figure 8(a)**), locating at 0.461, 0.465 and 0.477 THz, respectively, for the hydrochloric acid, acetone and ammonia vapors. Trying to match the spectral dip position, the effective inner core refractive indices can be calculated by FDTD method (**Figure 8(b)**), which are 1.016, 1.035 and 1.102, respectively, for the vapors of hydrochloric acid, acetone and ammonia. The relation is then summarized in **Figure 8(c)** between the spectral dip frequencies and the effective core-indices of the glass-pipe fiber, which is filled with different vapors. Those inner core refractive indices are distinct due to the different vapors even though the same liquid volume is inserted inside the pipe hollow core. The relation between the effective inner core refractive index and resonance frequency in **Figure 8(c)** can be fit with one polynomial to evaluate any slight index variation inside the pipe hollow core. For the example of water sample, the relating spectral dip is consistent with and without inserting the vapor because of the considerably low density, comparing to those volatile liquids. It means the effective inner core index of water case approximates to 1.0.

Different volatile liquids have individual vapor pressures at normal atmospheric pressure and temperature. There are different quantities of vaporized molecules in the same hollow core volume. Vapor pressures of volatile liquids are proportional to the quantities of vaporized molecules in the same chamber. High vapor molecule density in the hollow core eventually results in the higher effective core-index than the air core, performing apparent blue shift at the resonant spectral dip at 0.452 THz. When we consider the vapor pressures in literatures for water, hydrochloric acid, acetone and ammonia at 1 atmosphere and 20°C, their saturated vapor pressures are, respectively, around 17 mm-Hg [25], 38 mm-Hg [26], 202 mm-Hg [26] and 308 mm-Hg [25]. Thereby, the effective core-index is proportional to the increase of the sample vapor pressure. When the vaporized molecular density is estimated on the basis of the ideal gases, the corresponding densities are illustrated in **Figure 8(d)**. The relation is fit with one nonlinear curve, representing the blue shift level would cease approaching the value of 10 nano-mole/mm^3. The minimum detectable molecular quantity would be approximately 7.8 micromole in the glass pipe, corresponding to a molecular density of 1.6 nano-mole/mm^3 when the system spectral resolution is about 4 GHz.

5. Conclusion

In this chapter, THz fiber sensors are successfully demonstrated in a PS-plastic wire, a PE-plastic ribbon integrated with a metal grating, a PP- and glass pipes. The presented fiber sensors are highly sensitive to the existence and variation of minute molecules and enable sensing analytes with different formations, including the liquids, powder grains, vapor gases, thin films and overlayers. A subwavelength plastic wire fiber has large percentages of evanescent field surrounding the fiber cores and there is negative waveguide dispersion sensitive to the refractive index variation in the air cladding. Fluidic channel is successfully integrated to detect minute concentration of melamine grains dissolved in the ethanol solutions ranged from 20 to 80 ppm. A plastic ribbon fiber also has the extended evanescent waveguide modes, but it is further advantageous to integrate a metal grating becoming a hybrid plasmonic waveguide based on the optical properties of a THz ribbon fiber, including the TM polarized wave guidance and the broadband transmission. When the optimal overlapping between the metal grating and ribbon waveguide modes, subwavelength confined surface waves can be excited and used for sensing powder grains or thin-film analytes. When the spectral resolution of THz system is reduced down to 1 GHz, the nano-scaled OPD detection could be realized by the plastic ribbon fiber sensor. The dielectric pipe fiber with a hollow air core is also considered as one important THz fiber sensor because of the built-in resonator along the pipe wall. To achieve highly sensitive detection, the applicable resonant modes for thin-film and vapor detections are absolutely different because of the different cross locations of the target analytes. Using the pipe fiber to adsorb liquids with high viscosities as one thin film layer on the inner pipe wall, the low frequency resonance modes or small pipe wall thicknesses are required to achieve the best sensitivity due to the much extended evanescent resonance field. For sensing the inserted vapors at the pipe hollow core, the sensitive THz waves contribute at the high order resonant modes because their resonant field well confined at the inner core efficiently interacting the vapors. In the sensing experiments, the detectable overlayer thickness and vaporized molecular density via THz pipe fiber sensors can be reduced to 2.9μm (~λ/225)and 1.6 nano-mole/mm^3, respectively.

Author details

Borwen You[1] and Ja-Yu Lu[2*]

*Address all correspondence to: jayu@mail.ncku.edu.tw

1 Division of Applied Physics, Faculty of Pure and Applied Sciences, University of Tsukuba, Tsukuba, Japan

2 Department of Photonics, National Cheng-Kung University, Tainan, Taiwan

References

[1] L. J. Chen, H. W. Chen, T. F. Kao, J. Y. Lu and C. K. Sun. Low-loss subwavelength plastic fiber for terahertz waveguiding. Opt. Lett. 2006;31:308–310.

[2] L. Tong, R. R Gattass, J. B Ashcom, S. He, J. Lou, M. Shen, I. Maxwell and E. Mazur. Subwavelength-diameter silica wires for low-loss optical wave guiding. Nature. 2003;426:816–819.

[3] J. Lou, L. Tong and Z. Ye. Modeling of silica nanowires for optical sensing. Opt. Express. 2005;13:2135–2140.

[4] L. Tong, J. Lou and E. Mazur. Single-mode guiding properties of subwavelength-diameter silica and silicon wire waveguides. Opt. Express. 2004;12:1025–1035.

[5] H. W. Chen, Y. T. Li and C. L. Pan, J. L. Kuo, J. Y. Lu, L. J. Chen and C. K. Sun. Investigation on spectral loss characteristics of subwavelength terahertz fibers. Opt. Lett. 2007;32:1017–1019.

[6] B. Ferguson and X. C. Zhang. Materials for terahertz science and technology. Nat. Mater. 2002;1:26–33.

[7] B. You, T. A. Liu, J. L. Peng, C. L. Pan and J. Y. Lu. A terahertz plastic wire based evanescent field sensor for high sensitivity liquid detection. Opt. Express. 2009;17:20675–20683.

[8] R. E. N. Baozeng, L. I. Chen, Y. U. A. N. Xiaoliang and W. A. N. G. Fu'an. Determination and correlation of melamine solubility. Chin. J. Chem. Eng. 2003;54:1001–1003.

[9] C. Yeh, F. Shimabukuro and P. H. Siegel. Low-loss terahertz ribbon waveguides. Appl. Opt. 2005;44:5937–5946.

[10] W. L. Barnes, A. Dereux and T. W. Ebbesen. Surface plasmon subwavelength optics. Nature. 2003;424:824–830.

[11] L. Shen, X. Chen and T. J. Yang. Terahertz surface plasmon polaritons on periodically corrugated metal surfaces. Opt. Express. 2008;16:3326–3333.

[12] S. A. Maier, S. R. Andrews, L. M. Moreno and F. J. G. Vidal. Terahertz surface plasmon-polariton propagation and focusing on periodically corrugated metal wires. Phys. Rev. Lett. 2006;97:176805.

[13] B. You, J. Y. Lu, W. L. Chang, C. P. Yu, T. A. Liu and J. L. Peng. Subwavelength confined terahertz waves on planar waveguides using metallic gratings. Opt. Express. 2013;21:6009–6019.

[14] C. R. Williams, S. R. Andrews, S. A. Maier, A. I. F. Dominguez, L. M. Moreno and F. J. G. Vidal. Highly confined guiding of terahertz surface plasmon polaritons on structured metal surfaces. Nat. Photonics. 2008;2:175–179.

[15] J. F. O'Hara, R. Singh, I. Brener, E. Smirnova, J. Han, A. J. Taylor and W. Zhang. Thin-film sensing with planar terahertz metamaterials: sensitivity and limitations. Opt. Express. 2008;16:1786–1795.

[16] G. Liu, M. He, Z. Tian, J. Li and J. Liu. Terahertz surface plasmon sensor for distinguishing gasolines. Appl. Opt. 2013;52:5695–5700.

[17] C. H. Lai, Y. C. Hsueh, H. W. Chen, Y. J. Huang, H. C. Chang and C. K. Sun. Low-index terahertz pipe waveguides. Opt. Lett. 2009;34:3457–3459.

[18] N. M. Litchinitser, A. K. Abeeluck, C. Headley and B. J. Eggleton. Antiresonant reflecting photonic crystal optical waveguides. Opt. Lett. 2002;27:1592–1594.

[19] C.-H. Lai, B. You, J.-Y. Lu, T.-A. Liu, J.-L. Peng, C.-K. Sun and H. C. Chang. Modal characteristics of antiresonant reflecting pipe waveguides for terahertz waveguiding. Opt. Express. 2010;18:309–322.

[20] A. M. Zheltikov. Ray-optic analysis of the (bio)sensing ability of ring-cladding hollow waveguides. Appl. Optics. 2008;47:474–479.

[21] B. You, J. Y. Lu, J. H. Liou, C. P. Yu, H. Z. Chen, T. A. Liu and J. L. Peng. Subwavelength film sensing based on terahertz anti-resonant reflecting hollow waveguides. Opt. Express. 2010;18:19353–19360.

[22] J. O. Carnal and M. S. Naser. The use of dilute solution viscometry to characterize the network properties of carbopol microgels. Colloid Polym. Sci. 1992;270:183–193.

[23] N. Kinrot. Analysis of bulk material sensing using a periodically segmented waveguide Mach–Zehnder Interferometer for biosensing. J. Lightwave Technol. 2004;22:2296–2301.

[24] B. You, J. Y. Lu, C. P. Yu, T. A. Liu and J. L. Peng. Terahertz refractive index sensors using dielectric pipe waveguides. Opt. Express. 2012;20:5858–5866.

[25] Chap. 5. In: J. A. Dean, editor. Lange's handbook of chemistry. McGraw-Hill; 1999.

[26] E. W. Washburn, editor. International critical tables of numerical data, physics, chemistry and technology, volume IV. Knovel; 2003.

Chapter 5

Terahertz Detectors (THzDs): Bridging the Gap for Energy Harvesting

Kapil Bhatt, Shilpi Shriwastava, Sandeep Kumar, Sandeep and Chandra Charu Tripathi

Additional information is available at the end of the chapter

http://dx.doi.org/10.5772/66347

Abstract

It is indispensable to integrate electronics with environment for better lives. Huge amount of solar energy, dark energy, and unused microwave energy is untapped till now due to insufficient availability of high frequency THz detectors. The difference between THz wave detection and THz electric field detection must be clear. THz wave detection connects the detection of explosives, drugs, astronomy, metals, and imaging applications, etc. On the other hand, THz electric field detection involves the conversion of electromagnetic (EM) radiations to usable DC power. The optimum choice of detectors for energy harvesting is a highly diverse area. The latter part is concentrated on the non-linear behavior of the incoming radiations and has been highlighted also. In this chapter, metal-insulator-metal (MIM) diode detectors have been explored to become a best choice for high frequency detectors.

Keywords: metal-insulator-metal (MIM) diodes, terahertz detectors (THzD), rectenna, tunneling

1. Introduction

In order to fulfill the world energy demands increasing day by day, it is perquisite to have some alternate for the conventional energy resources and energy harvesting methods. Since last few decades, advancement in the field of the THz rectification reveal another way to develop highly sensitive uncooled THz detector. Undoubtedly, a terahertz measurement requires a highly sensitive detector to obtain distinct spectra. One problem faced by the detection of THz waves is low photon energy of milli-electron volts (meV), making the development of a high-performance terahertz detector a difficult task. By serendipity, accurate modeling and applications of the nanoscale materials and devices have conquered these difficulties.

There is a fine line between the egoistical requirement and legitimate demand of energy around the world. By 2030, the demand for energy is projected to increase by 55%, while requirement of oil is increased from 11.4 to 17.7 billion tons [1]. Between 2005 and 2030, energy consumption is expected to increase by 50%, with the bulk of the demand coming from developing countries. The energy provided by the Sun at the Earth's surface on 1 day is more than enough to fulfill the needs of the whole Earth for more than 25 years, if utilizes efficiently. But due to low conversion efficiency, a prevalent commercial usage of solar energy is hardly seen. A portion of the increasing developments, utilizing the enhancement in nanotechnology may be able to change this situation. During the clear climate conditions, the Earth gets 1000 W/m^2 solar irradiation from the Sun [2]. This enormous energy received from the Sun has opened the ways to look for novel systems to convert unused energy radiated continuously from the Sun into usable electricity. The continuous research results in environmentally friendly and clean energy sources and will replace our conventional sources of energy such as coal, gas, oil, hydro, combustible, petroleum energy, nuclear energy, etc. By-products of these sources lead to greenhouse effects and the solar energy is considered to be the most excellent substitute for it as compared to other energy resources. There is a good scope for renewable/green energy corner as its share is less than 1% as compared to present energy resources.

The requirements of efficient energy utilization could be estimated from the fact that during the 2012 Olympics held in London, organizer had managed new facilities with energy efficient, sustainable, and recyclable designs and they cut down the watts to maintain the games clean, green, and energy efficient. The London games had left a benchmark as the most energy-efficient Olympics to till date. A similar kind of legacy is being followed by Rio Olympics 2016. This scenario has motivated the researchers toward exploring renewable energy and unused energy resources (dark energy, unused heat and radiations).

Rectenna is one of the energy harvesting devices, which is a combination of rectifying high frequency diode and an antenna. Rectenna (Rectifier antenna) is a system that is capable for harvesting the infrared and visible wavelengths. The heart of the rectenna system is a metal-insulator-metal (MIM) device that can convert such an energy collected by nanoantenna into usable electric energy.

2. Why THz region?

A lower electromagnetic (EM) region has been occupied for mobile communication systems, and the higher EM region has been occupied for optical communication. However, the THz region is still untapped and has not been completely in use, and this regime has become of colossal importance in extensive disciplines. This region embraces numerous aspects of space research, medical studies, atmospheric studies, natural science, plasma physics, defense and security, process control, etc. The close correlation of electrons and THz radiation is also reflected in the use of THz spectroscopy, which is a contemporary field for the researcher to attract toward it. THz spectroscopy is publicized to provide insight into the electron mobility in semiconductive and conductive nanostructures (like in MIM nanostructures and tunneling spectroscopy) that are suitable for fast electronics systems, which are used to convert solar power into electrons. MIM diodes are able to work in semiclassical and quantum region (**Figure 1**) further broader this window.

It is also applicable for nonlinear optics and image spectroscopy. THz imaging has become a significant function in this region. The initial development of the premature terahertz time domain spectroscopy was the innovation that a range of materials, which are opaque in the visible and near-IR region together with clothing and packaging materials, are reasonably transparent over a great extent of the THz range [3]. Room temperature operation of the infrared and far infrared frequencies detectors too increased the interest toward this region.

The popularity of the study of this region can be estimated by the fact that about 50% of the luminosity and most of the photons energy emitted since Big Bang of the space are contained by the THz radiations [4]. Every technology has dark side as well as brighter one; research in THz range has been obstructed by the absorption of the Earth's atmosphere amongst various main factors is the water vapors present there.

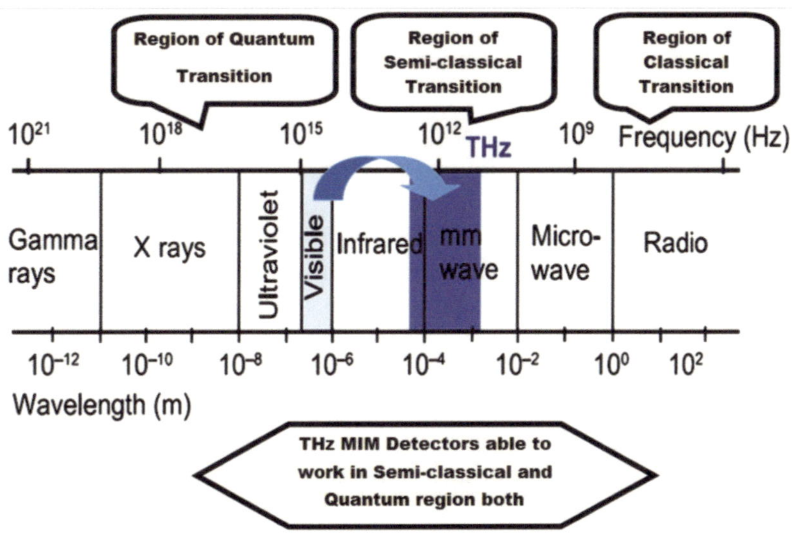

Figure 1. Schematic of the EM spectrum shows the working region for MIM devices.

3. Rectenna made easy

We are more than 7 billion today and multiplying every second around the globe. Specialists believe that further 50% extra energy will be required by 2050 to prolong humanity.

Nevertheless, it does not mean that population is the only stumbling block, but the ever inflating ego and diminishing humanity is also responsible for more energy needs. There is a fine line between the egoistical requirement and legitimate demand of energy around the world. Among various developed countries, Americans build up no more than 4.5% of the world's population and yet consume almost 20% of its energy. Its energy consumption is truly extraordinary [5]. In order to cope up with these problems, appropriate measures intended at reducing the dependence on the present fuels are needed, and the exploration for clean and renewable unconventional energy resources is one of the most urgent confront to the sustainable development of creature evolution.

There are several unconventional sources of energy like thermal energy, nuclear, vibration, wind, water, solar energy, etc. Solar energy is estimated to provide a significant contribution for the solution of energy problem today. Just a small portion of the Sun's energy that hits the Earth is sufficient to meet all our power needs. A little amount of the solar energy reaching the Earth is utilized for producing electricity by using present solar systems. Still the energy demand met by using solar energy is very less. Solar cells based on photovoltaic technology have many applications in solar ponds, vehicles based on solar energy, water pumps, lights, satellites calculators, solar telephones, and many more. But there are several shortcomings of this technology in terms of bulky solar panels, costly installation, low conversion efficiency, and dependencies on the weather conditions. The photovoltaic technology used for energy harvesting has limitation of maximum conversion efficiency by the Shockley-Queisser limit. However, in comparison the rectenna system is able to provide monospectral radio frequency-DC (RF-DC) with a conversion efficiency of 100% [6]. Moreover, photovoltaic technology is expensive and some poisonous by-products are produced during the manufacturing process. Therefore, in this perspective to obtain high efficiency, a green source of renewable energy, rectenna, has emerged as one of the best solutions for meeting the growing energy demand. A system that converts unused EM radiations and radiations from the Sun into usable DC is a rectenna system. It consists of a high frequency antenna to capture the incoming signal and provides the signal to prerectification block. The band-limited signal from prerectification block has been converted to DC which after sufficient filtration applied to output load, as shown in **Figure 2**.

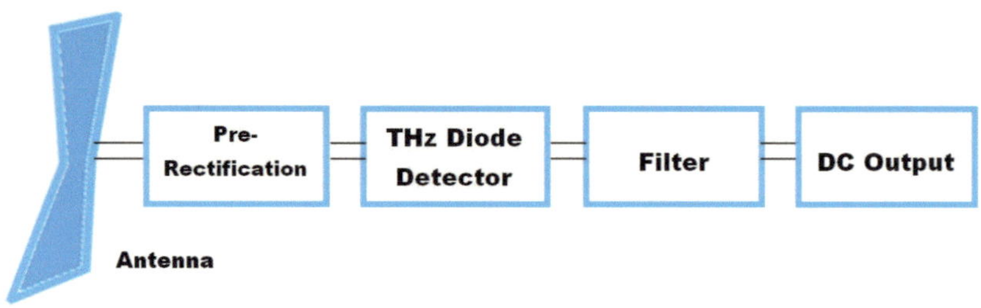

Figure 2. Block diagram of rectenna system.

The term rectenna, i.e., rectifying antenna, was coined by William C. Brown, he used an array of diodes for conversion of incoming microwave power into DC [7]. However, the rectenna

http://dx.doi.org/10.5772/66347

system in contrast also suffered from limited multispectral efficiency. The progress in the computer-aided design structure has currently cut down the design processes, in which complex electronic circuits such as high-frequency filters, antennas, and rectenna can be designed with a high level of buoyancy and reliability at very low cost and in small period of time.

3.1. Nanoscale antennas

For the collection of the EM field which is a combination of electric and magnetic fields, optimized antenna design and analysis must be taken care. Maxwell's field equations describe the full apparatus of EM fields. Antenna's geometric structure to associated wavelength and its polarization play a very important role. The receiving antenna must be located in the plane of polarization antenna for highest absorption of the electromagnetic fields. Several THz detectors and emitters establish the polarization with associated spectral response, e.g., dipole antennas are mainly based on linear polarization, whereas systems in incoming fields can either respond circularly, elliptically, or linearly polarized light [8].

Several nano-sized antennas, available [9–11] to capture the small wavelength signals, are half-wave dipole, spiral-type, Bow-Tie, Yagi-Uda, etc. The shape and size of the antenna is very important for THz energy collection. The μm-shaped antenna is required to collect the small wavelength signals. Latest nanotechnology for terahertz detection has now demanded for diode-integrated antennas or antenna-coupled diode detectors. The schematic diagram of integrated antenna with diode is shown in **Figure 3**.

Baily gave the concept of capturing the solar power with antennas for the first time employing the wave nature of incoming power [12]. Antenna-coupled detectors with micro-bolometer, metal-insulator-metal diode, and Schottky diodes have been used for detection in the sub-mm-wave and mm-wave regions with various antenna configurations [13, 14]. The surface plasmons are generated on the antenna due to incident EM radiations. Due to the appropriate impedance matching between the antenna geometry and the coupled MIM diode, these plasmons transfer to diode and it generates output DC.

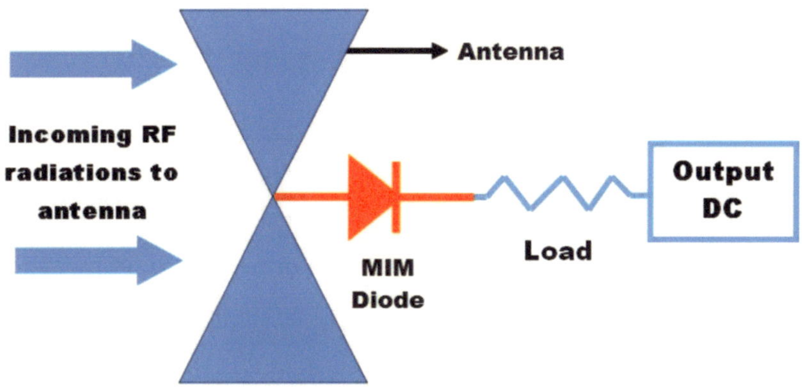

Figure 3. A typical schematic of antenna coupled diode detector.

3.2. Potential detectors available

Commonly available detectors for high-frequency (HF) range are Schottky diode, geometric diode, and MIM/MIIM. High challenges are involved in designing of these detectors due to complex geometries involved, HF challenges and increased fabrication cost. Nevertheless, due to the advancement in technology, highly optimized diode detectors are available nowadays which perform well up to several hundred THz.

3.2.1. Schottky diode detector

For THz detection, a long time battle has been continued between Schottky and MIM diodes [15]. The *n*-doped GaAs semiconductor is the key contender for THz applications, used in the metal-semiconductor structure. Schottky diode's series resistance is high-frequency-dependent parameter and essentially becomes a complex function of device geometry and material conductivity [16]. The diode conversion efficiency η_d [17] is the key to determine the system's performance given by

$$\eta_{\text{diode}} = \frac{\text{DC Output power}}{\text{RF Power incident on diode}} \tag{1}$$

Let

$$\eta_{\text{diode}} = \frac{1}{A+B} \tag{2}$$

The low-frequency term is given by

$$\begin{aligned} A = 1 + \frac{R_L}{\pi R_S}\left(1+\frac{V_{bi}}{V_D}\right)^2 \left[\theta_{ON}\left(1+\frac{1}{2\cos^2\theta_{ON}}\right) - \frac{3}{2}\tan\theta_{ON}\right] \\ + \frac{R_L}{\pi R_S}\left(1+\frac{V_{bi}}{V_D}\right)\frac{V_{bi}}{V_D}\left(\tan\theta_{ON} - \theta_{ON}\right) \end{aligned} \tag{3}$$

And high-frequency term is given by

$$B = \frac{R_S R_L c_j^2 \omega^2}{2\pi}\left(1+\frac{V_{bi}}{V_D}\right)\left(\frac{\pi-\theta_{ON}}{\cos^2\theta_{ON}} + \tan\theta_{ON}\right) \tag{4}$$

where c_j is the diode's junction capacitance given by

$$c_j = c_{j0}\sqrt{\frac{V_{bi}}{V_{bi}+V_D}} \tag{5}$$

Also

$$\tan\theta_{ON} - \theta_{ON} = \frac{\pi R_S}{R_L\left(1+\frac{V_{bi}}{V_D}\right)} \tag{6}$$

The cut-off frequency of Schottky is given by

$$fc = \frac{1}{2\pi R_s c_j} \tag{7}$$

where R_L is the DC load resistance and R_S is the diode's series resistance, c_j is the junction capacitance, V_{bi} is the diode's built-in voltage in the forward bias region, V_D is the self-bias voltage due to rectification across the terminals of the diode, θ_{ON} is forward-bias turn-on angle depends on diode input power, ω is the angular frequency. For higher frequency operation, the diode series resistance and junction capacitance must be lower. Planar Schottky diodes are able to operate at higher frequency with additional cooling mechanism required. Moreover, Schottky diodes cannot operate at zero bias; hence, MIM diodes are the best choice.

3.2.2. MIM diodes

The MIM diode is a combination of two metal plates separated by a very thin insulating layer of the order of few nanometers. The contact area between upper and lower layer determines the active device area. The area may vary from few μm^2 to several hundred μm^2 depends upon the operating frequency. There are various parameters needed to be optimized for high conversion efficiency like metal work function, insulator electron affinity, and its thickness, which are responsible for the best detector. The dependencies of these parameters can be estimated from the fact that the small deviation (in the range of Å) of an insulating layer and diode area can cause a significant change in the tunneling characteristics, and *I-V* responses [18]. Depending upon the application involved, the type of detector will be used. Comparison of different parameters has been shown in **Table 1**. Also, the fabrication of geometric diodes and other geometric field enhancement diodes is much more complex and expensive in comparison with MIM diodes. Detailed comparative analysis of these diodes has been discussed in our previous publication [19].

3.2.3. Geometric diode

It is the geometry of the device, which allows a preferential motion of charge carriers in a direction defined by its geometry. The asymmetric structure of the device forces to flow the charge carriers in one direction only, and it rectifies an alternating current as a result and hence gives the diode like actions. The diode acts as a funnel for the flow of carriers moving from left to right or right to left (depends on the design of geometry), with restricted flow in one direction and ease flow in other direction as shown in **Figure 4**. These diodes are able to convert from few THz to 1000 THz into usable DC [20].

Parameters	**Detectors**		
	Schottky	**MIM**	**Geometric**
Max frequency of operation	Up to 30 THz [21]	Up to 150 THz [22]	Up to 28 THz [23]
Minimum size of active area	NA	1 μm^2 [24]	1 nm (neck length) [23]
Zero bias resistance	NA	Few 100 kΩ [25]	Few 10 kΩ [26]
Diode capacitance	Few fF [27]	1.8×10^{-15} F [28]	1.8×10^{-17} F [28]
Applications	For few THz detection	For few 100 THz detection	For few 100 THz detection
Fabrication complexity	Small	High	Very high

Table 1. Comparison of High high-fFrequency detectors.

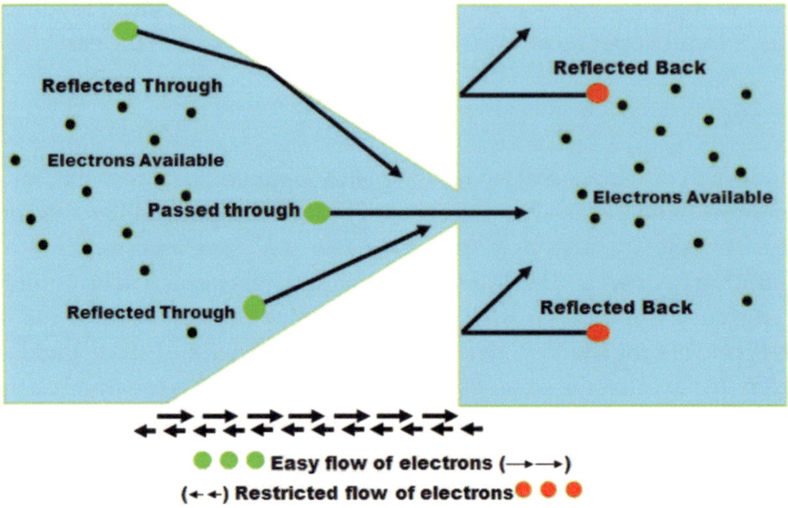

Figure 4. Schematic diagram of geometric diode.

4. Metal-insulator-metal (MIM) diodes

MIM diodes are the successor of Schottky diodes. The Schottky diodes have been routinely used for high-frequency applications like rectification and mixing. MIM diodes are not influenced by parasitic capacitors as compared to Schottky diodes. Extremely broader bandwidth, minute size, room temperature operation, effortlessly integrability with CMOS innovation [28], and zero bias voltage requirements are different encouraging factors for MIM diodes to be investigated with increased interest from last two decades. It is an exceptionally encouraging innovation for terahertz regime. The femtosecond quick exchanging time of these diodes makes them valuable in numerous low power and ultra-fast applications going from rectenna sunlight-based cells as an energy harvesting device [29] to imaging, sensing, and hot electron transistors, also in numerous refined microelectronic products such as switching memories, display devices (LCD backplanes), field emission cathodes [30], etc. MIM diode is a thin-film device in which the electrons tunnel through the insulator layer from the first metal layer to the second metal. **Figure 5** shows the schematic representation of the MIM diode.

Charge transport across the insulator occurs due to quantum-mechanical tunneling of electrons. A similar concept applies to MIM diodes, the process in which a particle penetrates or tunnel via the energy barrier in place of jump over it in contrary to classical mechanics. When the electron reaches the barrier, the insulator in between the metal electrodes absorbs some part of the energy passes through the barrier reaching to the second electrode, but another part reflects back due to some irregularities in thee insulator. How much the energy absorbs and reflected originally depends upon the different characteristics of material used such as work-function difference of metal electrodes, electron affinity of the insulator material, thickness of the barrier, applied potential energy, etc.

http://dx.doi.org/10.5772/66347

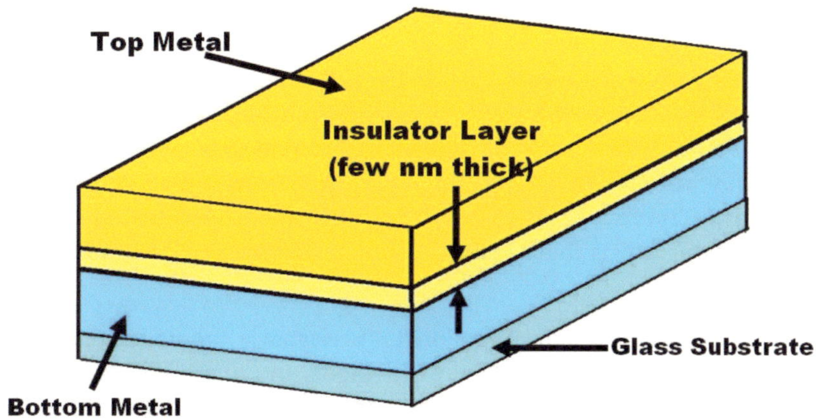

Figure 5. MIM diode schematic diagram.

Due to the tunneling phenomena involved, the energy beam absorbed makes it travel at ultra-high speeds. The speed is much higher than that on a normal conductor or a semiconductor like silicon. Tunneling is a highly nonlinear phenomenon, and hence these diodes are exploited for the rectification, mixing, and detection of alternating currents at very high frequencies even in the THz spectral region. **Figure 6** shows the typical schematic diagram of the MIM band and Fowler-Nordheim tunneling method.

The frequency response of the diode is governed by the resistor-capacitor (RC) time constant. Higher the cutoff frequency, lower the capacitance and resistance. The impedance of the diode must be close enough to the impedance of the antenna for maximum power transfer, and the capacitance must be minimized for operation at THz frequencies. Thickness cannot be decreased to facilitate tunneling, hence only area is left to be altered. Krishnan et al. report

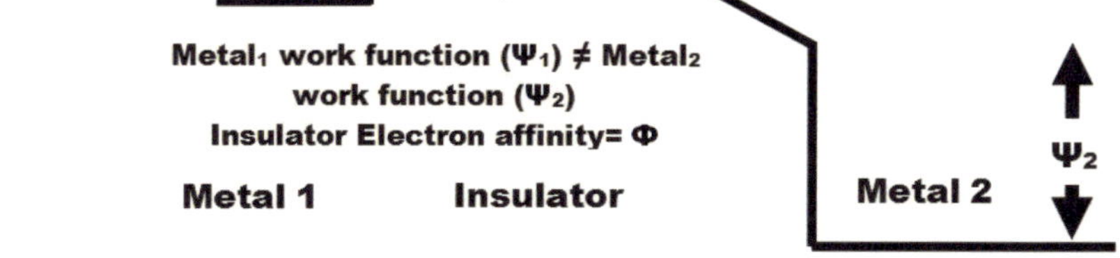

Figure 6. A typical schematic of MIM band diagram with FN tunneling.

the fabrication of terahertz MIM diode, the atto-farad capacitance (10^{-18}) is required, which further needs the device areas less than 100 nm × 100 nm for typical material systems which require e-beam lithography. However, this challenge of fabricating such nanoscale geometry has been overcome to a large extent by using modern lithographic techniques. Miniature-size diodes can be fabricated with the modern ultra-fine deposition techniques, such as e-beam lithography, reactive sputtering, etc., with small size, higher cutoff frequency, and thin insulating layer. **Table 2** lists these features of the MIM devices fabricated using different techniques.**Table 2** lists these features of the MIM devices fabricated using different techniques.

Trade-off between asymmetry and diode junction resistance can be overcome using multi-layer structure [34]. In comparison to the MIM diode, the tunneling current can be easily engineered in such multi-insulator MIM diodes by simply altering the type and number of insulators. This changes the number of barrier heights present at each interface and consequently the *I-V* characteristics. Reasonably good results for asymmetry and nonlinearity have been obtained by means of such diodes [35]. Among multi-insulator MIMs, the double insulator (MIIM) is a common preference [36].

4.1. Tunneling probability

Probability of electrons penetrating the insulator layer is called as tunneling probability. It is also called as transport probability, it is the ratio of the number of electrons incident on the interface with the electrons that reflected back. In order to tunneling take place, the two metals having insulators in between must have different work functions. When a bias voltage is applied between the metal electrodes of MIM diode, the potential barrier is reduced and charge carriers start tunneling between the two electrodes through the thin-film. Tunneling occurs in either direction, but by proper choice of metal electrodes with different work functions, the magnitude of the tunneling current will become larger in one direction than in the other direction. The tunneling current is calculated using Simmon's method [37]. This method assumes low-temperature operation hence neglecting thermal currents and incorporating tunneling as the main electron transport mechanism.

MIM combination	Active area	Insulator thickness	Patterning technique	Cut-off freq.	Ref.
Ni-NiO-Ni	110 × 110 nm^2	3.5 nm	E-beam lithography	30 THz	[31]
Ni-NiO-Cr	1.6 μm^2	3 nm	E-beam lithography	94 GHZ	[32]
Ni-NiO-Ag	3.1 × 10^{-4} μm^2	6 nm	E-beam lithography	343 THz	[33]
Cu-CuO-Cu	2 × 2 μm^2	2 nm	E-beam/reactive sputtering	–	[34]

Table 2. Comparison of MIM diodes fabricated using latest deposition methods.

Different algorithms such as TMM (transmission matrix method), WKB (Wentzel, Kramers, and Brillouin) approximation, and NEGF (nonequilibrium green function method) have been developed for calculating the tunneling probability. Simmon provides an approximation of a finite probability of electron penetration in the course of the insulator using WKB method. But this method overvalued the transport current. NEGF formulism is the method used to calculate the tunneling transmission probability. Using NEGF approximation, the total number of electrons emitted from left electrode and right electrode can be taken into consideration by defining two quantities, namely, left electrode and right electrode coupling functions, **Figure 7** shows the application of NEGF formulism to MIM diode [38]. TMM is also a powerful fast method for the analysis of transport properties of nonhomogeneous systems [32]. These algorithms have been compared for the calculation of transport probability in [39].

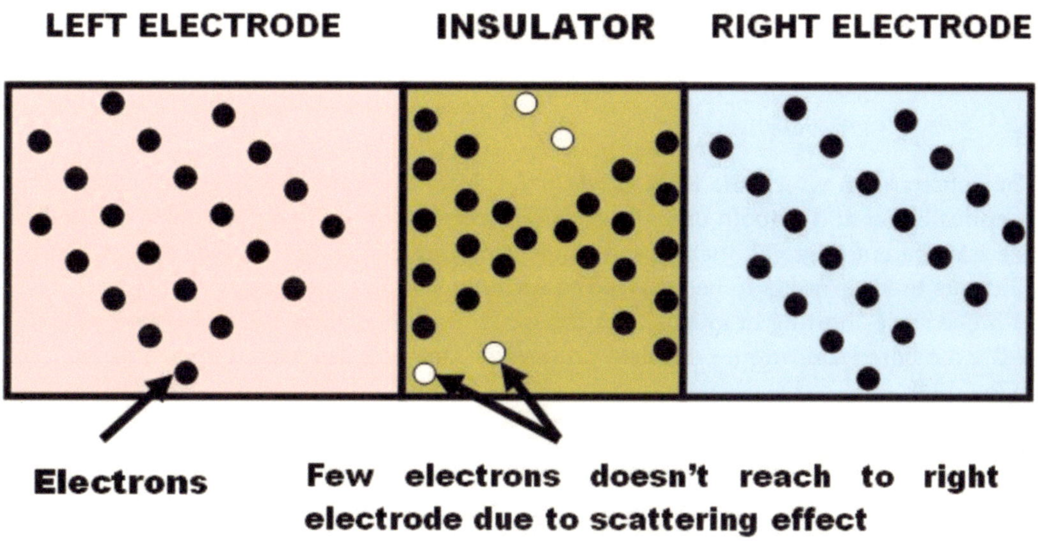

Figure 7. A typical schematic diagram showing the application of NEGF to MIM diode.

5. Modeling, fabrication and characterization of MIM diodes

An efficient modeling of MIM diode enables an investigator to predict the optimum material combination and device geometry for desired current-voltage characteristics. The accuracy of modeling depends on correct estimation of modeling parameters such as applied bias voltage, barrier thickness, dielectric constants of the insulator layers, and effective mass (assumed to be unity).

Various research groups are working to design an accurate model for rectenna system using different software such MATLAB, PSpice, Mathemaica, ADS, etc. [32, 34, 40]. We had not combined MIM diodes with antennas but modeled and fabricated MIM diodes for using two combinations (aluminum and chromium) of materials and assume the incoming energy to diode to be 100%.

5.1. Simulation

The diodes are simulated using model [35] and fabricated diode's *I-V* characteristics have been verified also. This simulator models the current density of MIM diode assuming a perfect insulator for a solely tunneling based analysis of current-voltage characteristics. We simulated the MIM diode for (Al-Al_2o_{3-} Cr) in MATLAB for current density and nonlinearity.

5.2. Fabrication

For the fabrication of MIM diode, the aluminum (work function of 4.28 eV) is used for the preparation of bottom electrode, and aluminum dioxide Al_2o_3 (having electron affinity of 1.25 eV) is used as a barrier layer after plasma oxidation of aluminum. For top electrode, chromium (work function of 4.5 eV) is used having sufficiently higher work function comparative to other materials with similar work function like titanium, niobium, and silver [41]. The basic steps involved in the fabrication of MIM diodes are explained below.

5.2.1. Substrate preparation

The substrate on which the MIM diode is fabricated is a microscopic glass slide, which is optically flat and smooth on both sides. The reason for using a glass substrate is that it is a transparent material, inexpensive, and readily available. Yet another reason is that, if silicon is used, it needs to be oxidized to form SiO_2 before the metals could be deposited to prevent the shorting of metals with the substrate. Since glass as a substrate needs to be cleaned before fabrication. Complete process for cleaning has been followed as explained in Ref. [34].

5.2.2. Metal deposition

The metal deposition for MIM diode is done by thermal evaporation. Two fundamental requirements for the metal to be deposited are the adhesion to the substrate and a smooth uniform surface. The fabricated MIM diode with the base metal made of aluminum and top metal made of chromium. The metal has been deposited using smart coat 3.0 by the thermal evaporation method. This method involves heating a material to the point of evaporation, usually at a pressure of around 10^{-6} mbar. The smart coat is cleaned properly by IPA or acetone. Using an automatic mode, the materials are loaded in the chamber and the deposition has been done. Thickness is accumulated at the program deposition rate. Basically, the potentiometer is controlling the current supply providing to the filament and accordingly controlling the rate of thickness Å/s, more the current more the rate of thickness. The rate of deposition can be monitored on the display panel. Mainly the current value is dependent on the material that are depositing, for example, for Al (aluminum) it will have 10—15 A current value and for Au (gold) it required > 70 A. Vibrations on the crystal holder produced due to external factors or due to internal vacuum pumps inside the smart coat 3.0 may vary the readings on thickness monitor. For the desired thickness (kÅ) continuous monitoring is required. **Figure 8** shows the smart coat 3.0 used for metal deposition.

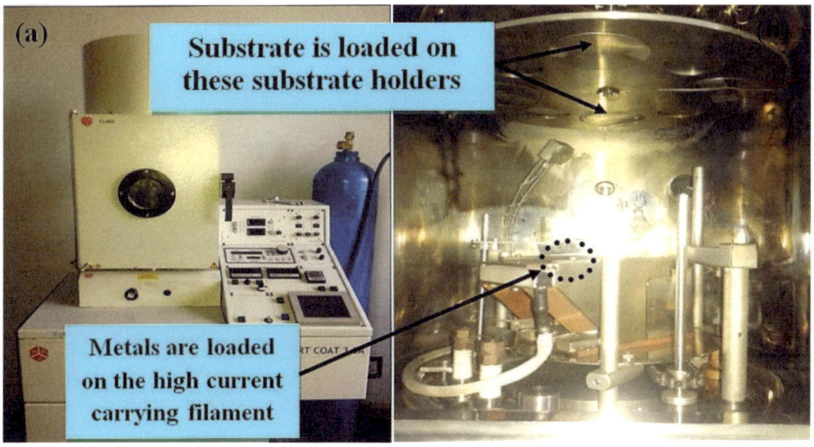

Figure 8. (a) Smart coat machine for thin-film deposition (b) Inside smart coat chamber.

5.2.3. Oxide formation

After loading the material in the smart coat chamber, the insulator layer is formed (using an aluminum mask) by the exposure of the oxygen (or argon) gas applied at a rate of 20–30 psi for a period of 10–12 minutes to form the plasma oxides on the base metal. Comparatively, thin layer with low impedance is formed in comparison with other deposition techniques. Similar procedure is used for the formation of top metal electrode as used for bottom electrode. After following the procedure completely, the final device is formed. **Figure 9** shows the fabricated diode and the mask used.

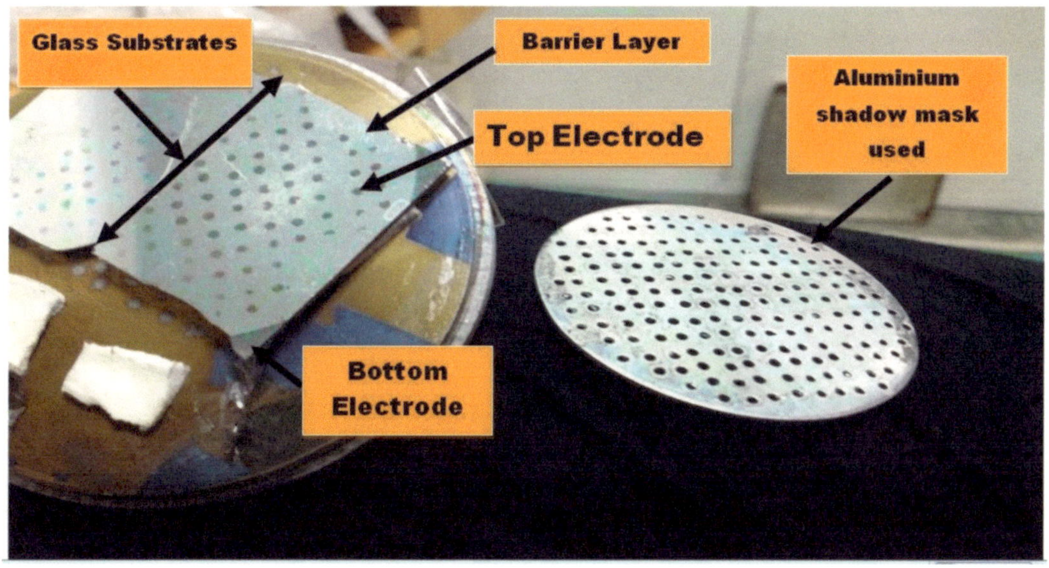

Figure 9. Fabricated MIM diode and mask used.

5.3. Electrical characterization

5.3.1. Experimental setup

The setup to measure the *I*(*V*) characteristics of the fabricated MIM tunnel diode is shown in **Figure 10**. It includes a probe station and a source meter. The source meter used is Model 2450 from Keithley instruments, which can vary the voltage from ±5 μV to ±220 V and can measure current from ±10 pA to ±1 A. The source measurement unit is connected to a probe station, and the measurements were taken so that any interference affecting the device could be avoided. This way a test bed was set up in which the device testing has been conducted. **Figure 10** shows the DC setup used for characterization of diode. The voltage sweep range set to be in low voltage in the values of few millivolts and the maximum sweep was selected from 1 to +1 V in order to get a proper *I-V* characteristics.

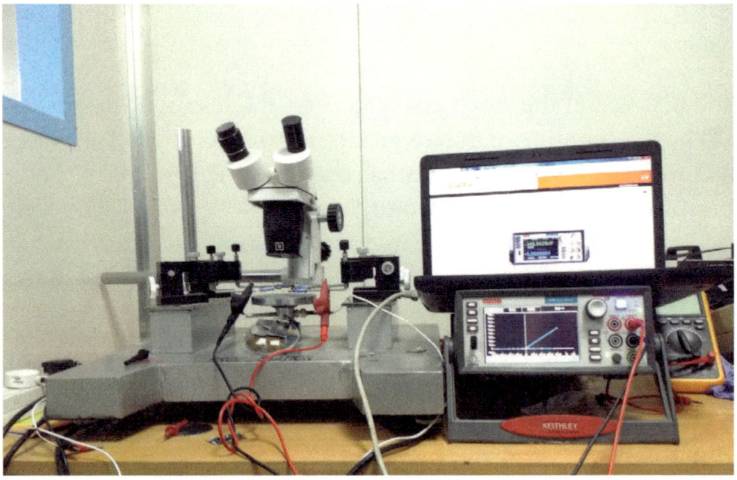

Figure 10. Setup for DC characterization.

Chromium is supplied with a negative bias and aluminum is supplied with a positive bias. The probes are lowered very slowly, so that the probe making contact with the top electrode is just in contact and the second probe on the bottom electrode is scraped a bit to ensure the proper contact with the aluminum.

5.3.2. DC characteristics

The DC characteristics of the fabricated diode MIM (Al-Al_2o_3-Cr) include the *I*(*V*) characteristics and the nonlinear characteristics. The responses of the diode are obtained with a biased voltage of ±1 V. The *I*(*V*) characteristics of the fabricated MIM diode which are compared with the simulated one and are in good agreement with is shown in **Figure 11**. Maximum current density observed is 2×10^{-4} A/cm^2 at 0.8 V which is less in comparison with the theoretical

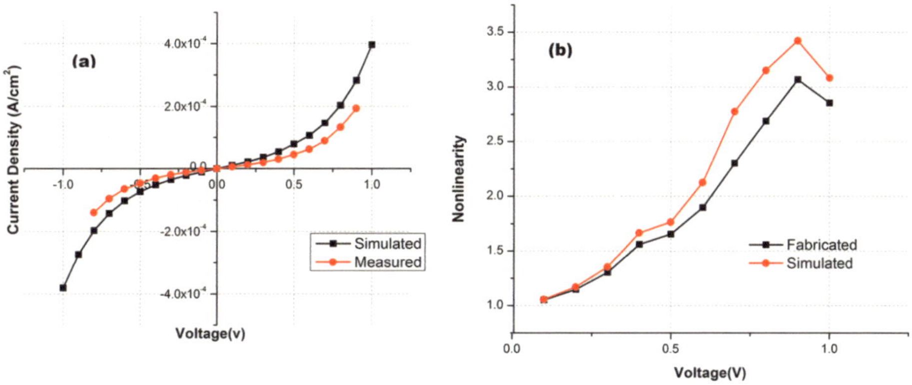

Figure 11. Measured DC characteristics of (Al-Al_2o_{3-} Cr) MIM diode, (a) current versus voltage relationship (b) nonlinearity versus voltage.

value due to some imperfections in fabrication processes. However, the nonlinearity is approximately 3.1, which is close to the modeled value.

6. Conclusion

Accurate detection and modeling of transport characteristics for any energy harvesting system are the main key for the calculation of *I-V* characteristics of MIM diode for THz energy harvesting applications. Due mainly to some poor characteristics of detector system the most of the harvesting systems suffered a lot. The MIM diode is one of the most suitable candidates for terahertz energy harvesting applications. Most of the innovations are the results of anterior modeling and fabrication of the device that result in better performance of the detectors. Since the future MIM detectors may incorporate thin insulating layers, which may result in structural irregularities; therefore, its meticulous modeling and optimum fabrication methods are essential before the actual implementation of the device. Potential approaches and factors responsible for designing most critical rectifying diodes operating at terahertz frequencies are discussed, and one of the material combinations for MIM diode has been modeled and characterized. It will augment the capability of naive researchers in identifying the latest THzDs to be used, their limitations and solutions to derive new paths for energy harvesting.

Acknowledgements

The author is grateful for providing research facilities in support of this work by Director of University Institute of Engineering & Technology, Kurukshetra University, Kurukshetra. The author acknowledges the help of Prof. R.K. Aggarwal, Computer Engineering Department, National Institute of Technology, Kurukshetra for various supports.

Author details

Kapil Bhatt*, Shilpi Shriwastava, Sandeep Kumar, Sandeep and Chandra Charu Tripathi

*Address all correspondence to: rai1kapil@gmail.com

University Institute of Engineering & Technology, Kurukshetra University, Kurukshetra, Haryana, India

References

[1] BBC Reasons for increase in demand for energy. www.bbc.co.uk/education/guides/zpmmmp3/revision.

[2] Nunzi J.M. Requirements for a Rectifying Antenna Solar Cell Technology. SPIE Photonics Europe, International Society for Optics and Photonics. 2010; 7712, 771204(1–7). DOI: 10.1117/12.855825

[3] Hu B.B. and Nuss M.C. Imaging with terahertz waves. Optics Letters. 1995; 20(16). DOI: 0146-9592/95/161716-03

[4] Leisawitz D. Scientific Motivation and Technology Requirements for the Spirit and Specs Far-Infrared/Submillimeter Space Interferometers. Proceedings of the SPIE, 2000; 4013, Munich, Germany, Mar. 29–31, 2000, pp. 36–46. DOI: 10.1117/12.393957

[5] World Population Balance. Population and Energy Consumption. http://www.worldpopulationbalance.org/population_energy.

[6] Joshi S., and Moddel G. Efficiency Limits of Rectenna Solar Cells: Theory Of Broadband Photon-Assisted Tunneling. Applied Physics Letters. 2013; 083901(1023), 1–5. DOI: 10.1063/1.4793425

[7] William C. B. The History of Development of Rectenna. Proceedings In Space Power, Symposium at JSC-NASA, Jan. 15-18, 1980, pp. 271–280.

[8] Fumeaux C, Herrman W., Kneubühl, F. K., and Rothuizen H. Nanometer Thin Film Ni-NiO-Ni Diodes for Detection and Mixing of 30 THz Radiation. Infrared Physics and Technology. 1998; 39, 123–183.

[9] Technical News Letter. http://ids.nic.in/Tnl_Jces_May%202012/PDF1/pdf/6.Nanteena.pdf.

[10] Kotter R. C., Green M.A., and Puzzer T. Theory and Manufacturing Processes of Solar Nanoantenna Electromagnetic Collectors. Solar Energy. December 2002; 73(6), 395.

[11] Corkish R., Green M.A., and Puzzer T. Solar Energy Collection by Antennas. Solar Energy. December 2002; 73(6), 395–401.

[12] Baily R. L. A Proposed New Concept for a Solar-Energy Converter. Journal of Engineering Power. 1972; 94(2), pp. 73–77.

[13] Gonzalez F. J., Abdel-Rahman M., and. Boreman G. D. Antenna-Coupled Vox Thin-Film Microbolometer Array. Microwave and Optical Technology Letters. 2003; 38, 235–237.

[14] Yang X., and Chahal P. Large-Area Low-Cost Substrate Compatible Cnt Schottky Diode for Thz Detection. Electronic Components and Technology Conference (ECTC), IEEE 61st.2011; 2158–2164.

[15] Acef O. et al. Comparison between Mim And Schottky Diodes As Harmonic Mixers For Visible Lasers and Microwave Sources. Optics communication. July 1994; 109(5–6), 428–434.

[16] Alain M. et al. Schottky Diode-Based Terahertz Frequency Multipliers and Mixers. Terahertz Electronic and Optoelectronic Components and Systems, 2010;11, pp. 480–495. DOI:10.1016/j.crhy.2010.05.002.

[17] Yoo T., and. Chang K. Theoretical and Experimental Development of 10 and 35 Ghz Rectenna's. IEEE Transactions on Microwave Theory and Technology. 1992; 40(6), 1259–1266.

[18] Krishnan S., Stefanakos E., and Bhansali S. Effects of dielectric thickness and contact area on current–voltage characteristics of thin film metal–insulator–metal diodes. Thin Solid Films. 2008; 516, 2244–2250.

[19] Bhatt K., and Tripathi C. C. Comparative Analysis of Efficient Diode Design for Terahertz Wireless Power Transmission System. Indian Journal of Pure & Applied Physics. 2015; 53, 827–836.

[20] Moddel G. Geometric Diode, Applications and Method. US Patent Application 20110017284, 2009.

[21] Hubers H. W., Schwaab G. W., and Roser H. P. Video Detection and Mixing Performance of Gaas Schottky-Barrier Diodes at 30 Thz and Comparison with Metal-Insulator-Metal Diodes. Journal of Applied Physics. 1994; 75, 4243–4248.

[22] Grossman E., Harvey T., and Reintsema C. Controlled Barrier Modification in Nb/Nbox/Ag Metal Insulator Metal Tunnel Diodes. Journal of Applied Physics. 2002; 91, 10134–9.

[23] Zhu Z., Joshi S., Grover S., and Moddel G. Graphene Geometric Diodes for Terahertz Rectennas. Journal of Physics D: Applied Physics. 2013; 46, 185101.

[24] Krishnan S., Rosa H. L., Stefanakos E., Bhansali S., and Buckle K. Design and Development of Batch Fabricatable Metal–Insulator–Metal Diode and Microstrip Slot Antenna as Rectenna Elements. Sensors and Actuators A: Physical. 2008; 142, 40–47.

[25] Krishnan S. Thin Film Metal-Insulator-Metal Tunnel Junctions for Millimeter Wave Detection, PhD Thesis, University of South Florida, 2008.

[26] Joshi S., Zhu Z., Grover S., and Moddel G. Infrared Optical Response of geometric diode rectenna solar cells. 38th IEEE Photovoltaic Specialists Conference. 2012; 002976–002978.

[27] Aik Y. T. Impact of Eddy Currents and Crowding Effects on High-Frequency Losses in Planar Schottky Diodes. IEEE Transactions on Electron Devices, 2011; 58 (10), 3260–3269.

[28] Grover S. Diodes for Optical Rectennas. PhD Thesis, University of Colorado, Boulder, 2011.

[29] Mohammad S. Novel Rectenna for Collection of Infrared and Visible Radiation. PhD Thesis, University of South Florida, Canada, 2005.

[30] Roy D. K. Quantum Mechanical Tunnelling and Its Applications. Philadelphia: World Scientific, 1986.

[31] Berland B. Photovoltaic Technologies beyond the Horizon Optical Rectenna Solar Cell, Final Report, ITN Energy Systems, INC Littleton, Colorado.

[32] Fu-Chien C. A Review on Conduction Mechanisms in Dielectric Films. Advances in Materials Science and Engineering. 2014, Article ID 578168, 18 pages, doi:10.1155/2014/578168.

[33] Islam E. H., and Ezzeldin A. S. Theoretical Study of Meal-Insulator-Metal Tunneling Diode Figures of Merit. IEEE, 2013; 49, pp. 72–79.

[34] Krishnan S., Bhansali S., Stefanakos E., and Goswami Y. Thin Film Metal-Insulator-Metal Junction for Millimeter Wave Detection. Procedia Chemistry. 2009; 1, 409–412.

[35] Eliasson B.J. Metal-Insulator-Metal Diodes for Solar Energy Conversion. PhD Thesis. Boulder: University of Colorado at Boulder, 2001.

[36] Grover S., and Moddel G. Engineering the Current–Voltage Characteristics of Metal–Insulator–Metal Diodes Using Double-Insulator Tunnel Barriers. Solid-State Electronics. 2012; 67, 94–99.

[37] Simmons J. G. Generalized Formula for the Electric Tunnel Effect Between Similar Electrodes Separated By a Thin Insulating Film. Journal of Applied Physics. 1963; 34 (6), 1793–803.

[38] Datta S. Quantum transport: atom to transistor. Cambridge University Press, Cambridge, UK, 2005.

[39] Bhatt K., and Tripathi C. C. Comparative Study of the Models Used for Calculation of Transport Characteristics in Metal-Insulator-Metal/MIIM Diodes. Advanced Science Letters. 2015: 21(8), 2570–2573.

[40] O'Regan T., Chin M., Tan C., and Birdwell A. Modeling, Fabrication, and Electrical Testing of Metal-Insulator-Metal Diode. 2011, ARL-TN -0464.

[41] Michaelson H. B, The work function of the elements and its periodicity. Journal of Applied Physics. 1977; 48, 4729–4733.

Section 2

THz Time Domain Spectroscopy

Chapter 6

THz Metamaterial Characterization Using THz-TDS

Christopher H. Kodama and Ronald A. Coutu, Jr.

Additional information is available at the end of the chapter

http://dx.doi.org/10.5772/67088

Abstract

The purpose of this chapter is to familiarize the reader with metamaterials and describe terahertz (THz) spectroscopy within metamaterials research. The introduction provides key background information on metamaterials, describes their history and their unique properties. These properties include negative refraction, backwards phase propagation, and the reversed Doppler Effect. The history and theory of metamaterials are discussed, starting with Veselago's negative index materials work and Pendry's publications on physical realization of metamaterials. The next sections cover measurement and analyses of THz metamaterials. THz Time-domain spectroscopy (THz-TDS) will be the key measurement tool used to describe the THz metamaterial measurement process. Sample transmission data from a metamaterial THz-TDS measurement is analyzed to give a better understanding of the different frequency characteristics of metamaterials. The measurement and analysis sections are followed by a section on the fabrication process of metamaterials. After familiarizing the reader with THz metamaterial measurement and fabrication techniques, the final section will provide a review of various methods by which metamaterials are made active and/or tunable. Several novel concepts were demonstrated in recent years to achieve such metamaterials, including photoconductivity, high electron mobility transistor (HEMT), microelectromechanical systems (MEMS), and phase change material (PCM)-based metamaterial structures.

The views expressed in this paper are those of the authors and do not reflect the official policy or position of the United States Air Force, Department of Defense, or the U.S. Government.

Keywords: terahertz, THz, time domain spectroscopy, TDS, metamaterials

1. Introduction to metamaterials

In the field of metamaterials, "meta-atoms" are designed to resonate and respond to stimuli in a similar manner as normal atoms [1]. From a macroscopic perspective, a matrix of meta-atoms can be viewed as a single homogeneous material—a metamaterial—with altered material properties. Meta-atoms designed to have electric and magnetic resonances are the base units of electromagnetic metamaterials. These have altered values of electric permittivity and magnetic permeability, which in turn results in a modified refractive index. Acoustic metamaterials, with altered mechanical material properties such as Poisson's ratio, have also been demonstrated [2]. In order to rightfully be considered a homogenous material, the meta-atoms constituting the metamaterial must be sufficiently small. For electromagnetic metamaterials, the meta-atoms need to be at least an order of magnitude smaller than the wavelengths of the desired operating frequencies [1].

Metamaterials are appealing because they offer the ability to artificially create custom material characteristics without having to depend on the fixed characteristics of existing materials. These metamaterials can further be designed to have exotic properties that are not seen in any naturally existing material. V.G. Veselago, in a paper in 1968, first considered and described the physical properties that would result from a material with simultaneously negative values of permittivity and permeability. Such materials are known as left-handed media [1].

In left-handed media, electromagnetic waves have a negative phase velocity; the waves seemingly follow the left-handed rule instead of the normal right-handed one. In addition, negative refraction occurs on a boundary between an ordinary and a left-handed medium. Negative refraction also causes left-handed concave and convex lenses to have reversed behavior: the concave lens converges light, and the convex lens diverges it. Surprisingly, a flat slab of left-handed material can even be used as a focusing lens [3]. Other exotic effects include inverse versions of the Doppler effect, Cerenkov radiation (i.e., radiative shock waves induced by high-speed particles traveling through a dielectric medium), and the Goos-Hänchen shift (i.e., a finite lateral shift between an incident beam of finite extent and its reflected counterpart) [1].

Although Veselago had established theoretical groundwork for understanding the nature negative-index materials, experimental demonstration of these materials would not occur for a few decades [3, 4].

J.B. Pendry in the late 1990s was the first to examine the fabrication of materials with customizable permeabilities and permittivities. In 1998, he demonstrated that a thin wire mesh would excite low frequency plasmons, meaning that the mesh would have customized, effective homogeneous permittivity values at low frequencies; these values could even be negative [4]. Soon after, he proposed a split-ring resonator (SRR) structure, which could analogously produce a desired effective permeability (including negative values) in the GHz range [5]. The structures he proposed, especially the split-ring resonator structure, are now ubiquitous in metamaterial devices at all frequencies.

The unique properties of metamaterials make them prime targets for research in the RF and photonics fields, where metamaterials can be used for enhanced lensing, beam steering, and phase shifting. Metamaterials have also greatly advanced the field of cloaking by freeing up the restriction of only being able to use naturally occurring materials and their material properties. Several different cloaking techniques have been demonstrated using metamaterials [6–9].

The THz spectrum has much potential for future applications, ranging from communications to bomb screening. Unfortunately, since most materials are transparent in the THz frequency range, controlling THz waves via modulation and other techniques is much more difficult when compared to electronic or optical devices and systems; metamaterials can help fill this technology gap. Since metamaterials can, in theory, be specifically designed to have certain material properties, they can interact with THz waves in ways that normal materials cannot [10, 11].

2. Metamaterials analysis and measurement

The split-ring resonator (SRR) is the fundamental unit cell used in a multitude of metamaterial designs. SRRs are split metallic rings meant to resonate with electromagnetic waves with wavelengths much larger than the SRR structure itself. The first SRRs proposed by Pendry consisted of two circular, concentric rings with oppositely oriented splits, as shown in **Figure 1(a)**. For the sake ease of fabrication, SRRs fabricated in the THz range typically have a much simpler geometry, like shown in **Figure 1(b)**.

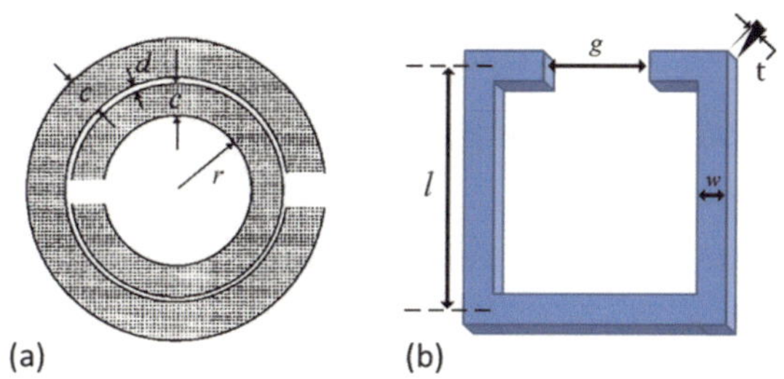

Figure 1. (a) Original SRR proposed by Pendry [5]; (b) a typical SRR shape used in THz research [11], [12].

SRRs are not the only method of generating resonant responses. In theory, pieces of metal in any geometrical shape can act as a resonator, albeit at frequencies determined by the geometry. For example, simple structures like rectangular hole arrays support surface wave polariton (SPP) resonant modes that can also strongly enhance incident radiation [13]. However, SRRs have two characteristics that make them desirable for use in metamaterials. SRRs couple well with the magnetic portion of electromagnetic waves due to its loop-like structure. This potentially allows SRR-containing metamaterials to have a frequency band with negative permeability. Second, the SRR is easily conceptualized as a simple LC resonator circuit, which allows one to quickly predict its lowest order resonant frequency.

As a basic approximation, the entire loop (ignoring the gap) can be viewed as a one-turn inductor and the gap as a parallel-plate capacitor. These quasi-static assumptions require that the dimensions of the SRR be much smaller than the wavelength of electromagnetic fields assumed to be passing through them (which conveniently is the same requirement needed to produce effective material properties from metamaterials). The wavelength of electromagnetic waves in free space at 1 THz is 300 μm, so metamaterial unit cells designed to operate in this range should have maximum dimensions of around 30 μm or less. The basic equations for the inductance, capacitance, and estimated resonant frequency of an SRR are

$$L \approx \mu_0 \frac{l^2}{t}, \quad C \approx \varepsilon_r \varepsilon_0 \frac{tw}{g}, \quad f_{\text{ideal}} \approx \frac{1}{2\pi}\sqrt{\frac{1}{LC}} = \frac{1}{2\pi}\frac{c}{\sqrt{\varepsilon_r}}\sqrt{\frac{g}{w\,l^2}}, \tag{1}$$

where t is the thickness of the SRRs, μ_0 is the vacuum permeability constant, l is the SRR side length, ε_r is the relative dielectric constant, ε_0 is the vacuum permittivity constant, w is the line width of the SRR, g is the gap width of the SRR, and c is the speed of light in vacuum [14–16].

The lowest order resonance of a SRR can be adequately described by a quasistatic model incorporating capacitors and resistors. The higher order resonances, however, cannot be modeled in such a manner. The original SRR in **Figure 1(a)** can be modeled as a transmission line due to the presence of two distinct conductors [17]. The single SRR in **Figure 1(b)** has only one conductor and thus cannot support any transmission line modes. Single SRRs can, however, produce evanescent modes confined to the surface, known as plasmon resonances. These plasmonic resonances occur at frequencies that have half-wavelengths, which are multiples m of the perimeter of the SRR:

$$f_{\text{dip}} = \frac{c}{\sqrt{\varepsilon_{r,\text{eff}}}}\frac{1}{(4l-g)}\frac{m}{2}, \tag{2}$$

where c is the speed of light, $\varepsilon_{r.eff}$ is the effective permittivity constant, l is the side length of the square SRR, g is the gap length of the SRR, and m is a positive integer greater than or equal to 1 [1]. The LC resonance can also be interpreted as the lowest order plasmon resonance, with $m = 1$ in the above equation [18].

These equations can only serve as basic approximations for the resonant frequency of a fabricated SRR. This is especially true in the THz range, where limits on feature size lead to nonidealities such as significant wire widths and gap widths. If more accuracy is desired, more precise equations need to be utilized for the inductance and capacitance terms for the LC resonant frequency, and the effective SRR perimeter seen by higher order resonances will differ from simply $(4l - g)$, depending on the current distribution through the width of SRR. As an alternative to an analytical solution, SRRs can instead be modeled using electromagnetic simulation software such as CST Microwave Studio, ANSYS HFSS, or COMSOL.

Fabricated THz SRRs can be measured and characterized using THz time-domain spectroscopy (THz-TDS). In THz-TDS, broadband THz pulses are passed through a sample and then are Fourier transformed to obtain spectral data. A simple planar SRR structure is shown in **Figure 2(a)**. The SRR is patterned in a layer of gold, atop an intrinsic silicon substrate; the alignment of the incident THz pulse is shown in the inset. The solid line in **Figure 2(b)** represents the transmission coefficient of the THz pulse passed through the sample. The LC resonance can be seen at approximately 0.8 THz, and a higher order resonance can be seen near 2.2 THz. The dashed orange line represents simulated data from models in CST Microwave Studio. The simulated data agree well with measured data, except in the region near 2.5 THz and above. In this region, the spectral power in the measured signal is not strong enough to fully resolve the sharp, higher order resonances of the SRR [19].

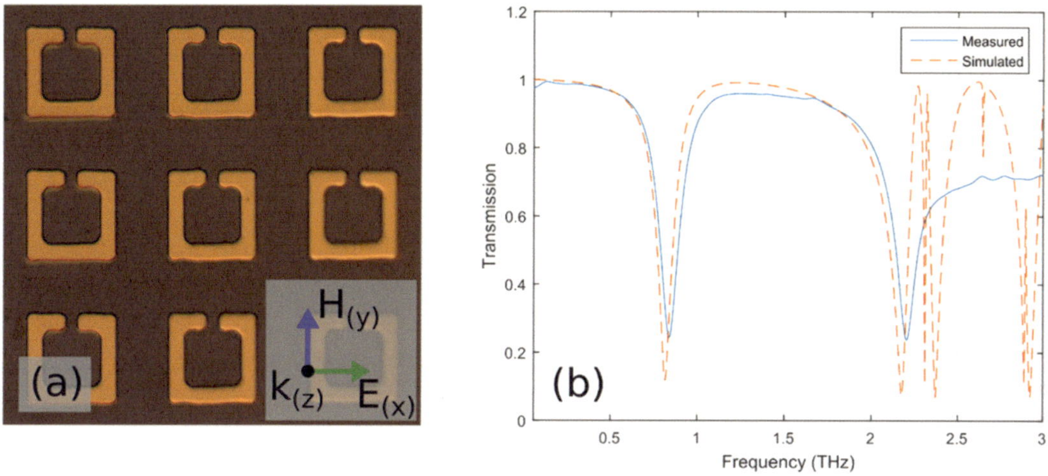

Figure 2. Gold SRRs on silicon substrate are shown in (a). The SRRs are 20-μm squares with 5-μm line widths, 3-μm gap widths, and a 39-μm periodicity. The plot in (b) displays the transmission response of the SRRs [19].

3. Fabrication processes

Since THz metamaterial unit cells need to be electrically small compared to THz wavelengths, these unit cells have widths of around 30 μm or less. This requirement makes microfabrication and micro-electromechanical systems (MEMS) processes suitable choices for creating THz metamaterials.

There are three main techniques that can be used to make microelectronic and MEMS devices: bulk micromachining, surface micromachining, and microforming. These techniques are often used interchangeably to create complete devices. In all three of these techniques, the process of photolithography plays an important part in creating devices out of designs.

In photolithography, a layer of photoresist—a polymer sensitive to certain wavelengths of light, typically in the ultraviolet range—is selectively exposed to light screened by a patterned mask. The exposed sections of photoresist undergo a chemical change, depending on the type of polymer. For "positive" photoresist, the exposed photoresist weakens and will dissolve in a developer solution, leaving behind a patterned layer of unexposed photoresist. This patterned photoresist can then act as a mask for other processing steps.

For example, the gold MMs shown in **Figure 2(a)** could have been fabricated using the process shown in **Figure 3**, if a suitably thick photoresist was used along with an evaporation deposition process, which would deposit highly nonconformal material to the edges of the resist. Otherwise, there will be difficulty stripping the photoresist in the fourth step. More advanced, dual-layer photoresist patterning techniques can also be used to create crisper patterns or to be used in conjunction with conformal depositions like sputter deposition. Processes involving the incremental addition and/or etching of thin layers of material on a substrate to create a device are known as surface micromachining techniques.

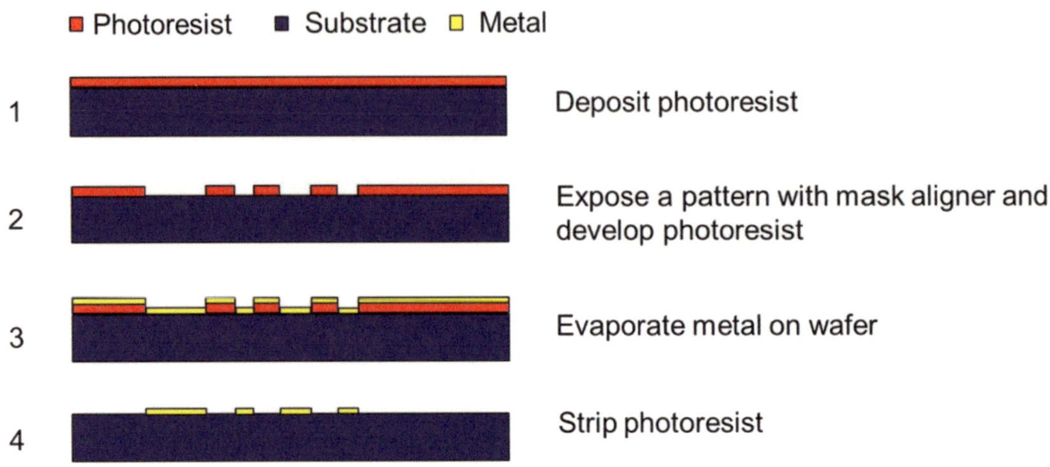

Figure 3. An example of the use of photolithography to deposit gold.

In bulk micromachining techniques, a device is fabricated primarily through etching processes into the bulk of a substrate. Depending on the crystallography of the substrate and the etchant type, etching processes can be isotropic or highly anisotropic in certain crystallographic directions. There are also more involved methods, such as deep reactive ion etching (DRIE), which can create high aspect ratio structures through sequential reactive ion etching (RIE) of a sidewall-protected cavity [20].

Microforming processes, also known as high aspect ratio micromachining (HARM) processes, use especially thick, patterned photoresist layers as molds for metal layers deposited via electroplating. In the LIGA microforming method, X-ray-based photoresist PMMA can be patterned in thick layers without worry of diffraction to create very high aspect ratio devices. LIGA is an acronym for the three German words lithographie, galvanoformung, and abformung, which, respectively, stand for lithography, electroplating, and polymer replication [21–23]. Other thick photoresists such as SU-8, or multiple layers of SF-11, can also be used as a lower cost alternative to create high aspect ratio structures.

In addition to micromachining and MEMS processes, additive manufacturing techniques show promise for THz metamaterials. One advantage to additive manufacturing is the ability to create devices without requiring labor-intensive cleanroom methods. However, the present capability of many of these techniques does not allow for the sub-millimeter resolution needed to create THz metamaterials, with a few exceptions. In inkjet printing, droplets of material are deposited onto a substrate via moving, microscopic nozzle heads. This method has been shown to feasibly fabricate layers of SRRs in the low THz range [24–26].

Inkjet-printed SRRs display resonances that are similar but slightly degraded compared to their cleanroom-fabricated counterparts; the degradation is largely caused by print variation among the individual printed metamaterial unit cells. This can be seen as a tradeoff to the advantages of less labor-intensive processes and lower cost manufacturing offered by inkjet printing. Another additive manufacturing method used to create THz metamaterials is laser decal transfer [27]. In this process, a laser pulse ejects a small amount of material from a donor substrate and transfers it onto an underlying substrate. The process is analogous to a pen (the laser) writing on top of a piece of carbon paper. Using a digital micromirror device (DMD), the laser can be spatially modulated to transfer more complex shapes onto a target substrate. With this process and an x-y stage manipulator, metal SRRs were printed with a minimum feature size of 6 μm [28]. The printed SRRs were highly uniform and produced a transmission response nearly identical to that of a comparable lithographically prepared sample. The main limitation of this method is that the deposited material needs to be cured at around 150°C after laser transfer, so it may not be suitable for some temperature-sensitive applications.

4. Active metamaterials

Metamaterials with magnetic resonances in the THz range were first demonstrated by Yen et al. in 2004 [29]. They were able to couple into the strong magnetic modes of the SRRs by sending incident waves at oblique angles in ellipsometry experiments. There is much ongoing work on developing active THz metamaterials through several methods [10, 30–32]. In order for metamaterials to be useful in dynamic applications like modulation, they need an incorporated active component that can modify its resonance [33].

An active metamaterial was first demonstrated by Padilla et al. in 2006 [34]. They successfully shunted the SRR response of single SRRs on a GaAs substrate by utilizing an optical pump. Without any optical excitation, the SRRs exhibited a typical LC resonance. When the optical pump was activated, photo-excited carriers spontaneously formed across the entire surface of the GaAs wafer, which made the entire surface metallic. This in turn nullified the effects of the SRRs and their LC resonance. Nonlinear response can also be achieved through SRRs fabricated on doped GaAs layers [35]. With this technique, incident THz radiation of zero to low levels causes a metallic response due to the doped GaAs layer. However, with increasing THz field strengths, inter-valley scattering becomes more frequent, causing decreased carrier mobility and decreased conductivity. With the disappearance of the suppressing GaAs conductivity, the SRR's LC resonance becomes more prominent, and a notch appears in the transmission. At even higher incident radiation strengths, impact ionization in the gaps of the SRRs causes the conductivity to increase again, suppressing SRR response.

Building on those results, Chen and Padilla demonstrated modulation through voltage-biased SRRs on n-GaAs, with the SRR-nGaAs junction acting as a Schottky barrier [30, 31]. This design is shown in **Figure 4**. Without the presence of a bias voltage, the doped layer effectively suppresses the resonant response. When a high enough voltage is applied between the SRRs and the n-GaAs layers, a depletion region will form underneath the SRRs. This depletion region both electrically isolates the SRRs from the n-GaAs layer and reduces the overall conductivity of the n-GaAs layer, which restores the suppressed LC resonance of the SRRs.

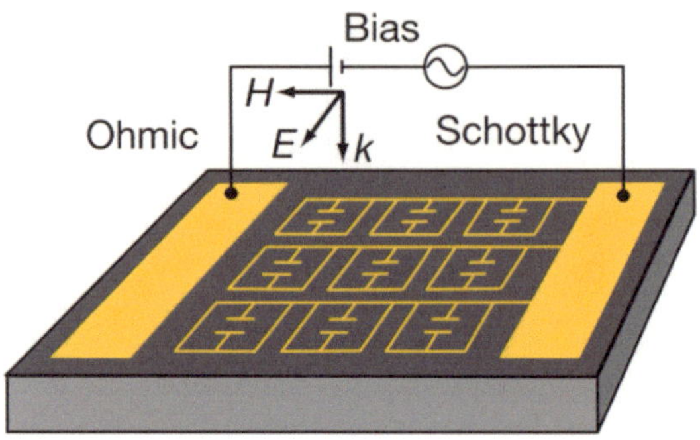

Figure 4. Schottky-based modulation of split-ring resonator resonances [31].

In 2007, Chen et al. demonstrated modulation via frequency shifting by using SRRs enhanced with silicon in the capacitor gaps, as shown in **Figure 5** [36]. When the SRRs are exposed to laser pulses, photoexcited carriers in the silicon layer effectively lengthen the internal SRR capacitor, which alters the resonant frequency of the structure.

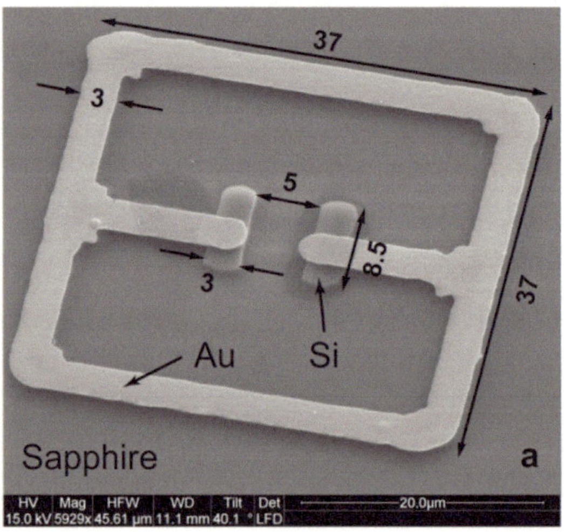

Figure 5. Split-ring resonator modulated through photoexcited silicon [36].

Shrekenhamer and Rout created a novel THz metamaterial design that incorporated high electron mobility transistors (HEMTs) in the gaps of SRRs, as shown in **Figure 6** [37]. At zero volts gate bias, a 2D electron gas channel naturally forms between the source and drain of the pHEMT, which is connected to opposite sides of the SRR gap. The electron channel shorts the SRR gap and removes the LC resonance from the structure's response. When a negative bias is applied to the gate, the channel is eliminated, and the SRR LC resonance is restored.

http://dx.doi.org/10.5772/67088

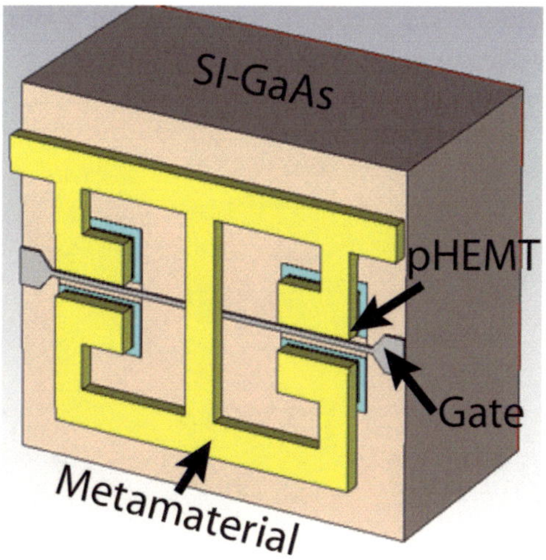

Figure 6. High electron mobility transistor–based modulation of split-ring resonators [37].

The four previously mentioned active metamaterial structures used electrical properties, namely semiconductors and semiconductor junctions, to achieve tunability in the THz frequency range. Mechanically tunable SRRs have also been demonstrated using MEMS processes. For example, Coutu et al. used MEMs cantilever beams arrays to create electrostatically tunable meta-atoms for the RF frequency range [38]. There have since been several demonstrations of THz-range meta-atoms using MEMS. In one such design, bi-material cantilevers are used to alter the resonance of SRRs, shown in **Figure 7(a)** [39]. In the rest state, the cantilevers are bent downward due to residual stresses intentionally introduced during the fabrication process. This shorts the SRR gaps and eliminates the LC resonance. When the device is heated, thermal expansion coefficient difference between the top and bottom half of the cantilever causes the beam to straighten out, which removes the electrical contact and restores the LC resonance. In [40], the capacitor arms themselves are treated as bi-material cantilevers, which are designed to have an initial deflection and can be lowered with an external voltage. This is shown in **Figure 7(b)**.

Bi-material cantilevers can also be used to actively change the in-plane orientation of SRRs, as shown in **Figure 7(c)** [42]. By changing the angle of incidence of the incoming electromagnetic wave with respect to the metamaterial unit cell, the magnetic coupling between the SRRs and the incident wave can be enhanced, strengthening the measured resonance.

Another method of modulating metamaterials is through the use of phase change materials (PCMs) and metal-insulator transition (MIT) materials. PCMs and MIT materials are types of materials that have vastly different electrical and/or optical properties based on their crystallographic alignment [43, 44]. Transitions between crystalline and amorphous states can be rapidly achieved, typically through thermal, optical, or electrical methods.

Bouyge et al. demonstrated reconfigurable bandpass filtering by using vanadium dioxide (VO_2) switches on microwave-range SRRs [45]. Vanadium dioxide is a MIT material with a transition temperature of 340 K and is volatile, meaning that it will revert back to its original phase after cooling back down [44]. With the VO_2 acting as a reconfigurable switch, patches of metal could be electrically connected or disconnected to the SRR, thus altering the resonance of the structure.

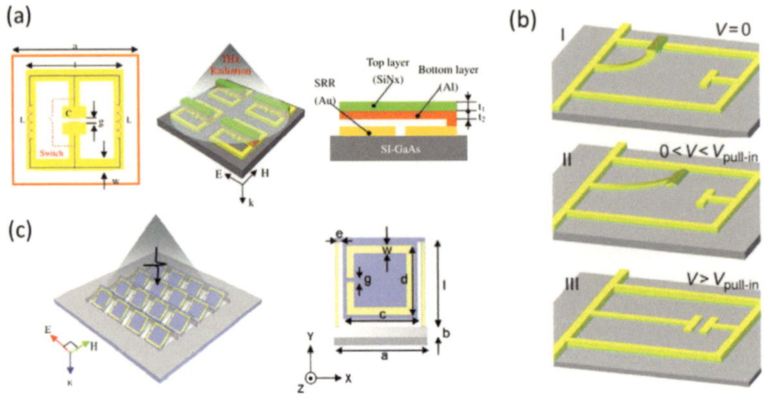

Figure 7. Various microelectromechanical systems (MEMS) incorporated, terahertz (THz) frequency range, tunable split-ring resonator (SRR) elements. In (a), THz SRRs are modulated using bi-material cantilevers [39]. The design in (b) uses bi-material cantilevers for the inner capacitive gap arms of THz SRRs, modulating the SRR resonances when these arms are deflected [40]. The third design in (c) shows structurally tunable THz SRRs which use bi-material cantilevers to bend the entire array in different directions [41].

Active metamaterial structures designed for the near-infrared range, like the one shown in **Figure 8**, have also been demonstrated using vanadium dioxide [46]. The planar hybrid structure shown can switch between electrically appearing as a normal SRR surface when the VO_2 is insulating or a full metal plane when the VO_2 has transitioned to the conductive state. The VO_2 in the self-aligned hybrid SRR causes a slight shift in the resonant frequency of the SRR before and after phase transition.

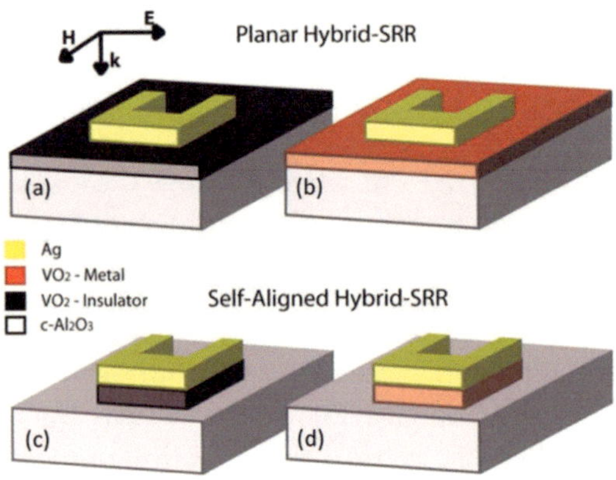

Figure 8. Near-infrared metamaterials with vanadium dioxide (VO_2). In (a) and (b), gold split ring resonators (SRRs) are fabricated on a layer of VO_2. In (a), before transitioning the VO_2 layer, the VO_2 acts like a dielectric and the SRR response is dominates the behavior of the device. In (b), while the VO_2 layer is transitioned, the VO_2 layer is metallic and the overall response of the material is that of a flat metal plane. In (c) and (d), the SRRs have two materials in their thickness, a metal layer and a VO_2 layer. Before transitioning, as shown in (c), the VO_2 is a dielectric, which only affects the response of the SRR with a small red-shift. In (d), during the VO_2 transition to a metallic phase, the effective metal thickness of the SRR doubles, which slightly increases the resonant frequency of the device [46].

http://dx.doi.org/10.5772/67088

Phase change materials have also been applied in some rudimentary metamaterial and plasmonics designs. Phase change materials differ from MIT materials in the fact that their phase transitions are nonvolatile, meaning that an amorphous PCM will crystallize on heating but would not reamorphize on slow cooling. To successfully reamorphize a PCM, a fast heat pulse (from a laser, via Joule heating, etc.) accompanied by a strong heat sink is needed to "melt-quench" the material back into an amorphous phase. A dynamic metamaterial switch using germanium antimony telluride (GST) is shown in **Figure 9**. [47]. This switch is controlled by a short laser pulse, which can transition the GST between the conductive and insulating states. In addition to GST, the chalcogenide gallium lanthanum sulfide (GLS) has also been shown to have potential for an active, plasmonics-based switch [48] and tunable layer for metamaterials in the infrared range [49].

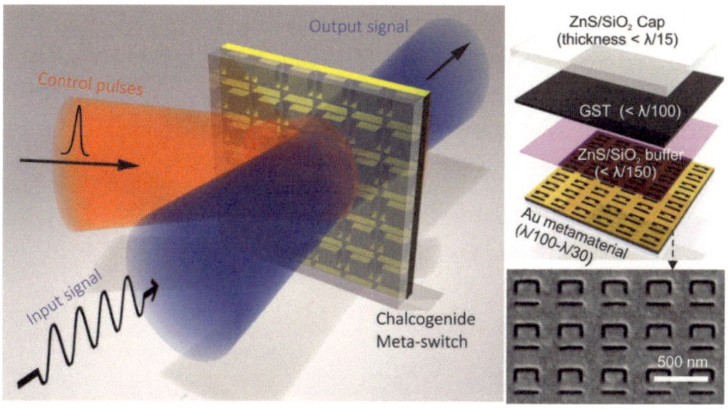

Figure 9. Metamaterial absorber with incorporated phase change material layer for dynamic applications [47].

Kodama et al. have demonstrated the use of germanium telluride (GeTe), another PCM, for achieving tunable SRRs. As shown in **Figure 10**, GeTe layer was placed in the gaps of SRRs. When the GeTe is amorphous and insulating, the SRRs exhibit their typical LC resonance; when the GeTe is crystalline and conductive, the SRR gap becomes electrically shorted, and the LC resonance is eliminated.

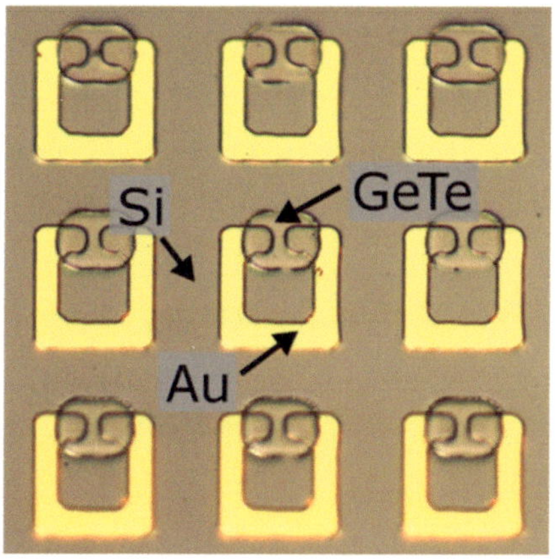

Figure 10. Tunable split-ring resonators using germanium telluride in the split gaps [19].

5. Conclusion

This chapter was designed to familiarize the reader with terahertz metamaterials. Starting with fundamental metamaterial unit, cell concepts were outlined, methods for producing metamaterial samples were detailed, and the latest developments in active THz metamaterial research have been reviewed. We hope we have been able to convey the potential that metamaterial research has to offer for terahertz technologies and systems.

Acknowledgements

The authors thank the Air Force Office of Scientific Research (AFOSR) for funding this effort (F4FGA06141J001) and the Air Force Research Laboratory (AFRL) Sensors Directorate for their assistance. The authors also thank the Air Force Institute of Technology (AFIT) clean-room staff, Mr. Richard Johnston and Mr. Adam Fritche.

Author details

Christopher H. Kodama and Ronald A. Coutu, Jr.*

*Address all correspondence to: Ronald.Coutu@marquette.edu

Marquette University, WI, USA

References

[1] R. Marqués, F. Martín, and M. Sorolla, *Metamaterials with Negative Parameters: Theory, Design and Microwave Applications*, 1st ed. Hoboken: Wiley, 2013.

[2] M.-H. Lu, L. Feng, and Y.-F. Chen, "Phononic crystals and acoustic metamaterials," *Mater. Today*, vol. 12, no. 12, pp. 34–42, Dec. 2009.

[3] V. G. Veselago, "The electrodynamics of substances with simultaneously negative values of ε and μ," *Sov. Phys. Uspekhi*, vol. 10, no. 4, p. 509, 1968.

[4] J. B. Pendry, A. J. Holden, D. J. Robbins, and W. J. Stewart, "Low frequency plasmons in thin-wire structures," *J. Phys. Condens. Matter*, vol. 10, no. 22, p. 4785, 1998.

[5] J. B. Pendry, A. J. Holden, D. J. Robbins, and W. J. Stewart, "Magnetism from conductors and enhanced nonlinear phenomena," *Microw. Theory Tech. IEEE Trans.*, vol. 47, no. 11, pp. 2075–2084, Nov. 1999.

[6] S. A. Cummer, B.-I. Popa, D. Schurig, D. R. Smith, and J. Pendry, "Full-wave simulations of electromagnetic cloaking structures," *Phys. Rev. E*, vol. 74, no. 3, p. 36621, Sep. 2006.

[7] P. Collins and J. McGuirk, "A novel methodology for deriving improved material parameter sets for simplified cylindrical cloaks," *J. Opt. A Pure Appl. Opt.*, vol. 11, no. 1, p. 15104, 2009.

[8] L. Pei-Ning, L. You-Wen, M. Yun-Ji, and Z. Min-Jun, "A multifrequency cloak with a single shell of negative index metamaterials," *Chin. Phys. Lett.*, vol. 28, no. 6, p. 64206, 2011.

[9] F. Bilotti, S. Tricarico, and L. Vegni, "Plasmonic metamaterial cloaking at optical frequencies," *Nanotechnol. IEEE Trans.*, vol. 9, no. 1, pp. 55–61, Jan. 2010.

[10] R. D. Averitt, W. J. Padilla, H. T. Chen, J. F. O'Hara, A. J. Taylor, C. Highstrete, M. Lee, J. M. O. Zide, S. R. Bank, and A. C. Gossard, "Terahertz metamaterial devices," in *Proc. SPIE*, 2007, vol. 6772, p. 677209.

[11] W. Withayachumnankul and D. Abbott, "Metamaterials in the terahertz regime," *Photonics J. IEEE*, vol. 1, no. 2, pp. 99–118, Aug. 2009.

[12] C. H. Kodama, "*Tunable Terahertz Metamaterials with Germanium Telluride Components*," Air Force Institute of Technology, Dayton, OH, 2016.

[13] A. K. Azad and W. Zhang, "Resonant terahertz transmission in subwavelength metallic hole arrays of sub-skin-depth thickness," *Opt. Lett.*, vol. 30, no. 21, pp. 2945–2947, Nov. 2005.

[14] O. Sydoruk, E. Tatartschuk, E. Shamonina, and L. Solymar, "Analytical formulation for the resonant frequency of split rings," *J. Appl. Phys.*, vol. 105, no. 1, p. 014903-1 - 014903-4, 2009.

[15] J. C. Maxwell, *A Treatise on Electricity and Magnetism*, vol. 1. Dover, Dover Publications, New York, 1954.

[16] E. B. Rosa and F. W. Grover, "Formulas and tables for the calculation of mutual and self-inductance," *Bull. Bur. Stand.*, vol. 8, pp. 1–287, 1912.

[17] M. Shamonin, E. Shamonina, V. Kalinin, and L. Solymar, "Resonant frequencies of a split-ring resonator: analytical solutions and numerical simulations," *Microw. Opt. Technol. Lett.*, vol. 44, no. 2, pp. 133–136, 2005.

[18] C. Rockstuhl, F. Lederer, C. Etrich, T. Zentgraf, J. Kuhl, and H. Giessen, "On the reinterpretation of resonances in split-ring-resonators at normal incidence," *Opt. Express*, vol. 14, no. 19, pp. 8827–8836, 2006.

[19] C. H. Kodama and R. A. Coutu, "Tunable split-ring resonators using germanium telluride," *Appl. Phys. Lett.*, vol. 108, no. 23, 2016.

[20] F. Laermer and A. Schilp, "*Method of Anisotropically Etching Silicon*." Google Patents, 1996.

[21] S. A. Campbell, *Fabrication Engineering at the Micro- and Nanoscale*, 4th ed. New York: Oxford University Press, 2013.

[22] V. Saile, *"Introduction: LIGA and Its Applications,"* in *LIGA and Its Applications*, Germany: Wiley-VCH Verlag GmbH & Co. KGaA, 2009, pp. 1–10.

[23] P. Rai-Choudhury, *Handbook of Microlithography, Micromachining, and Microfabrication: Micromachining and Microfabrication*. SPIE Optical Engineering Press, London, UK, 1997.

[24] K. Takano, T. Kawabata, C.-F. Hsieh, K. Akiyama, F. Miyamaru, Y. Abe, Y. Tokuda, R.-P. Pan, C.-L. Pan, and M. Hangyo, "Fabrication of terahertz planar metamaterials using a super-fine ink-jet printer," *Appl. Phys. Express*, vol. 3, no. 1, p. 16701, 2010.

[25] H. Teguh Yudistira, A. Pradhipta Tenggara, V. Dat Nguyen, T. Teun Kim, F. Dian Prasetyo, C. Choi, M. Choi, and D. Byun, "Fabrication of terahertz metamaterial with high refractive index using high-resolution electrohydrodynamic jet printing," *Appl. Phys. Lett.*, vol. 103, no. 21, 2013.

[26] M. Walther, A. Ortner, H. Meier, U. Löffelmann, P. J. Smith, and J. G. Korvink, "Terahertz metamaterials fabricated by inkjet printing," *Appl. Phys. Lett.*, vol. 95, no. 25, 2009.

[27] R. C. Y. Auyeung, H. Kim, S. Mathews, and A. Piqué, "Laser forward transfer using structured light," *Opt. Express*, vol. 23, no. 1, pp. 422–430, Jan. 2015.

[28] H. Kim, J. S. Melinger, A. Khachatrian, N. A. Charipar, R. C. Y. Auyeung, and A. Piqué, "Fabrication of terahertz metamaterials by laser printing," *Opt. Lett.*, vol. 35, no. 23, pp. 4039–4041, Dec. 2010.

[29] T. J. Yen, W. J. Padilla, N. Fang, D. C. Vier, D. R. Smith, J. B. Pendry, D. N. Basov, and X. Zhang, "Terahertz magnetic response from artificial materials," *Science (80-.).*, vol. 303, no. 5663, pp. 1494–1496, 2004.

[30] K. Fan and W. J. Padilla, "Dynamic electromagnetic metamaterials," *Mater. Today*, vol. 18, no. 1, pp. 39–50, Jan. 2015.

[31] H.-T. Chen, W. J. Padilla, J. M. O. Zide, A. C. Gossard, A. J. Taylor, and R. D. Averitt, "Active terahertz metamaterial devices," *Nature*, vol. 444, no. 7119, pp. 597–600, Nov. 2006.

[32] I. B. Vendik, O. G. Vendik, M. A. Odit, D. V Kholodnyak, S. P. Zubko, M. F. Sitnikova, P. A. Turalchuk, K. N. Zemlyakov, I. V Munina, D. S. Kozlov, V. M. Turgaliev, A. B. Ustinov, Y. Park, J. Kihm, and C.-W. Lee, "Tunable metamaterials for controlling THz radiation," *Terahertz Sci. Technol. IEEE Trans.*, vol. 2, no. 5, pp. 538–549, Sep. 2012.

[33] M. Rahm, J.-S. Li, and W. Padilla, "THz wave modulators: a brief review on different modulation techniques," *J. Infrared Millim. Terahertz Waves*, vol. 34, no. 1, pp. 1–27, 2013.

[34] W. J. Padilla, a. J. Taylor, C. Highstrete, M. Lee, and R. D. Averitt, "Dynamical electric and magnetic metamaterial response at terahertz frequencies," *Phys. Rev. Lett.*, vol. 96, no. 10, p. 107401, Mar. 2006.

[35] K. Fan, H. Y. Hwang, M. Liu, A. C. Strikwerda, A. Sternbach, J. Zhang, X. Zhao, X. Zhang, K. A. Nelson, and R. D. Averitt, "Nonlinear terahertz metamaterials via field-enhanced carrier dynamics in GaAs," *Phys. Rev. Lett.*, vol. 110, no. 21, p. 217404, May 2013.

[36] H.-T. Chen, J. F. O'Hara, A. K. Azad, A. J. Taylor, R. D. Averitt, D. B. Shrekenhamer, and W. J. Padilla, "Experimental demonstration of frequency-agile terahertz metamaterials," *Nat Phot.*, vol. 2, no. 5, pp. 295–298, May 2008.

[37] D. Shrekenhamer, S. Rout, A. C. Strikwerda, C. Bingham, R. D. Averitt, S. Sonkusale, and W. J. Padilla, "High speed terahertz modulation from metamaterials with embedded high electron mobility transistors," *Opt. Express*, vol. 19, no. 10, pp. 9968–9975, May 2011.

[38] R. A. Coutu, P. J. Collins, E. A. Moore, D. Langley, M. E. Jussaume, and L. A. Starman, "Electrostatically tunable meta-atoms integrated with in situ fabricated MEMS cantilever beam arrays," *Microelectromech. Syst. J.*, vol. 20, no. 6, pp. 1366–1371, Dec. 2011.

[39] H. Tao, A. Strikwerda, C. Bingham, W. J. Padilla, X. Zhang, and R. D. Averitt, "Dynamical control of terahertz metamaterial resonance response using bimaterial cantilevers," in *Progress in Electromagnetics Research Symposium*, Electromagnetics Academy, 2008, pp. 856–859.

[40] F. Ma, L. Yu-Sheng, Z. Xinhai, and C. Lee, "Tunable multiband terahertz metamaterials using a reconfigurable electric split-ring resonator array," *Light Sci. Appl.*, vol. 3, p. e171, May 2014.

[41] H. Tao, A. C. Strikwerda, K. Fan, W. J. Padilla, X. Zhang, and R. D. Averitt, "MEMS based structurally tunable metamaterials at terahertz frequencies," *J. Infrared Millim. Terahertz Waves*, vol. 32, no. 5, pp. 580–595, 2011.

[42] H. Tao, W. J. Padilla, X. Zhang, and R. D. Averitt, "Recent progress in electromagnetic metamaterial devices for terahertz applications," *Sel. Top. Quantum Electron. IEEE J.*, vol. 17, no. 1, pp. 92–101, Jan. 2011.

[43] C. H. Lam, "History of phase change materials," in *Phase Change Materials: Science and Applications*, 1st ed., S. Raoux and M. Wuttig, Eds. New York, NY: Springer, 2009, pp. 1–14.

[44] A. H. Gwin, "Materials Characterization and Microelectronic Implementation of Metal-Insulator Transition Materials and Phase Change Materials," Air Force Inst Tech, 2015.

[45] D. Bouyge, A. Crunteanu, J.-C. Orlianges, D. Passerieux, C. Champeaux, A. Catherinot, A. Velez, J. Bonache, F. Martin, and P. Blondy, "Reconfigurable bandpass filter based on split ring resonators and vanadium dioxide (VO2) microwave switches," in *Microwave Conference, 2009. APMC 2009. Asia Pacific*, 2009, pp. 2332–2335.

[46] M. J. Dicken, K. Aydin, I. M. Pryce, L. A. Sweatlock, E. M. Boyd, S. Walavalkar, J. Ma, and H. A. Atwater, "Frequency tunable near-infrared metamaterials based on VO2 phase transition," *Opt. Express*, vol. 17, no. 20, pp. 18330–18339, Sep. 2009.

[47] B. Gholipour, J. Zhang, J. Maddock, K. F. Macdonald, D. W. Hewak, and N. I. Zheludev, "All-optical, non-volatile, chalcogenide phase-change meta-switch," in *Lasers and Electro-Optics Europe (CLEO EUROPE/IQEC), 2013 Conference on and International Quantum Electronics Conference*, 2013, p. 1.

[48] Z. L. Sámson, S.-C. Yen, K. F. MacDonald, K. Knight, S. Li, D. W. Hewak, D.-P. Tsai, and N. I. Zheludev, "Chalcogenide glasses in active plasmonics," *Phys. Stat. Solidi—Rapid Res. Lett.*, vol. 4, no. 10, pp. 274–276, 2010.

[49] Z. L. Sámson, K. F. MacDonald, F. De Angelis, B. Gholipour, K. Knight, C. C. Huang, E. Di Fabrizio, D. W. Hewak, and N. I. Zheludev, "Metamaterial electro-optic switch of nanoscale thickness," *Appl. Phys. Lett.*, vol. 96, no. 14, 2010.

Chapter 7

Determining the Complex Refractive Index of Materials in the Far-Infrared from Terahertz Time-Domain Data

Maxime Bernier, Frédéric Garet and
Jean-Louis Coutaz

Additional information is available at the end of the chapter

http://dx.doi.org/10.5772/66348

Abstract

Terahertz time-domain spectroscopy is a well-established technique to study the far-infrared electromagnetic response of materials. Measurements are broadband, fast, and performed at room temperature. Moreover, compact systems are nowadays commercially available, which can be operated by nonspecialist staff. Thanks to the determination of the amplitude and phase of the recorded signals, both refractive index and absorption coefficient of the sample material can be obtained. However, determining these electromagnetic parameters should be performed cautiously when samples are more or less transparent. In this chapter, we explain how to extract the material parameters from terahertz time-domain data. We list the main sources of error, and their contribution to uncertainties. We give rules to select the most adapted technique for an optimized characterization, depending on the transparency of the samples, and address the case of samples with strong absorption peaks or exhibiting scattering.

Keywords: terahertz time-domain spectroscopy, transmission TDS, reflection TDS, attenuated total reflection, extraction precision, Kramers-Kronig relations, scattering

1. Introduction

Up to the end of the 1980s, the far-infrared electromagnetic response of materials was mostly investigated thanks to Fourier transform infrared (FTIR) spectroscopy, which exhibits several advantages. During one scan, the recorded time-equivalent waveform is built from information delivered by the entire spectrum, whereas other dispersive prism- or grating-based spectrometers receive at any time only signal from a narrow band, i.e., a weaker signal with a smaller signal-over-noise ratio (SNR). Second, and oppositely to dispersive spectrometers, the resolution of FTIR instruments is not limited by the size of the source. These are respectively known as the multiplex and *étendue* advantages [1]. Nevertheless, FTIR instruments display some drawbacks mostly due to the lack of efficient sources and detectors of far-infrared waves. The sources must be broadband in view of achieving narrow waveforms and thus performing broad spectral measurements. Generally, blackbody-like sources are implemented: they are rather powerless and deliver incoherent light. Thus, a long integration time is necessary to get a high SNR. Moreover, to obtain a high-frequency resolution, the waveform must be recorded over a long equivalent time window, during which noise is also recorded. Sensitive detectors, like Si bolometers, operate only at cryogenic temperatures. These experimental problems, i.e., long recording times and cryogenic temperature, were solved by Auston and Chueng [2] in 1985 who introduced and demonstrated the new concept of coherent time-domain far-infrared spectroscopy, known today as terahertz time-domain spectroscopy (THz-TDS). This initial work was completed by researchers at IBM Corp. [3, 4], who definitively installed THz-TDS as a very competing tool to study the far-infrared properties of materials and devices. Since these pioneering researches, a strong effort has been devoted by numerous laboratories worldwide to improve THz-TDS equipment and techniques. Today, several books describe this technology [5–9] and commercial systems are available [10]. THz-TDS typically permits to investigate the range 0.1–5 THz, but some recent systems allow one to reach the mid-infrared, i.e., frequencies larger than 10 THz [11–13]. In addition, the time-domain technique makes possible to perform optical-pump and THz-probe time-resolved experiment to study the carrier dynamics in semiconductors [14, 15] or the kinetics of photo-induced chemical reactions [16, 17]. Moreover, nonlinear THz effects can be observed, thanks to the huge THz peak power in THz-TDS systems fed by amplified mode-locked lasers [18].

In this chapter, we describe the principles of THz-TDS and we explain how to extract the refractive index and the coefficient of absorption of materials from THz-TDS data. We study the precision of this determination versus different error sources. We give rules to choose, depending on the samples under test, the most adapted THz-TDS technique among transmission or reflection ones. We also treat the case of materials that exhibit strong absorption bands, and heterogeneous materials that scatter and/or diffract the THz beam.

2. Principles and basics of THz-TDS

In THz-TDS setups, a train of ultra-short laser pulses excites a THz antenna, which converts each optical pulse into an electromagnetic (EM) burst and radiates it in free space. In other words, the carrier frequency of the optical pulse is rectified and only its envelope is saved. Because of a noninstantaneous response of the antenna, the conversion widens the EM pulse duration when compared to the optical one. Thanks to some THz optical system, the EM pulse is focused onto a receiving THz antenna, which is triggered by a delayed part of the laser beam. In the receiving antenna, a nonlinear process mixes both incoming EM pulse and laser pulse, giving rise to a signal integrated by the reading electronics, which is proportional to the convolution product of the laser pulse and the electrical field of the EM pulse. By varying the time delay between emission and detection, which corresponds to a time-equivalent sampling, the temporal waveform of the convolution product is obtained. Two major features should be noted: (i) emitter and receiver are enlightened by the same pulsed laser beam, thus they are perfectly synchronously excited; and (ii) because it is triggered by ultra-short laser pulses, the receiver records the EM pulse signal only when it is excited by the laser pulses (typically during a 1-ps time slot): noise in the time interval between two consecutive laser pulses (typically 10 ns) is not recorded. This amazing 1 ps/10 ns = 10^4 ratio, associated with the perfect synchronization between emission and detection, together with the high stability of mode-locked laser pulse comb, makes the dynamics of THz-TDS extremely high, usually larger than 60 dB in power. The spectrum of the signal is calculated through a numerical Fourier transform of the convolution trace. The Fourier transform supplies a complex value, with a modulus and a phase. The phase is related, in the time domain, to the relative origin of the time delay between emission and detection while the modulus spectrum depends strongly on (i) the spectral efficiency of the emitting and receiving THz antennas, and (ii) the adjustment of the quasi-optical THz-TDS system. The most common THz antennas are photo-conducting switches made from ultrafast semi-insulating semiconductors, like low-temperature grown GaAs (LTG-GaAs). Basically, a microstrip line with a narrower gap at its center is deposited over the semiconductor substrate. In emission, the structure is DC biased and the gap becomes conductive when illuminated by a laser pulse. It behaves as a dipole whose moment varies promptly due to the photoconduction process. This dipole radiates in the far-field region an EM signal proportional to the second time-derivative of the moment variation, i.e., to the first derivative of the current (conduction and displacement) surge flowing through the gap. In such antennas, detection occurs through a complementary effect. The gap is biased by the incoming THz field $E(t)$ that accelerates the free carriers synchronously generated by the triggering laser pulses. This current, proportional to the THz field, is read and time-integrated by the electronics, usually a lock-in amplifier and it writes:

$$S(\tau) \propto \int_{-\infty}^{+\infty} E(t)N(t-\tau)\,dt, \tag{1}$$

where τ is the time delay between emission and detection, and $N(t-\tau)$ is the number of photo-carriers resting at time t from a generation occurring a delay τ before. The spectrum of $S(\tau)$ is obtained by a Fourier transform of Eq. (1):

$$\tilde{S}(\omega) = \mathrm{TF}[S(\tau)] \propto \int_{-\infty}^{+\infty}\int_{-\infty}^{+\infty} E(t)N(t-\tau)e^{-j\omega\tau}\,dt\,d\tau \propto \tilde{E}(\omega)\tilde{N}^{*}(\omega) \tag{2}$$

ω is the angular frequency, $\tilde{E}(\omega)$ and $\tilde{N}(\omega)$ are the spectra of $E(t)$ and $N(t)$, and $\tilde{N}^{*}(\omega)$ is the complex conjugate of $\tilde{N}(\omega)$. From now on, a tilde denotes a complex variable.

THz emission can also be obtained either by illuminating the bare surface of an ultrafast semiconductor wafer, at which photo-generated carriers are accelerated inside the wafer by surface fields or/and by the Dember effect [19], or by optical rectification in an electro-optic (EO) crystal effect. Detection is also commonly performed by EO sampling. Characterization of a sample is mostly achieved in transmission by locating the sample in the THz beam, and recording the THz waveforms without (reference) and with the sample. Then, one Fourier transforms the waveforms, and the complex transmission coefficient $\tilde{T}_{\text{meas}}(\omega)$ of the sample is equal to the ratio of the signal and reference spectra:

$$\tilde{T}_{\text{meas}}(\omega) = \frac{\tilde{S}(\omega)}{\tilde{S}_{\text{ref}}(\omega)} = \frac{\tilde{E}(\omega)\tilde{N}^{*}(\omega)}{\tilde{E}_{\text{ref}}(\omega)\tilde{N}^{*}(\omega)} = \frac{\tilde{E}(\omega)}{\tilde{E}_{\text{ref}}(\omega)}. \tag{3}$$

The same procedure is performed in reflection, providing the experimental complex coefficient of reflection $\tilde{R}_{\text{meas}}(\omega)$. In this case, the reference signal is the THz waveform reflected by a perfect mirror placed exactly at the same position as the sample to be tested. To avoid the difficulty of the exact positioning of the reference mirror, attenuated total reflection (ATR) scheme is preferred [20]. ATR set-ups are especially dedicated to characterize liquids or powders since (i) THz radiations are strongly absorbed by classical liquids like water avoiding any transmission measurement, and (ii) ATR scheme is sensitive enough to characterize dilutions whose solute concentration could be as small as 1%. In ATR experiment, the THz beam is reflected against the base of a prism whose index of refraction is higher than that of the studied substance to achieve total internal reflection of the THz beam. The reference is recorded without the material placed upon the prism base to finally obtain the measured complex total reflection coefficient $\tilde{R}_{\text{ATR,meas}}(\omega)$.

The complex refractive index $\tilde{n} = n(\omega) + j\kappa(\omega)$ of the sample to be characterized is determined when the calculated transfer function (i.e., either $\tilde{T}(\omega,\tilde{n})$, $\tilde{R}(\omega,\tilde{n})$, or $\tilde{R}_{\text{ATR}}(\omega,\tilde{n})$), in which $\tilde{n}$ is the only adjustable variable, is equal to the experimental ones ($\tilde{T}_{\text{meas}}(\omega)$, $\tilde{R}_{\text{meas}}(\omega)$, or $\tilde{R}_{\text{ATR,meas}}(\omega)$, respectively). When dealing with nonmagnetic materials, analytical expressions of the transfer functions are rather simple as long as the samples are slabs with parallel and flat faces. At the sample location, the incoming THz beam must be a plane wave. In practice, this is verified even with focused THz Gaussian beams, as far as the sample, placed at the focal point, is thinner than the Rayleigh length of the beam. Thus, analytical expressions of $\tilde{T}(\omega,\tilde{n})$, $\tilde{R}(\omega,\tilde{n})$, and $\tilde{R}_{\text{ATR}}(\omega,\tilde{n})$ are as follows:

$$\tilde{T}(\omega,\tilde{n}) = \frac{4\tilde{K}}{(\tilde{K}+1)^2}\phi^{-1}FP(\omega,\tilde{n})e^{-j\frac{\omega}{c}\cos\theta\, d} \tag{4}$$

$$\tilde{R}(\omega,\tilde{n}) = \pm\frac{1-\tilde{K}}{1+\tilde{K}}(1-\phi^{-2})FP(\omega,\tilde{n}),\ \tilde{R}_{\text{ATR}}(\omega,\tilde{n}) = \pm\frac{1-\tilde{K}}{1+\tilde{K}} \tag{5}$$

with $\tilde{K}_{TE} = \tilde{n}^2\tilde{K}_{TM} = \frac{\sqrt{\tilde{n}^2-\sin^2\theta}}{\cos\theta}$, $\phi = e^{j\frac{\omega}{c}d\sqrt{\tilde{n}^2-\sin^2\theta}}$, and $FP(\omega,\tilde{n}) = \left[1-\left(\frac{\tilde{K}-1}{\tilde{K}+1}\right)^2\phi^2\right]^{-1}$.

θ is the angle of incidence, d is the sample thickness, and c the speed of the light in vacuum. In Eq. (5), the sign "+" is for the TE polarization case and "-" for the TM one. In the ATR case, $\tilde{K}$ takes similar expressions in which $\tilde{n}$ is substituted by $\tilde{n}/n_p$, where n_p is the refractive index of the prism. $FP(\omega,\tilde{n})$ is the Fabry-Perot term that accounts for the rebounds of the THz pulse in the sample. In ATR experiments, we suppose that d is much larger than the evanescent length of the THz beam in the sample, so rebounds are neglected as well as in reflection when the sample is opaque or strongly absorbing. In this case, for both reflection and ATR cases, solving $\tilde{R}(\omega,\tilde{n}) = \tilde{R}_{\text{meas}}(\omega)$ is easy because Eq. (5) can be analytically inversed:

$$\tilde{K} = \frac{1\pm\tilde{R}(\omega,\tilde{n})}{1\mp\tilde{R}(\omega,\tilde{n})} \tag{6}$$

Solving Eq. (4) (or Eq. (5) for thin samples) is more difficult because of the oscillatory complex exponential term ϕ. Consequently, the modulus and phase of $\tilde{T}(\omega,\tilde{n})$ or $\tilde{R}(\omega,\tilde{n})$ plotted in function of n and κ exhibit several minima (see **Figure 1**). Therefore, finding the right zero of $\tilde{T}(\omega,\tilde{n}) = \tilde{T}_{\text{meas}}(\omega)$ or $\tilde{R}(\omega,\tilde{n}) = \tilde{R}_{\text{meas}}(\omega)$ is not an easy task.

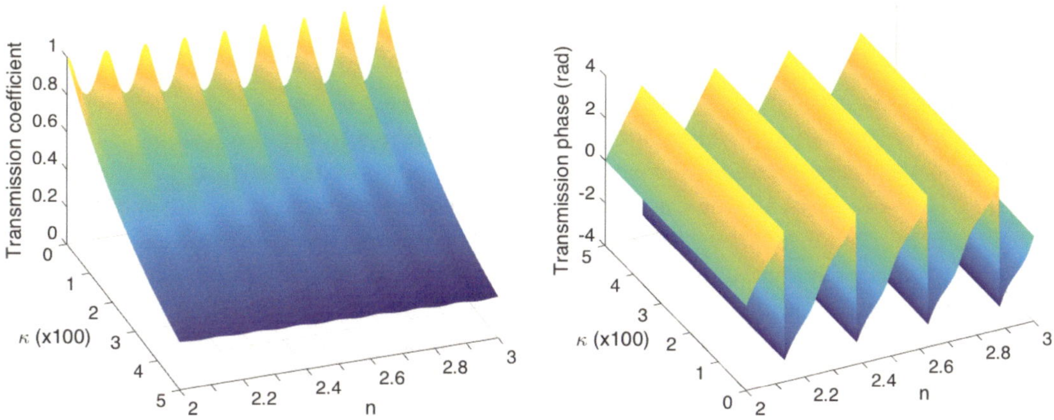

Figure 1. Map of the modulus (left) and the wrapped phase (right) of $\tilde{T}(\omega, \tilde{n})$ versus n and κ. The sample is 1.2-mm thick and the frequency is 1 THz.

To get rid of the oscillatory behavior, we proposed [21] to employ an error function $\delta(\omega, \tilde{n})$ that exhibits a monotonous shape with a single minimum that is quickly found with any numerical method:

$$\delta(\omega, \tilde{n}) = \left[\ln(|\tilde{X}(\omega, \tilde{n})|) - \ln\left(\left|\tilde{X}_{meas}(\omega)\right|\right)\right]^2 + \left[\arg(\tilde{X}(\omega, \tilde{n})) - \arg\left(\tilde{X}_{meas}(\omega)\right)\right]^2 \quad (7)$$

with $\tilde{X} = \tilde{T}$ or $\tilde{R}$. **Figure 2** shows $\delta(\omega, \tilde{n})$ in transmission versus n and κ for two different cases: (i) transmission modulus $T_{meas} = 0.01$, phase $\varphi_{T,meas} = 10$ rad, $\omega d/c = 3$ (e.g., a 180-μm thick sample studied at $f = 0.8$ THz), (ii) $T_{meas} = 0.003$, $\varphi_{T,meas} = 30$ rad, $\omega d/c = 30$ (e.g., a 2.4-mm thick sample studied at $f = 0.6$ THz). A single zero of $\delta(\omega, \tilde{n})$ is clearly seen, at which n and κ can be almost instantaneously extracted using a numerical routine.

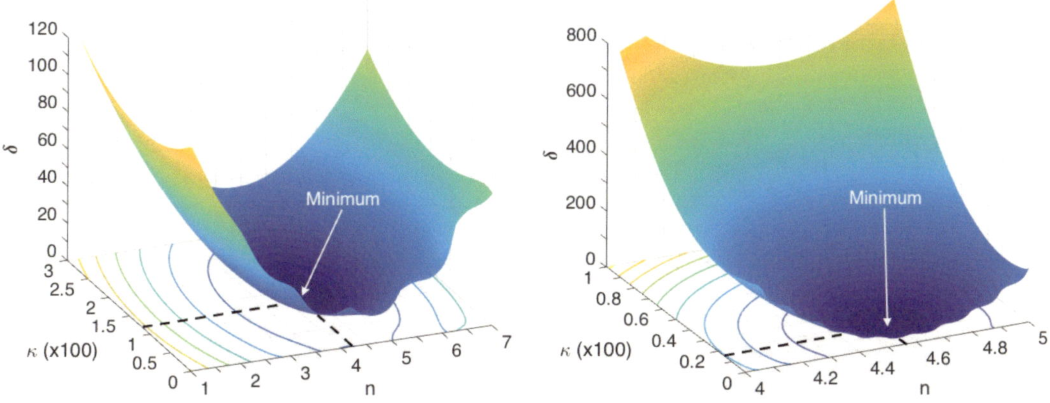

Figure 2. Maps of the error function δ versus n and κ, with T meas = 0.01, φ T,meas = 10 rad, $\omega d/c = 3$ (left), and T meas = 0.003, φ T,meas = 30 rad, $\omega d/c = 30$ (right).

3. Precision on the parameters determination

The material parameters (n and κ) determined by THz-TDS are obtained with a precision that is limited [22] by (i) a bad knowledge of the sample parameters (erroneous thickness, not parallel sides, surface roughness, inhomogeneity, etc.), (ii) a bad positioning of the sample, (iii) the fact that the THz beam is not a plane wave, and (iv) the experimental noise. Also, one must take into account errors and noises arising from a deficient delay line (mechanical drift, registration, etc.). In Ref. [22], Withayachumnankul et al. have given a complete study of the causes of uncertainty in THz-TDS in terms of variance of the recorded signals. Here our approach is simplified but leads to analytical expressions, which emphasize the role and the weight of each experimental parameter.

3.1. Effect of a bad value of the sample thickness

A bad value of the thickness d modifies the results in the case of transmission THz-TDS. The induced error Δn is calculated by setting the differential of $\tilde{T}(\omega, \tilde{n})$ equal to zero. We get:

$$\Delta\tilde{n} = \Delta d \frac{\partial \tilde{T}}{\partial d} / \frac{\partial \tilde{T}}{\partial \tilde{n}} \tag{8}$$

Let us suppose, for the sake of sim and reflection are obtained by differentiating plification, that the Fabry-Perot rebounds can be removed by a proper time-windowing of the THz waveform. Thus, Eq. (4) of $\tilde{T}(\omega, \tilde{n})$ becomes:

$$\tilde{T}(\omega, \tilde{n}) \approx \frac{4\tilde{K}}{(\tilde{K}+1)^2} e^{j\frac{\omega}{c}\left(cos\theta - \sqrt{\tilde{n}^2 - sin^2\theta}\right)d} \tag{9}$$

A simple calculation leads to:

$$\Delta\tilde{n}_{TE} = \frac{j\frac{\omega}{c\tilde{n}}(\tilde{X} - cos\theta)\tilde{X}^2}{\frac{\tilde{X} - cos\theta}{\tilde{X} + cos\theta} + j\tilde{X}\frac{\omega}{c}d}\Delta d, \quad \Delta\tilde{n}_{TM} = \frac{j\frac{\omega}{c\tilde{n}}(\tilde{X} - cos\theta)\tilde{X}^2}{\frac{\tilde{X} - \tilde{n}^2 cos\theta}{\tilde{X} + \tilde{n}^2 cos\theta}\left(1 - 2\frac{\tilde{X}^2}{\tilde{n}^2}\right) + j\tilde{X}\frac{\omega}{c}d} \tag{10}$$

with $\tilde{X} = \sqrt{\tilde{n}^2 - sin^2\theta}$. Usually, experiments are performed at normal incidence, for which Eq. (10) simplifies into:

$$\Delta\tilde{n} = j\frac{\omega}{c}\frac{\tilde{n} - 1}{\frac{\tilde{n}-1}{\tilde{n}+1} + j\frac{\omega}{c}\tilde{n}d}\Delta d. \tag{11}$$

Typically, an error of 1% on d leads to an error smaller than 1% on n and κ (see **Figure 3**).

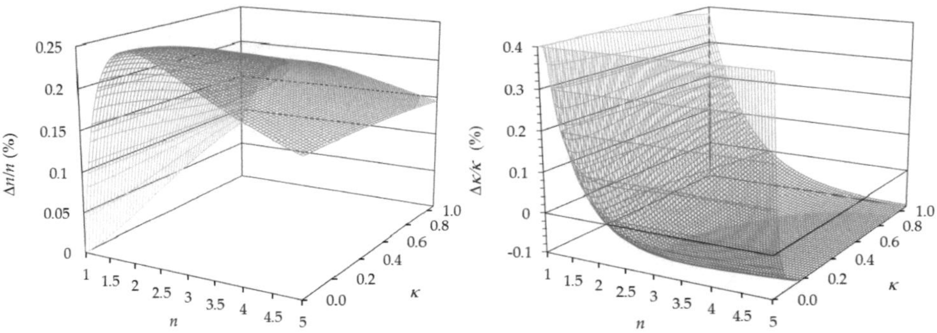

Figure 3. Δn (left) and $\Delta\kappa$ (right) in %, versus n and κ, induced by a Δd =10 μm thickness error. The sample is 1-mm thick ($\Delta d/d$= 1%) and f = 1 THz.

3.2. Effect of an angular tilt

Using the same procedure, we investigate the influence of a bad orientation of the sample. The derivation is done versus the tilt angle $\Delta\theta$. Under normal incidence, we get in transmission:

$$\Delta\tilde{n}_{TE} = \frac{j\frac{\omega}{c}\tilde{n}\,d}{\frac{\tilde{n}-1}{\tilde{n}+1}+j\frac{\omega}{c}d}\Delta\theta, \Delta\tilde{n}_{TM} = \frac{j\frac{\omega}{c}\tilde{n}\,d}{\frac{\tilde{n}-1}{\tilde{n}+1}+j\frac{\omega}{c}\tilde{n}d}\Delta\theta. \quad (12)$$

A slight difference is obtained if the angular tilt is along the direction of the E-field (TE) or perpendicular to it (TM). As shown in **Figure 4**, an angular tilt of 1° induces typically an error Δn much smaller than 1%, except for small values of n and κ, for which the error is of the order of 1%. However, as imaging the THz-TDS beam is almost impossible, adjusting precisely the orientation of the sample in the beam is a difficult task and angular error tilts of several degrees are possible, which leads to errors larger than 1%.

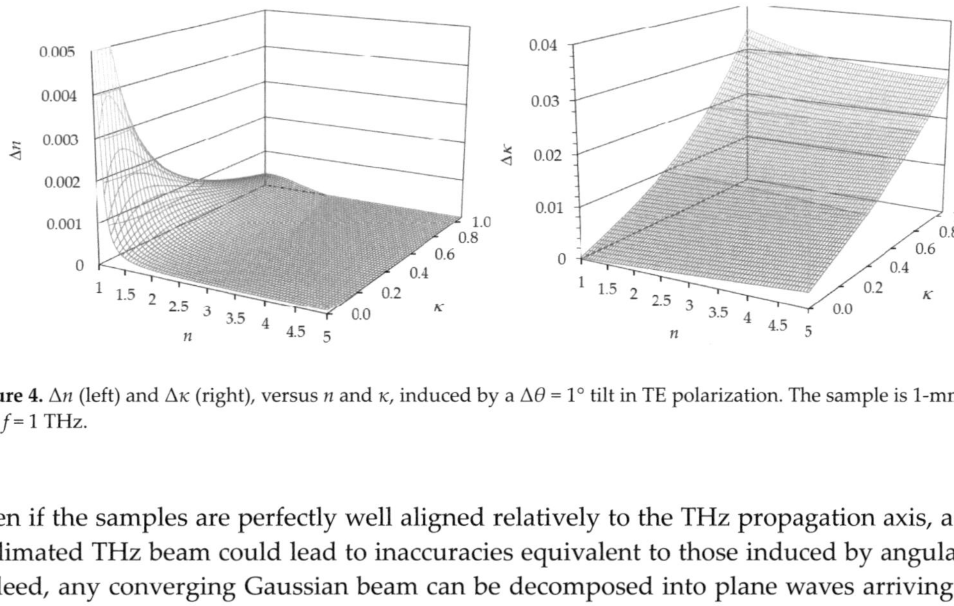

Figure 4. Δn (left) and $\Delta\kappa$ (right), versus n and κ, induced by a $\Delta\theta = 1°$ tilt in TE polarization. The sample is 1-mm thick and $f = 1$ THz.

Even if the samples are perfectly well aligned relatively to the THz propagation axis, a non-collimated THz beam could lead to inaccuracies equivalent to those induced by angular tilt. Indeed, any converging Gaussian beam can be decomposed into plane waves arriving onto the sample under different incidence angles, from 0° (along the propagation axis) up to the diffraction angle whose value depends on the frequency. Referring to **Figure 4**, this could induce an error of about 1%. This unwanted effect gets even worse as THz converging Gaussian beam probes thick samples, because it defocuses the THz beam that impinges the receiver. Consequently, the detected signal can be respectively larger or weaker than expected, which leads to an over- or underestimation of the sample losses.

In reflection, a bad orientation (angular tilt $\Delta\theta$) of the tested sample leads to an additional error given by:

$$\Delta\tilde{n}_{TE} = \Delta\tilde{n}_{TM} = \frac{\tilde{n}^2-1}{2\tilde{n}}(\Delta\theta)^2 \quad (13)$$

By comparing expressions (12)—transmission—and (13)—reflection—it appears that the error Δn induced by an angular tilt is weaker (typ. 1/1000) in reflection than in transmission: this is due to the additional path of propagation in the material induced by the tilt in the transmission scheme, while the effect on the Fresnel coefficients at the sample interfaces is smaller. Nevertheless, it should be noticed that an angular tilt, in reflection THz-TDS, is more perturbing, as all the energy of the deviated THz beam may not reach the detector. This error, which depends on each set-up design, could be quite large in reflection THz-TDS, but it could be wiped out by carefully aligning the set-up and the sample.

3.3. Effect of a bad positioning of the reference mirror in reflection THz-TDS

In the case of reflection THz-TDS, the main geometrical error source is the misalignment of the sample as compared to the reference mirror. For the sake of simplicity, we suppose that the sample is thick enough to neglect the Fabry-Perot rebounds. Let δ be the difference in position, which induces an erroneous phase difference $2\delta\,\omega/c$:

$$\tilde{R}_{\text{meas}}(\omega,\tilde{n}) \;=\; \tilde{R}(\omega,\tilde{n})\; e^{j2\frac{\omega}{c}\delta} \;=\; \frac{\tilde{n}-1}{\tilde{n}+1}\, e^{j2\frac{\omega}{c}\delta} \;=\; \tilde{R}(\omega,\tilde{n}+\Delta\tilde{n}) \;=\; \frac{\tilde{n}+\Delta\tilde{n}-1}{\tilde{n}+\Delta\tilde{n}+1} \tag{14}$$

This phase difference leads to an error $\Delta\tilde{n}$:

$$\Delta\tilde{n} \;\approx\; j2\tfrac{\omega}{c}(\tilde{n}^2-1)\delta \tag{15}$$

Because of the j term, the error Δn depends on the imaginary part of $\tilde{n}^2$, i.e., it is proportional to $n\kappa$. For transparent materials ($\kappa \approx 0$), the error is negligible, but it becomes large for opaque materials that are usually characterized in reflection THz-TDS. The error $\Delta\kappa$ that varies as $n^2-\kappa^2$ is small for opaque materials and large for transparent materials. Therefore, THz-TDS in reflection is a well-adapted and precise technique for the determination of the absorption of opaque materials. **Figure 5** shows the effect of a 1-µm position shift versus n and α at $f = 1$ THz. The error Δn is of the order of a few percent, while the error $\Delta\alpha$ is larger especially for transparent materials ($\Delta\alpha$ ~100 cm^{-1} for $n = 2.6$).

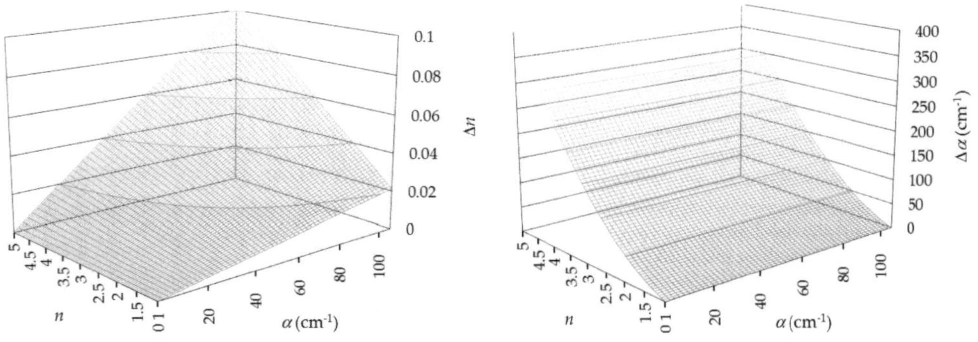

Figure 5. Δn (left) and $\Delta\alpha$ (right), versus n and α, induced by a $\delta = 1$-µm shift of the mirror position in reflection THz-TDS for $f = 1$ THz.

These rather large errors are induced by very small shifts, here $\delta = \lambda/300$, and get even larger at higher frequencies, according to Eq. (15). Therefore, in reflection THz-TDS, a great attention must be paid to position the sample at the exact location of the reference mirror, or, if not possible, to either control, measure, or correct the induced phase difference. For that purpose, it exists several experimental [23–26] or numerical [27] solutions.

3.4. Effect of a noise with photo-conducting THz antennas

Noise makes uncertain the measured values of the magnitude and the phase of the THz signals. Let us treat here only the case of photo-conducting antennas. We call σ^2 the noise power, i.e., the square of the variance of the THz signals. A first noise σ_E^2 is generated by the emitter: shot noise due to the random arrival of the pump laser photons, fluctuation of the laser intensity, etc. Drift and mechanical vibrations of the optical delay line as well as random fluctuations of the laser beam direction add a noise-equivalent contribution. It depends strongly on the equipment: Withayachumnankul et al. [22] have measured an amplitude variance of the order of 10^{-3} that was mostly due to delay-line registration and mechanical drift. This noise is of the same order as σ_E^2 defined previously and can be included in the emitter noise. The THz beam, together with its noise, is then reflected or transmitted by/through the sample toward the receiving antenna. At the receiver, two additional noises perturb the recorded signal, namely, the shot noise σ_{sh}^2 that is proportional to the recorded current S_{meas} ($\sigma_{sh}^2 = 2e\Delta f\, S_{meas}$, Δf is the detection bandwidth and e the electron charge), and a noise σ_D^2 that is independent of the current (Johnson noise, amplification noise, thermal noise, etc.). The total noise is the sum of all these contributions:

$$\sigma^2 = \left|\frac{S_D}{S_E}\right|^2 \sigma_E^2 + 2e\,\Delta f\; S_D + \sigma_D^2, \tag{16}$$

where S_E is the current recorded without sample in the set up (reference current), whereas S_D is the current recorded with the sample. Taking into account the small value of the noise when compared to the signal (perturbation approach), we get:

$$\sigma_X^2 = A(\omega)X^2 + B(\omega)X + C(\omega). \tag{17}$$

Here, X is the modulus of the complex coefficients of reflection $\tilde{R}$ or transmission $\tilde{T}$, depending on the type of performed THz-TDS measurements. A, B, and C are parameters specific to each THz-TDS set-up and are given by:

$$A(\omega) = \frac{2\,\sigma_E^2 + 2e\,\Delta f\,|S_E| + \sigma_D^2}{|S_E|^2}, \quad B(\omega) = \frac{2e\,\Delta f}{|S_E|}, \quad C(\omega) = \frac{2e\,\Delta f\,\sigma_D^2}{|S_E|^2} \tag{18}$$

Figure 6 shows the A, B, and C coefficients of our homemade THz-TDS set up, built around a mode-locked laser (Tsunami Spectra-Physics, 50 fs pulse duration at the antennas, 82 MHz repetition rate, pumped with a CW Millenia laser) and LTG-GaAs dipole-like THz antennas. The coefficients were determined as follows. Several samples of different thicknesses, and thus of different transmission coefficients, were made from the same material (Stycast glue). For each sample, several transmission THz-TDS data were recorded, and the noise was deduced from the standard deviation of the transmission spectra. Then, the noise was plotted at any given frequency as a function of T and fitted with expression (17). Typically, at the maximum of sensitivity of the LTG-GaAs antennas, i.e., around 0.5 THz, $A \approx 15 \times B \approx 200 \times C$. It follows that the main source term is the noise in the emitter (σ_E^2), which is larger than the noise in the detector (σ_D^2). As the main difference in receiving and emitting antennas is the photocurrent (both antennas are similar and excited by the same laser power), we conclude that the optical shot noise and laser intensity fluctuations in the emitter are the major source of noise.

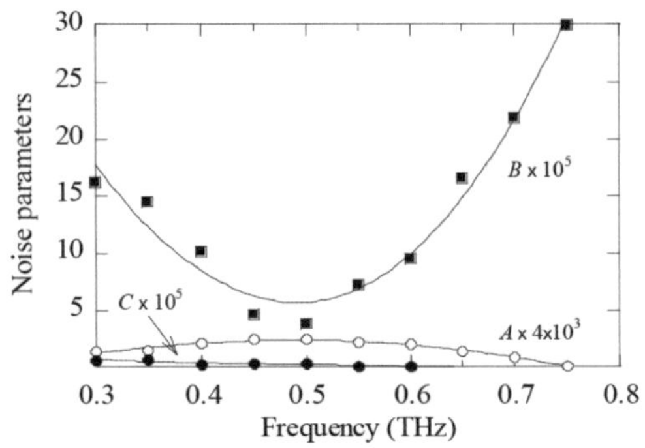

Figure 6. *A, B,* and *C* (SI unit) coefficients experimentally determined from transmission THz-TDS of several samples of Stycast having different thicknesses. The data have been recorded with a THz-TDS system equipped with LT-GaAs antennas. The lines are a guide to the eye.

Relations (17) and (18) indicate that the standard deviation is proportional to the detection bandwidth Δf. Thus, reducing Δf, i.e., increasing the integration time of a lock-in amplifier, decreases the error as well. On the other hand, the *A, B,* and *C* coefficients vary inversely to the available incoming THz signal. Increasing the THz power of the emitting antenna reduces the uncertainties. The influence of noise on the measurement precision is derived as follows. One writes any experimental spectral component, in terms of magnitude and phase, as the sum of the actual value and noise:

$$\tilde{X}_{meas}(\omega, \tilde{n}) = X_{meas}\, e^{j\varphi_{meas}} = X\, e^{j\varphi} + X_{noise}\, e^{j\varphi_{noise}} \tag{19}$$

Assuming the noise is much smaller than the signal, one easily gets:

$$X_{meas} = X + \cos\left(\varphi - \varphi_{noise}\right) X_{noise}, \qquad \tan\varphi_{meas} = \frac{X \sin\varphi + X_{noise}\sin\varphi_{noise}}{X\cos\varphi + X_{noise}\cos\varphi_{noise}} \tag{20}$$

With a fully random phase noise, the related standard deviations of the modulus and phase of the measured signals are:

$$\sigma_X = \overline{\left(X_{meas} - X\right)^2}^{1/2} = \sqrt{\frac{1}{2}\overline{X^2_{noise}}}, \qquad \sigma_\varphi = \frac{\sigma_X}{X}. \tag{21}$$

The validity of our analysis is experimentally verified as depicted in **Figure 7**. The standard deviations of the modulus *R* and phase φ of the signal reflected by a HR-Si wafer is plotted versus frequency. The dashed lines represent the standard deviation as obtained from averaging eight different measures, whereas the continuous lines are calculated with Eq. (21). Between 0.2 and 1.2 THz, SNR is large and the agreement between calculation and experiment is good. Outside this range, the noise is too large and thus our perturbation approach is no more valid.

Finally, one should derive the experimental standard deviation from the actual measured signals. The *X* value is the ratio of the transmitted or reflected signal S_X over reference signal S_{Ref}:

$$X_{meas} = \frac{S_X + S_{X,noise}}{S_{Ref} + S_{Ref,noise}} \approx X + \frac{S_{X,noise} - X\, S_{Ref,noise}}{S_{Ref}}. \tag{22}$$

The standard deviation is calculated from Eq. (22):

$$\sigma_X^2 = \frac{\sigma_X^2 + X^2\, \sigma_{Ref}^2}{S_{Ref}^2} \tag{23}$$

http://dx.doi.org/10.5772/66348

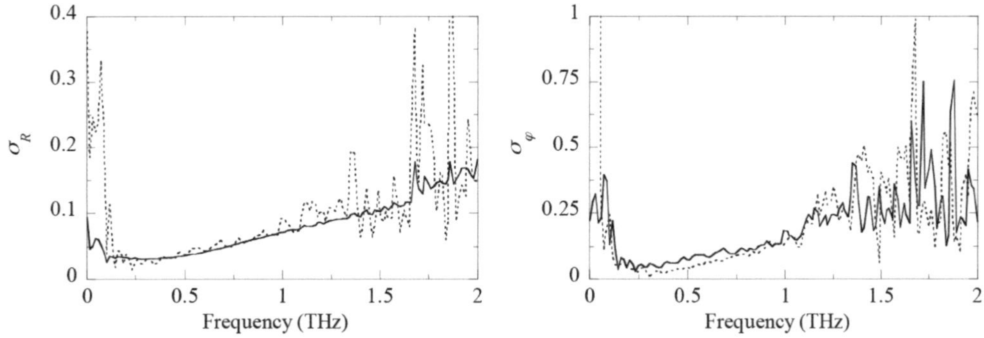

Figure 7. Standard deviations σ_R and σ_φ versus frequency for the THz signal reflected by a HR-Si wafer. The dashed line is obtained from eight different measurements, whereas the continuous line is calculated.

The noise-induced errors Δn_{T_R} and $\Delta \kappa_{T_R}$ in transmission and reflection are obtained by differentiating Eq. (4) or Eq. (5), respectively. In the general case of oblique incidence, the differential expressions are huge and complicated but get simpler in the case of normal incidence, when rebounds can be filtered. In this latter case one obtains:

$$\Delta n_T \simeq (1+n)^3 \frac{n\beta(1+n)+n-1}{(n-1)^2+n^2\beta^2(1+n)^2} e^{\kappa\beta} \frac{\sigma_T}{4}, \; \Delta \kappa_T \simeq \frac{\beta(1+n)-n+1}{n\beta(1+n)+n-1} \Delta n_T \tag{24}$$

$$\Delta n_R = \Delta \kappa_R = 2\frac{\left|(1+R^2)\cos\varphi - 2R\right| + \left|\sin\varphi\right|(1-R^2)}{\left(1-2R\cos\varphi + R^2\right)^2}\sigma_R. \tag{25}$$

$\beta = \frac{\omega d}{c}$. Examples of so-determined error Δn_{T_R} are presented in **Figure 8**. For the Stycast samples, THz-TDS experiments performed in transmission leads to the amazing precision $\Delta nT/n < 0.05\%$ between 0.6 and 1 THz. However, this value is only the noise-induced error and the actual total imprecision (~1%) is mainly due to a bad position of the sample or a bad knowledge of the sample thickness, as explained above. Conversely, characterizing HR-Si in reflection is definitively not the best way, as the uncertainty is never below 10%. We explain below how to choose the optimized THz-TDS technique in order to get a more precise determination of the material parameters. In the case of oblique incidence and of ATR measurements, one cannot get analytical expressions like Eqs. (24) and (25), and thus the errors should be numerically estimated.

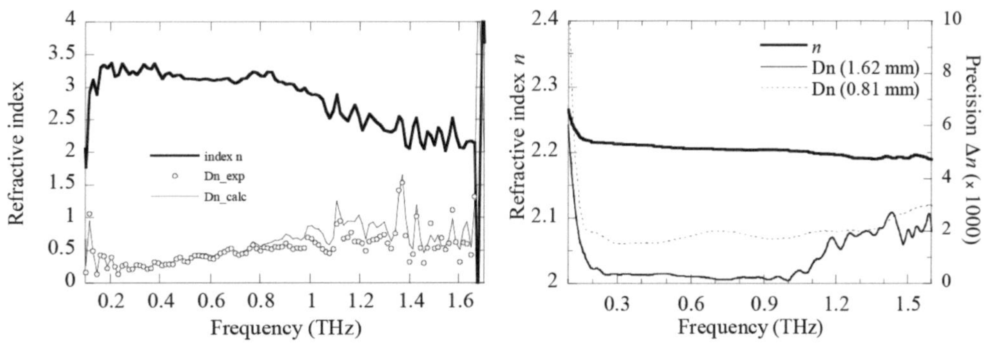

Figure 8. Spectra *n(f)* together with the error *Δn(f)*; left: HR-Si measured in reflection. Continuous line and open circles are calculated using Eq. (25) in which σR is estimated using expression (17) with A, B and C given in Figure 6, and measured, respectively; right: Stycast measured in transmission (sample thickness 0.81 and 1.62 mm).

Figure 9 summarizes the influence for the different error sources in transmission THz-TDS. The sample is a 1-mm thick slab, with $n = 2$ and $\kappa = 0.01$. The errors Δn and $\Delta\kappa$ are plotted versus frequency in the case of a thickness error $\Delta d = 0.01$ mm, of an angular tilt $\Delta\theta = 1°$, and a noise defined by Eq. (24) in which σ_T is substituted by expression (17) with $A = 10^{-3}$, $B = 10^{-5}$, and $C = 10^{-6}$.

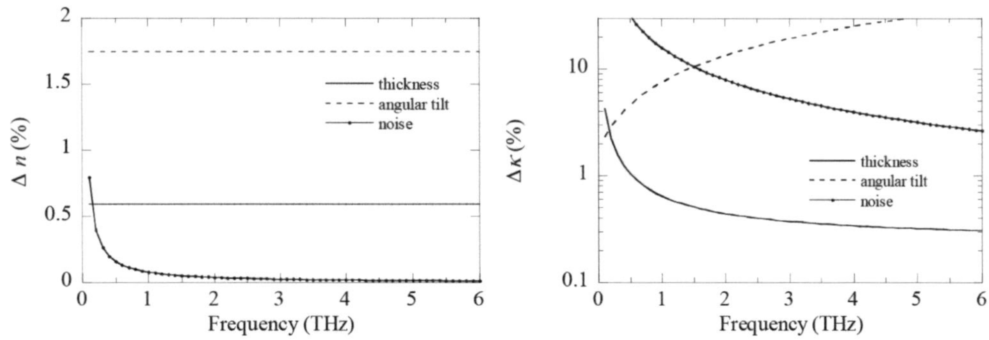

Figure 9. Δn (left) and $\Delta\kappa$ (right) versus frequency as induced by a 1% thickness error (thick continuous line), by a 1° tilt (thin continuous line) and by the noise (dotted line). Thes estimation is made in transmission for a 1-mm thick sample with $n = 2$ and $\kappa = 0.01$.

The angular tilt is by far the largest source of error for the refractive index n. κ is mostly affected by the noise at lower frequencies, while the angular tilt effect is predominant at higher frequencies. In any case, $\Delta\kappa$ is larger than Δn, especially at lower frequencies, where $\Delta\kappa$ in percent tends toward infinity because of the $1/\kappa$ term. The sum of all these errors is typically 1–2% for n and about 10% for κ. Because of the Fourier transform properties, the precision on the frequency is simply related to the recording time window, i.e., in most of THz set ups, to the length of the mechanical delay line. If the THz waveform is recorded over a time window $\Delta\tau$, the frequency resolution is $\Delta f = 1/\Delta\tau$. Common delay lines are 25–50 mm long, thus $\Delta\tau = 150 \sim 300$ ps, and $\Delta f = 3 \sim 6$ GHz. Using longer delay lines [28] to achieve a better frequency resolution is made difficult because of long-term fluctuations of the laser power and possible weak deviations of the direction of the laser beam.

4. Transmission or reflection THz-TDS?

For opaque samples, it is compulsory to perform THz-TDS measurements in reflection because no THz signal is transmitted. For low-index transparent samples, transmission scheme is preferable, as the induced phase variation is integrated over the whole sample length, while the Fresnel phase change in reflection are weaker. For samples with a moderate absorption coefficient, the choice of the best experimental scheme is not obvious. However, the error study presented in the previous paragraph helps in selecting the optimized THz-TDS characterization technique, i.e., either in reflection or in transmission. For the sake of simplicity, we address

http://dx.doi.org/10.5772/66348

here only the case of normal incidence. We suppose that the samples are perfectly placed in the THz beam (no angular tilt) and that the sample thickness is perfectly known. We suppose also that the sample is thick enough to permit a record of only the directly reflected or transmitted THz pulse, by a proper time-windowing of the other rebounds. Thus, the only source of inaccuracy is induced by intrinsic noises, given by relation (18), resulting on the errors expressed by Eqs. (24) and (25), where σx is expressed using Eq. (17). In reflection, Δn and $\Delta \kappa$ do not depend on the phase term $\beta = \omega d/c$, whereas they do in transmission. Indeed, the transmitted THz signal encounters a phase increase due to propagation through the sample. Thus, we perform our numerical study with β as parameter. **Figure 10** (left) shows $\Delta n_R/\Delta n_T$ versus n and κ calculated for $\beta = 20$ and $A = 10^{-3}$, $B = 10^{-5}$, and $C = 10^{-6}$.

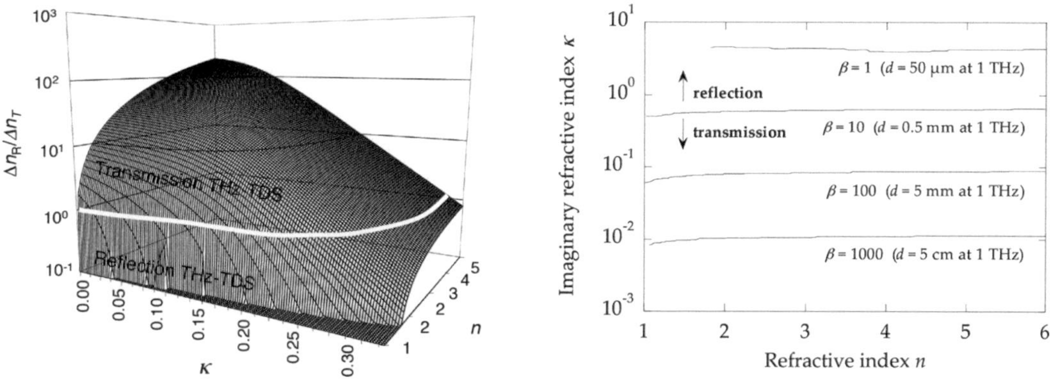

Figure 10. 3D map of the ratio $\Delta n R/\Delta n T$ versus n and κ for $\beta = 20$ (left); limit between the reflection and transmission schemes for an optimized parameter extraction (right).

The white curve indicates the limit $\Delta n_R = \Delta n_T$. Above this limit, $\Delta n_R > \Delta n_T$ and thus transmission TDS is more precise than reflection TDS. This happens when absorption is rather small ($\kappa < 0.3$). At higher absorption, reflection is preferable because the transmitted signal becomes weaker. The limit between both techniques is plotted in the (n,κ) space on **Figure 10** (right) for different β. For a given set-up (given A, B, C values), the technique to be selected depends mostly on κ. For example, a 0.5-mm thick sample studied at 1 THz ($\beta = 10$) should be characterized in transmission as soon as $\kappa < 0.6$, i.e., $\alpha < 12$ cm^{-1}.

5. Combined transmission and reflection THz-TDS

In the far infrared, some materials exhibit strong absorption lines due to molecular resonances or/and due to collective excitations. In gases, narrow molecular resonances correspond to the excitation of mechanical vibrations of the whole molecule structure. In liquids and solid materials, the molecular resonances are coupled and broadened by thermal and density inhomogeneity at the molecular scale, resulting in wider absorption bands. In crystals, excitation of phonons leads also to a strong absorption of the THz waves. When characterizing such materials in transmission THz-TDS, there could be no signal detected in transmission within the absorption bands. The phase is lost in these spectral regions, and thus extraction of the material parameters at higher frequencies is no more possible using common THz-TDS extraction procedures as $2m\pi$ phase jumps (m is an unknown integer) of the transmitted phase φ_T occur at the absorption peaks. On the other hand, reflection THz-TDS is applicable even in the absorption regions, but its precision is inferior to that of transmission THz-TDS, especially because of a possible shift in position with regards to a reference mirror. Thus, even if R is measured over the whole achievable THz range, the precision Δn_R of the refractive index obtained in reflection THz-TDS is worse than the one Δn_T obtained in transmission, excepted in the absorption bands, as shown in **Figure 11** (left). Both transmission and reflection techniques may be combined to precisely evaluate $\tilde{n}(\omega)$ of such materials over the whole experimental bandwidth [29]. Basically, the procedure consists in determining with a great precision $\tilde{n}_T$ from transmission data in the first region of transparency (below 1.4 THz for maltose, see **Figure 11**, left). Thus, $\tilde{n}_R$ is extracted, with a minor precision, over the whole spectrum from the reflection data. The possible bad position δ of the sample as regards to the reference minor is corrected by adding the necessary frequency-dependent phase term $\Delta \varphi_R = 2\delta\omega/c$, so that $\tilde{n}_R$ equalizes $\tilde{n}_T$ in the first spectral region and thus $\tilde{n}_R$ is corrected over the whole spectrum. In the second region of transparency (1.6–2 THz), the possible $2m\pi$ step of the phase φ_T occurring at the resonance is retrieved by forcing $\tilde{n}_T$ obtained in transmission to be equal to the corrected $\tilde{n}_R$.

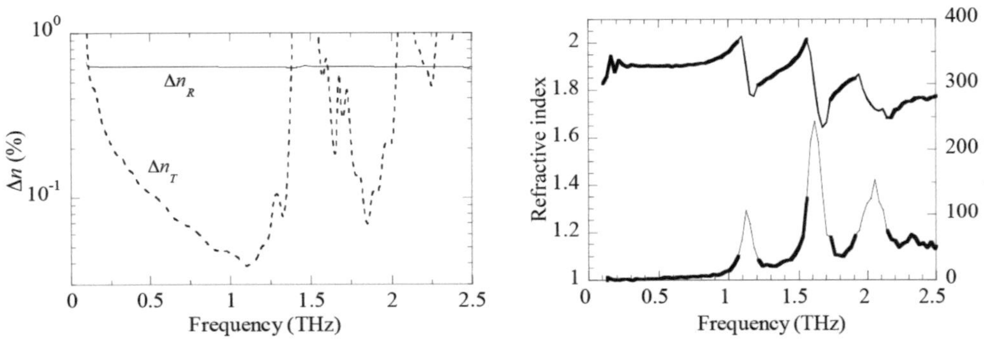

Figure 11. Left: Relative noise-induced error on the refractive index of maltose (2-mm thick) obtained in transmission (Δn_T) and in reflection (Δn_R) THz-TDS. Right: Refractive index n and coefficient of absorption α of maltose, as extracted from THz-TDS measures with an 890-μm thick sample. The thick curves are from corrected transmission data, while the thin ones are from corrected reflection data.

The same procedure is repeated for each observed saturated resonant peak. Finally, $\tilde{n}$ is accurately obtained from corrected transmission in all regions of transparency, while in the absorption peaks, we set $\tilde{n} = \tilde{n}_R$. An example is given in **Figure 11** (right). An 890-μm thick pellet of pure maltose has been characterized by THz-TDS. In the vicinity of the absorption peaks at 1.15, 1.65, and 2.05 THz, transmission of the sample is below the noise level. In these spectral ranges, we save the value of $\tilde{n}_R$ extracted from corrected reflection data, while in the other regions of transparency, we retain the phase-corrected transmission values $\tilde{n}_T$. Absorption peaks as high as 250 cm^{-1} can be precisely evaluated.

6. Combined transmission THz-TDS and Kramers-Kronig analysis

Reflection schemes are sometimes not available with commercial systems, preventing any phase correction procedure as detailed in the previous paragraph. Fortunately, phase jumps can be corrected from transmission measurements only. The causality of THz response of natural materials makes possible to calculate $n(\omega)$ from $\alpha(\omega)$ with the Kramers-Kronig (KK) relations. In fact, the phase jumps, due to saturation of transmission in the absorption peaks, impact mainly the extraction of $n(\omega)$ and very slightly $\alpha(\omega)$ outside the absorption bands. Thus, the idea [30] here is to extract $\alpha(\omega)$ from TDS data in the transparency spectral regions and to perform a KK calculation to get $n(\omega)$. Missing $\alpha(\omega)$ values (in the absorption peak) induce an error on $n(\omega)$, which is quite small because it is spread over the spectrum thanks to the integral KK calculation. Comparing, in the transparency regions, $n(\omega)$ determined by the KK transformation and the one extracted from THz-TDS data permits to know and correct the phase jumps occurring at each resonance, if any. A last extraction, with the phase corrected transmission TDS data, leads to a very precise determination of $n(\omega)$ between the absorption peaks. In fact, THz-TDS data are obtained over a limited spectral range, while KK calculation should be completed from 0 to infinity. The resulting error is minimized by performing a singly subtractive Kramers-Kronig (SSKK) transform [31]:

$$n(\omega) = n(\omega_a) + \frac{c}{\pi} PP \int_0^{\infty} \frac{\alpha(\omega')\left(\omega^2 - \omega_a^2\right)}{\left(\omega'^2 - \omega_a^2\right)\left(\omega'^2 - \omega_a^2\right)} d\omega' \tag{26}$$

PP stands for "principal part." The SSKK transformation requires knowing the refractive index at a given angular frequency, namely, the anchorage angular frequency ω_a. When dealing with THz spectra exhibiting saturated peaks, employing SSKK has two advantages: to minimize errors due to a limited experimental bandwidth, and to take advantages of the precisely known value $n(\omega_a)$ as long as ω_a is chosen before the first saturated peak. This technique combining transmission THz-TDS and KK analysis has been used to retrieve the refractive index of a 790-μm thick DAST sample. The refractive index n_{SSKK} has been calculated from α_T with an anchorage frequency $f_a = 0.5$ THz, and compared to the optical parameters extracted from a 235-μm thick DAST sample (dashed curves), which do not suffer from any saturation effects. Because of the large absorption peaks of DAST centered at 1.1 and 3.1 THz, no transmitted signal was detected around these peaks. As seen in **Figure 12**, this results in a saturation of the absorption peaks together with an offset of $n_T(\omega)$, determined by a classical THz-TDS extraction, after the low transmission bands. The SKKK method leads to a corrected spectrum $n_{SSKK}(\omega)$, which however suffers from a small discrepancy due to missing $\alpha(\omega)$ data. But $n_{SSKK}(\omega)$ is sufficiently close to the actual $n(\omega)$, which permits to find the phase jump value by comparing $n_{TDS}(\omega)$ and $n_{SSKK}(\omega)$, and then to perform again a classical extraction with the phase corrected over the saturated peak frequencies.

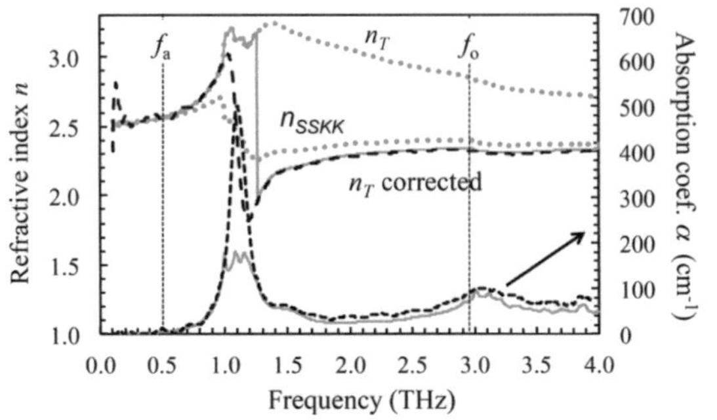

Figure 12. n_T and α_T of DAST extracted from transmission THz-TDS measurements of a 790-μm thick sample, and the refractive index corrected by the SSKKR method (solid line), by evaluating the $2m\pi$ phase jump at $f_0 = 2.95$. n_{SSKK} (attention n et pas eta) has been calculated from α_T with an anchorage frequency $f_a = 0.5$ THz. These results are compared with the ones obtained with a 235-μm thick DAST sample (dashed curves) that does not suffer from any saturation effect.

7. Scattering effect

When dealing with chemical compounds, many samples are pellets fabricated by pressing a powder mixture of the material to be studied and of a transparent hosting matrix. This technique permits to characterize materials (named hereafter "inclusion" material) that are extremely absorbent in the THz range, or that cannot be manufactured as bulky or film samples, or that could be dangerous at a high concentration. The optical parameters (refractive index and absorption coefficient) of the inclusion material, diluted in the hosting powder, are retrieved through a model describing the electromagnetism response of the mixture. If this mixture can be considered as homogeneous, i.e., if powder grains and more generally inhomogeneity are smaller than the wavelength, there is no scattering: the electromagnetic response of the material can be described by effective medium theories. Among many theories, the most popular models are the Maxwell-Garnett one (MG) [32] and the Bruggmann (BG) one [33]: the inclusions are supposed to be spheroids or ellipsoids embedded in the hosting matrix. According to these models, the dielectric constant of the mixture respectively obeys:

$$\text{MG}: \varepsilon_{\text{mixture}} = \varepsilon_{\text{host}} \frac{2(1-p)\varepsilon_{\text{host}} + (1+2p)\varepsilon_{in}}{(2+p)\varepsilon_{\text{host}} + (1-p)\varepsilon_{in}}, \tag{27}$$

$$\text{BG (spheroids)}: \quad (1-p)\frac{\varepsilon_{\text{host}} - \varepsilon_{\text{mixture}}}{\varepsilon_{\text{host}} + 2\,\varepsilon_{\text{mixture}}} + p\,\frac{\varepsilon_{in} - \varepsilon_{\text{mixture}}}{\varepsilon_{in} + 2\,\varepsilon_{\text{mixture}}} = 0. \tag{28}$$

p is the volume concentration of the inclusions. Basically, MG models allow one to nicely fit experimental data when both p and the contrast $\varepsilon_{in}/\varepsilon_{host}$ remain small. For larger contrasts, the Bruggeman's model is preferable. **Figure 13** (left) shows the real part of the permittivity, at 0.6 THz, of a mixture of HDPE and fructose powder versus the volume fraction of fructose. Both MG (continuous line) and Bruggeman (dashed line) models well fit the experimental data (open circles), because scattering effects do not affect so much the time of flight of the THz beam across the sample, i.e., the real refractive index n. On the other hand, the imaginary part of the permittivity (**Figure 13**, right) cannot be fitted by a MG model because scattering losses, which are not taken into account in the MG theory, are misinterpreted as absorption.

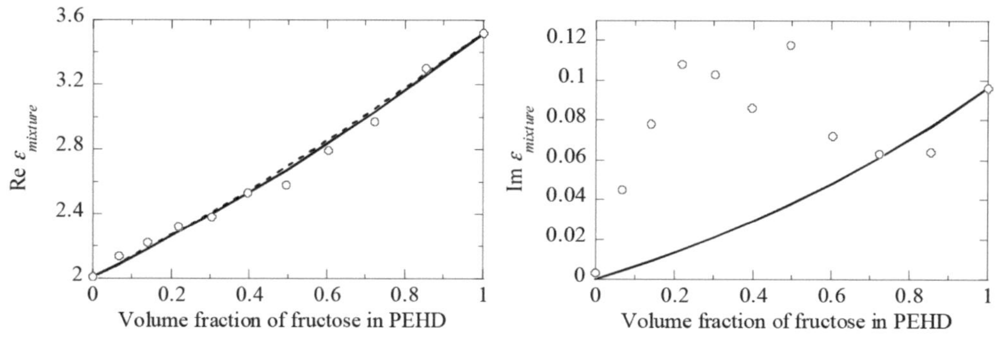

Figure 13. Real part (left) and imaginary part (right) of the permittivity $\varepsilon_{\text{mixture}}$ of a mixture of HDPE and fructose powders versus the volume fraction of fructose at 0.6 THz. Open circles: experimental data; continuous line: MG model; dashed line: BG model.

Thus, in many heterogeneous samples, scattering occurs and depends on the relative size of the scattering particles as compared to the wavelength [34]. If the particle size is smaller than about $\lambda/20$, scattering may be neglected while, if it is bigger than about 10 × λ, rays are geometrically deviated by powder grains. In between these limits, when particles are smaller than the wavelength, the scattering process is well described by the Rayleigh theory, while Mie scattering occurs for grain size comparable or bigger than the wavelength. Because of the typical size of the grains in common powders, Rayleigh scattering is almost negligible at THz frequencies and Mie scattering [35] model or even more complicated theories must be employed. As there is no analytical expression available to render the scattered amplitude, numerical codes are used. Nevertheless, Mie theory can be approximated at low scatterer concentration by simplified models, like the one proposed by Raman [36] to explain the Christiansen effect [37]. In this case, the equivalent absorption is expressed as:

$$\alpha_{\text{scattering}} = \Delta K^2 f^2 \left(n_{in} - n_{\text{host}}\right)^2. \tag{29}$$

Δ is the average size of the scattering grains and K is a coefficient that depends on the shape, the concentration, and the distribution of the grains. Such model is established on the probability, at one dimension, that an incoming photon is scattered by a grain. It shows a loss variation proportional to the square frequency. Multiple scattering is not entered into the theory, which means that the absorption coefficient of a sample does not depend on its thickness. **Figure 14** (left) illustrates the scattering effect in mixtures of HDPE and sugars (fructose, maltose, glucose). At the lowest frequencies, absorption obeys a f^2-law (dashed lines). Let us point out that the f^2-dependence is also the behavior of Mie-type scattering, which should be considered. **Figure 14** (right) shows the absorption of a HDPE-maltose mixture corrected from scattering to be compared to absorption obtained with a pure maltose sample.

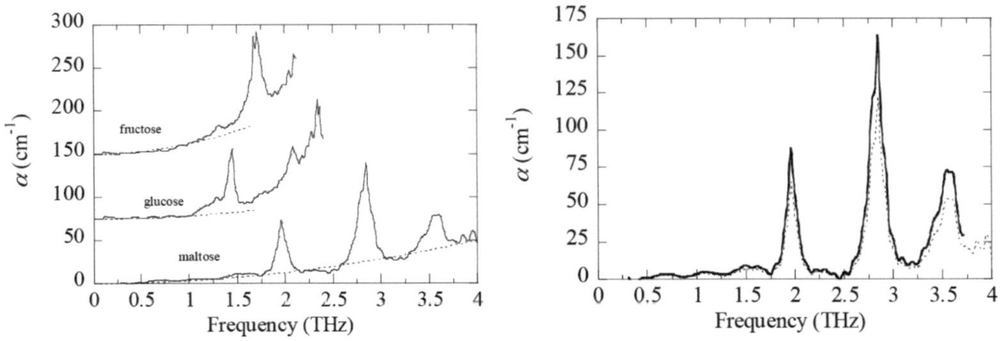

Figure 14. Left: Absorption of a mixture of powders of 25% of HDPE and 75% of sugars (maltose, glucose, and fructose) versus frequency. For the sake of visibility, the glucose and fructose curves are vertically shifted of 75 and 150 cm^{-1}, respectively. The dashed lines are parabolic fitting curves; (right) absorption of maltose (continuous) and HDPE-maltose mixture (dashed) corrected from scattering.

The limit of validity of the Raman model is also shown in **Figure 15**. The measured absorption of the mixture is plotted versus the volume fraction p of fructose (**Figure 15**, left).

Figure 15. Left: Measured absorption of a mixture HDPE-fructose powder mixture versus the volume fraction p of fructose at different frequencies. The lines are a guide to the eye; (right) measured absorption of a mixture HDPE-fructose powder mixture ($p = 30\%$) versus the size Δ of the fructose grain at different frequencies. The lines are parabolic fits.

At different frequencies, the experimental data are well fitted by a linear curve for $p < 20–30\%$ and for $p > \sim 50\%$, (28). In between, the transition fructose-in-HDPE to HDPE-in-fructose cannot be described by the Raman model. **Figure 15** (right) presents the measured absorption

http://dx.doi.org/10.5772/66348

versus the size Δ of the fructose grains, for a constant volume ratio $p = 30\%$. The experimental points are fitted by parabolic curves, which do not obey to relation (29), except at the limit $\Delta \rightarrow 0$. A more rigorous modeling of the scattering process in such mixtures is based on the Mie scattering theory. Several numerical codes are freely available to estimate the Mie scattering, like the one from the Oregon Medical Laser Center [38]. **Figure 16** exhibits the extinction coefficient, taking into account both absorption and scattering losses, at 1.5 THz of a HDPE-fructose mixture ($p = 30\%$). Mie modeling gives a curve (dotted line) below the experimental data (dots), when the bulk value $n_{HDPE} = 1.51$ is entered in the calculation. A better fit is obtained with $n_{HDPE} = 1.41$. Such a result could be explained by residual air inclusions (20%) in the pellet which decrease the effective refractive index of the host substance [39]. The agreement between experimental data and calculated ones is good, especially as no adjustable parameter is input in the calculation. Moreover, some geometrical factors cannot be taken into account, like for example the shape of the fructose grains, which looks like cubes instead of spheres as required to apply Mie hypothesis.

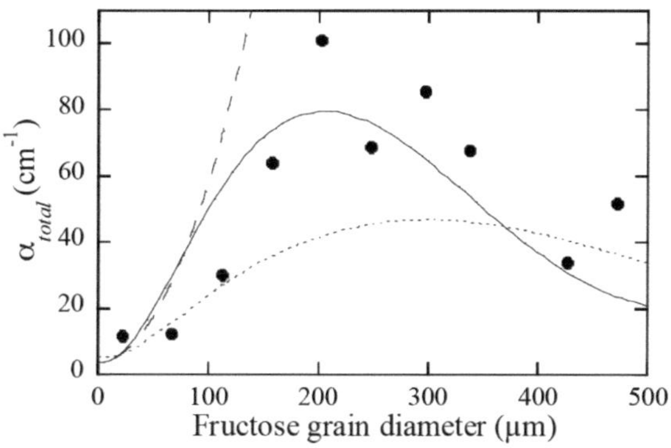

Figure 16. Absorption, including scattering, of a mixture of HDPE-fructose ($p = 30\%$) determined from transmission THz-TDS (dots). The lines are calculated with a Mie scattering code [39]: with $n_{HDPE} = 1.51$ (dotted), with $n_{HDPE} = 1.41$ (continuous). The dashed line is a parabolic dependence at small grain sizes.

8. Conclusion

We have described here the main features of THz-TDS. This is a unique technique to quantitatively and precisely characterize the electromagnetic response of materials and devices over a broadband in the far-infrared domain. However, because of limited space, we did not address in this chapter some additional possibilities of THz-TDS in terms of material characterization:

- Anisotropic materials: Their characterization by THz-TDS is not tricky as most of THz antennas (photo-conducting switches or electro-optic crystals) are polarization-sensitive, with a rather good rejection level (a few percent). Even cross-polarization effects can be investigated by rotating the receiving antenna around the optical axis of the THz system.

Moreover, THz-TDS supplies the phase of the transmitted signals with respect to the involved THz beam polarization, from which the anisotropic parameters of a sample can be deduced.

- Metamaterials: They are quite easy to manufacture for the THz range, as they require repetitive features whose size is smaller than the wavelength, i.e., in the range below a few tens of microns. This is achieved with MEMS or/and microelectronics technologies. Such THz metamaterials [40] show amazing properties, like negative refractive index (left-handed property), chirality, and so on. THz-TDS is especially well adapted to characterize these metamaterials because it delivers both amplitude and phase of the reflected or/and transmitted signal. Thus, for example, a negative phase, due to propagation in a left-handed material, is clearly observed in the THz-TDS experimental spectra.
- Time-resolved THz-TDS [41]: This technique can be applied if the sample under test is sensitive to light. In this case, a third part of the pulsed laser beam illuminates the sample with an adjustable delay as compared to the impinging THz pulse. The light-induced modification of the sample properties, mostly by photo-generation of free carriers, changes the transmission of the THz beam. By varying the optical-THz time-delay, the photo-induced excitation in the material is time resolved. This method works very well to study the carrier dynamics in semiconductors and in superconductors, as far as the carrier lifetime is not too short as compared to the THz pulse duration. The main limitation of the technique is the rather long duration of the THz pulse, i.e., few hundreds of femtoseconds, as compared to very fast phenomena in matter. Investigating short events requires a proper deconvolution of the temporal records, which is not an easy task.

The list of applications of THz-TDS is for sure quite long and giving it exhaustively is almost impossible. Moreover, the continuous progress of technology makes the technique really easy to use, especially with commercial systems, and numerous new scientific results are regularly published.

Author details

Maxime Bernier*, Frédéric Garet and Jean-Louis Coutaz

*Address all correspondence to: maxime.bernier@univ-savoie.fr

Laboratory IMEP-LAHC, UMR CNRS 5130, University of Savoie Mont-Blanc, Savoy, France

References

[1] Bell R. J.; Introductory Fourier Transform Spectroscopy, 1972, Academic Press, New York.

[2] Auston D. H., Chueng K. P.: Coherent time-domain far-infrared spectroscopy. J. Opt. Soc. Am. B. 1985; **2**: 606–612.

[3] van Exter M., Fattinger Ch., Grischkowsky D.: Terahertz time-domain spectroscopy of water vapor. Opt. Lett. 1989; **14**: 1128–1130.

[4] Pastol Y., Arjavalingam G., Halbout J.-M., Kopcsay G. V.: Coherent broadband microwave spectroscopy using picosecond optoelectronic antennas. Appl. Phys. Lett. 1989; **54**: 307–309.

[5] Mittleman D., editor, Sensing with terahertz radiation, Springer Series in Optical Sciences, 2003 (Vol. 85).

[6] Sakai K., editor, Terahertz Optoelectronics, Springer Topics in Applied Physics, 2005 (Vol. 97).

[7] Zhang X.-C. and Xu J., Introduction to THz Wave Photonics, Berlin: Springer 2009.

[8] Lee Y.-S., Principles of Terahertz Science and Technology, Berlin: Springer 2009.

[9] Bründermann E., Hübers H.-W., and Kimmitt M., THz techniques, Springer Series in Optical Sciences, 2012 (Vol. 151).

[10] China: Oplan (www.oplanchina.com); Germany: Batop (www.batop.de), Menlo Systems (www. menlosystems.com), Accurion (www.accurion.com), Toptica (www. toptica.com), LaserQuantum/GigaOptics (www.laserquantum.com); Japan: Advantest (www.advantest.co.jp); Lithuania: Ekspla-Teravil (www.ekspla.com); Switzerland: Rainbow Photonics AG (www.rainbowphotonics.com); UK: Bruker-TeraView (www. teraview.com); USA: Zomega (www.zomega-terahertz.com), Del Mar Photonics (www.delmarphotonics.com), Picometrix (www.picometrix.com), Emcore (www. emcore.com), ARP (http://arphotonics.net), MicroTech Instruments Inc. (www.mtinstruments.com).

[11] Ho I.-C., Guo X., Zhang X.-C. : Design and performance of reflective terahertz air-biased-coherent-detection for time-domain spectroscopy. Opt. Expr. 2010; **18**: 2872–2883.

[12] Vieweg N., Fischer B. M., Reuter M., Kula P., Dabrowski R., Celik M. A., Frenking G., Koch M., Jepsen P. U.: Ultrabroadband terahertz spectroscopy of a liquid crystal. Opt. Expr. 2012; **20**: 28249–28256.

[13] Matsubara E., Nagai M., Ashida M.: Coherent infrared spectroscopy system from terahertz to near infrared using air plasma produced by 10-fs pulses. J. Opt. Soc. Am. B. 2013; **30**: 1627–1630.

[14] Nemec H., Kadlec F., Kuzel P.: Methodology of an optical pump-terahertz probe experiment: An analytical frequency-domain approach. J. Chem. Phys. 2002; **117**: 8454–8466.

[15] Ulbricht R., Hendry E., Shan J., Heinz T. F., Bonn M.: Carrier dynamics in semiconductors studied with time-resolved terahertz spectroscopy. Rev. Mod. Phys. 2011; **83**, 543–586.

[16] Beard M. C., Schmuttenmaer C. A.: Using the finite-difference time-domain pulse propagation method to simulate time-resolved THz experiments. J. Chem. Phys. 2001; **114**: 2903–2909.

[17] Beard M. C., Turner G. M., Schmuttenmaer C. A.: Terahertz spectroscopy. J. Phys. Chem. 2002; **106**: 7146–7159.

[18] Leitenstorfer A., Nelson K. A., Reimann K., Tanaka, K.: Focus on nonlinear terahertz studies. New Journal of Physics. 2014; **16**, 045016.

[19] Krotkus A., Coutaz J.-L.: Non-stoichiometric semiconductor materials for terahertz optoelectronics applications. Semicond. Science Techn. 2005; **20**: S142.

[20] Hirori, H., Yamashita K., Nagai, M., Tanaka K.: Attenuated total reflection spectroscopy in time domain using terahertz coherent pulses. Japanese journal of applied physics. 2004; **43**: L1287.

[21] Duvillaret L., Garet F., Coutaz J.-L.: A reliable method for extraction of material parameters in terahertz time-domain spectroscopy. IEEE J. Sel. Top. Quant. Electron. 1996; **2**: 739–746.

[22] Withayachumnankul W., Fischer B. M., Lin H., Abbott D.: Uncertainty in terahertz time-domain spectroscopy measurement. J. Opt. Soc. Am. B. 2008; **25**: 1059–1072.

[23] Pashkin A., Kempa M., Nemec H., Kadlec F., Kuzel P.: Phase-sensitive time-domain terahertz reflection spectroscopy. Rev. Sci. Instrum. 2003; **74**: 4711–4717.

[24] Khazan M., Meissner R., Wilke I.: Convertible transmission-reflection time-domain terahertz spectrometer. Rev. Sci. Instrum. 2001; **72**: 3427–3430.

[25] Nagashima T., Hangyo M.: Measurement of complex optical constants of a highly doped Si wafer using terahertz ellipsometry. Appl. Phys. Lett. 2001; **79**: 3917–3919.

[26] Thrane L., Jacobsen R. H., Jepsen P. U., Keiding S. R.: THz reflection spectroscopy of liquid water. Chem. Phys. Lett. 1995; **240**: 330–333.

[27] Lucarini V., Ino Y., Peiponen K.-E., Kuwata-Gonokami M.: Detection and correction of the misplacement error in terahertz spectroscopy by application of singly subtractive Kramers-Kronig relations. Phys. Rev. B. 2005; **72**: 125107.

[28] Hoshina H., Seta T., Iwamoto T., Hosako I., Otani C., Kasai Y.: Precise measurement of pressure broadening parameters for water vapor with a terahertz time-domain spectrometer. J. of Quant. Spectr. Rad. Transfer. 2008 ; **109**: 2303–2314.

[29] Bernier M., Garet F., Coutaz J.-L.: Precise Determination of the Refractive Index of Samples Showing Low Transmission Bands by THz Time-Domain Spectroscopy. IEEE Trans. THz Sci. Technol. 2013 ; **3**: 295–301.

[30] Bernier M., Garet F., Coutaz J.-L.: Accurate characterization of resonant samples in the terahertz regime through a technique combining time-domain spectroscopy and Kramers-Kronig analysis. IEEE Trans. THz Sci. Technol. 2016 ; **6**: 442–450.

[31] Lucarini, V., Saarinen, J. J., Peiponen, K. E., & Vartiainen, E. M.: Kramers-Kronig relations in optical materials research. Springer Science & Business Media 2005 (Vol. 110).

[32] Garnett J. C. M. : Colours in metal glasses and in metallic films. Phil. Trans. R. Soc. Lond. 1904; **203**: 385.

[33] A. G. Bruggeman: Berechnung verschiedener physikalischer Konstanten von heterogenen Substanzen. Ann. Phys. (Leipzig). 1935; **24**: 636–679.

[34] Menzel, W. P. Remote sensing applications with meteorological satellites. Online lecture of University of Wisconsin-Madison, Space Science and Engineering Center, Cooperative Institute for Meteorological Satellite Studies; 2005.

[35] Mie G.: Annalen der Physik. 1908; **330**: 377.

[36] Raman C. V., Viswanathan K. S. Proc. Indian Acad. Sci. A.1955; **41**: 55.

[37] Franz M., Fischer B.M., Walther M.: The Christiansen effect in terahertz time-domain spectra of coarse-grained powders. Appl. Phys. Lett. 2008; **92**: 021107–01.

[38] Available from: http://omlc.ogi.edu/software/mie/

[39] Garet F., Hofman M., Meilhan J., Simoens F., Coutaz J.L.: Evidence of Mie scattering at terahertz frequencies in powder materials. App. Phys. Lett. 2014; **105**: 031106.

[40] Yen T. J., Padilla W. J., Fang N., Vier D. C., Smith D. R., Pendry J. B., Basov D. N., Zhang X.: Terahertz magnetic response from artificial materials. Science. 2004; **303**,1494–1496.

[41] Groeneveld R. H. M., Grischkowsky D.: Picosecond time-resolved far-infrared experiments on carriers and excitons in GaAs-AlGaAs multiple quantum wells. J. Opt. Soc. Am. B. 1994; **11**: 2502–2507.

Chapter 8

Terahertz Waveform Measurements Using a Chirped Optical Pulse and Terahertz Spectroscopy of Reverse Micellar Solution: Towards Time-resolved Terahertz Spectroscopy of Protein in Water

Hiroshi Murakami

Additional information is available at the end of the chapter

http://dx.doi.org/10.5772/67195

Abstract

One challenging research target using terahertz spectroscopy is time-resolved terahertz spectroscopy of protein molecules to clarify the relationship between protein's functions and the low-frequency collective motion within the molecule. Our results on two research topics necessary for this target are described. One is single-shot measurements of terahertz waveform that has large advantages in time-resolved terahertz spectroscopy. We examined experimentally and theoretically single-shot measurements using chirped optical pulses. The other is terahertz spectroscopy of reverse micellar solutions in which nanometer-sized water droplets are formed. Protein powder is usually used as a sample for terahertz spectroscopy because the absorption of terahertz waves by solvent water is very strong in aqueous solutions of protein, although protein molecules work in water. The absorption of terahertz waves by the nonpolar oil solvent in reverse micellar solution is considerably weak compared with that by water. We demonstrated that terahertz absorption spectra of protein in liquids are obtained by the use of protein-containing reverse micellar solution. On the other hand, a nanometer-sized water droplet in reverse micellar solutions is a promising candidate for studies of supercooled water. We made temperature-dependent terahertz spectroscopy of the water droplet to study collective water motions due to hydrogen bond networks.

Keywords: terahertz spectroscopy, time-resolved, single-shot measurement, biomolecules, reverse micelle, water, chirped pulse, nanoconfinement, temperature dependence

1. Introduction

1.1. Terahertz time-domain spectroscopy

At present, terahertz time-domain spectroscopy (THz-TDS) using a femtosecond laser is a popular method to measure the optical constant of materials in the terahertz frequency range, and its applications for medical science and industry are being investigated [1, 2]. A schematic diagram of THz-TDS using two photoconductive antennas is depicted in **Figure 1(a)**. A femtosecond laser pulse is divided into two beams. One is used to generate a terahertz electromagnetic wave by accelerating photoelectrons due to ultrafast laser irradiation onto the DC-biased photoconductive antenna. The other is used to obtain the temporal waveform of the terahertz electronic field that biases the other photoconductive antenna, where the photocurrent induced by the laser pulse is proportional to the amplitude of the electric field and is measured with a lock-in amplifier. The terahertz wave exhibits roughly a picosecond monocyclic pulse (**Figure 1(b)**) and a very small part of its waveform can be sampled by the femtosecond probe pulse. The temporal overlapping between the two pulses is sequentially varied by use of an optical delay line, and the whole terahertz waveform is obtained. The temporal waveform of the terahertz field pulse transmitted through a sample material is converted to the spectrum in the frequency domain by the Fourier transform (**Figure 1(c)**) and then the optical constant of the material is derived using the spectrum obtained in the same way from the terahertz field pulse transmitted through a reference material (e.g., solvent liquid for a solution sample). One excellent advantage of this method is to obtain the complex permittivity (or refractive index) of materials directly from the measurement because one measures the electric field amplitude, and not the intensity, as seen in **Figure 1(b)**.

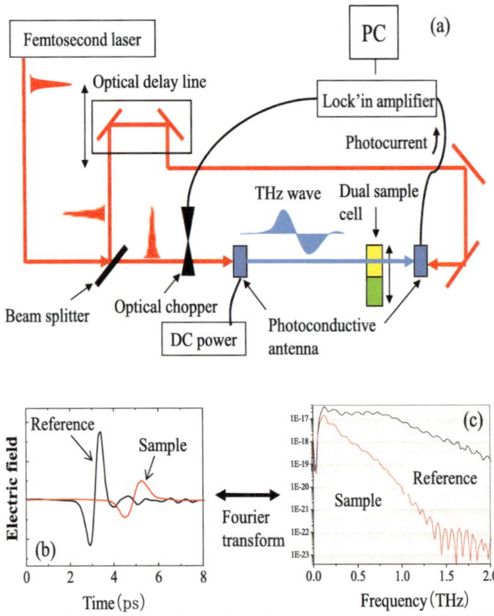

Figure 1. (a) Schematic diagram of terahertz time-domain spectroscopy (THz-TDS). (b) Terahertz waveforms after transmission through reference and sample solutions. (c) Frequency spectra obtained from the Fourier transform of the terahertz waveforms.

http://dx.doi.org/10.5772/67195

1.2. Optical-pump terahertz-probe time-resolved terahertz spectroscopy

The next step of THz-TDS is optical-pump terahertz-probe (OPTP) spectroscopy using pulsed terahertz waves; that is, one examines the time-dependent complex permittivity of materials in the terahertz frequency range after abrupt change in the electronic state of the material induced by an optical laser pulse [3–5]. Further, an enhancement in the peak intensity of the terahertz pulse wave has opened up terahertz-pump terahertz-probe spectroscopy and nonlinear terahertz spectroscopy [6–8]. These time-resolved terahertz spectroscopies have been performed largely for solids such as semiconductor and graphene. OPTP spectroscopy requires two optical delay-line systems if THz-TDS is employed for the terahertz wave probe; one is used for THz-TDS, and the other controls the time difference between the optical and terahertz pulses at the material. One issue of this OPTP spectroscopy is that it is difficult to apply it to sample materials that undergo an irreversible process or show a relaxation process with a time constant longer than the period of pulse repetition of the femtosecond laser because the terahertz waveform measured using THz-TDS is deformed. One method to overcome this issue is a single-shot measurement of terahertz waveforms; accordingly, this measurement needs only one optical delay-line system for control of pump-probe delay time.

1.3. Single-shot measurements of terahertz waveforms using a chirped optical pulse

Single-shot measurements of a terahertz waveform can be achieved by a combination of electro-optic detection using a chirped optical pulse (EODCP) and measurements of the chirped pulse with a spectrometer. This was first proposed and accomplished by Jiang and Zhang [9]. The result was discussed on the basis of a theoretical expression approximately derived [10]. We examined the method in detail as functions of the chirp rate and spectral resolution of the spectrometer and analyzed the experimental results using the theoretical expression analytically derived [11]. Moreover, on the basis of EODCP, we measured the complex refractive index of a material in the terahertz range and conducted OPTP spectroscopy. These results show that this method is applicable to terahertz spectroscopy and a promising method for OPTP spectroscopy. These results are described in Section 2.

1.4. Relationship between terahertz collective motions within a protein molecule and its function

Biomolecules, such as protein and DNA, play a fundamental role in the biological function of living cells. Further, it is believed that low-frequency collective motions in a protein molecule in the terahertz range play an important role in protein's functions [12–15]. Hence, terahertz spectroscopy of biomolecules has been extensively conducted [16–21]. Moreover, OPTP spectroscopy allows us to observe the collective motions during the protein function. Time-dependent change in the terahertz motions of a protein is measured after an optical pulse initiates a chemical reaction in the molecule, such as enzyme reaction and protein folding, by OPTP spectroscopy. If a specific collective motion is involved in the reaction, spectral change is expected to be observed at the frequency characteristic of the collective motion. However, there is a serious issue for terahertz spectroscopy of proteins. Although proteins work in water, protein powders are usually used as a sample for terahertz spectroscopy because absorption of terahertz wave by solvent water is very strong in aqueous solutions of protein and because it is difficult to obtain the terahertz signal due to the protein molecule. To overcome this problem, we employ reverse micelles.

1.5. Reverse micelle

A reverse micelle is formed by self-assembly of surfactant molecules in a nonpolar oil solvent and becomes a nanometer-scale spherical cage filled with water, as shown in **Figure 2** [22, 23]. The size of reverse micelles can be experimentally controlled by the water-surfactant molar ratio (w_0). Water-soluble molecules, such as proteins and DNA, can be dissolved in the reverse micelle. Because absorption of terahertz waves by nonpolar solvents is considerably weak compared with polar solvents such as water, the absorption background signal due to the solvent was very low for reverse micellar solutions. We demonstrated that terahertz signal due to protein molecules is obtained by use of protein-containing reverse (PCR) micellar solution [24], as described in Section 4.

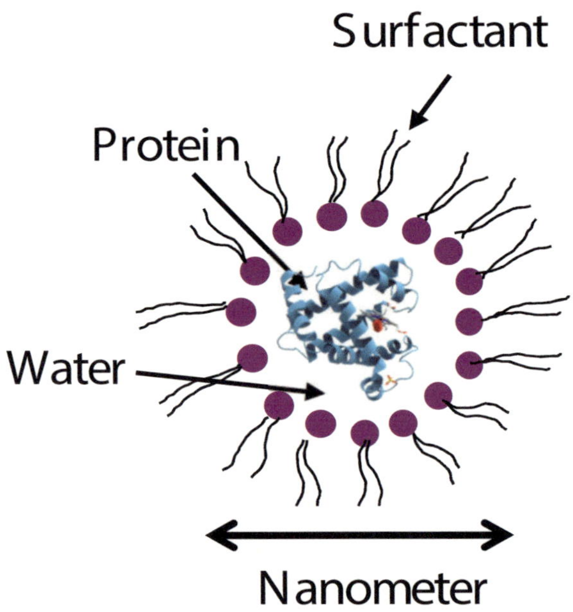

Figure 2. Schematic cross-section of a protein-containing reverse micelle.

A reverse micelle is a potential candidate for studying biomolecules and their surrounding waters under the condition similar to those in living cells. Cells are crowded with many kinds of molecules, and so the surroundings of biomolecules are different from those in the dilute aqueous solution. This is referred to as a macromolecular crowding effect [25, 26]. Therefore, the effect needs to be studied to clarify biomolecular function. In particular, the difference in the state of the surrounding waters of biomolecules will have a significant effect on their function [27, 28]. Nevertheless, biomolecules are usually studied in the dilute aqueous solutions.

1.6. Terahertz spectroscopy of nanoconfined water for studies on thermodynamic anomaly of water at 228 K

Water has the density maximum at 4°C. This is a well-known anomalous property of water, which is responsible for the fact that ice floats on the top of liquid water. Further, most thermodynamic properties of supercooled water display strong anomalies, for example, the isobaric specific heat and thermal expansion appear to diverge at 228 K [29]. Since the discovery of the anomalies, much attention has been paid to the properties of supercooled water. Several hypotheses, such as the liquid-liquid phase transition hypothesis, were proposed for a unified understanding of water [30, 31]. As for the experimental studies of supercooled water, it is necessary to use mesoporous materials such as MCM-41 because water confined in such materials does not freeze far below the melting point [32–35]. However, the interaction between confined water and the internal surface of the cage used for the confinement could affect the properties of water [31]. This is a serious issue for supercooled water studies.

A reverse micelle is a promising candidate for studies of supercooled water distinguished from the mesoporous material by the following reasons. Reverse micelles in liquids will provide soft confinement compared with the nanopore in solid materials. Further, we have recently shown by use of a molecular probe introduced into reverse micelles that water shedding from reverse micelles with an aqueous cavity radius of ~1 nm occurs below the melting point of water and that the water droplets extracted are dispersed in the solution down to the melting point of the oil solvent (~170 K) [36]. This indicates that it is possible to make temperature-dependent measurements of nanometer-sized water droplets over a wide temperature range. Moreover, the water shedding allows us to use a nanometer-sized water droplet free from the cage.

There are studies on the temperature-dependent properties of confined-water using reverse micelles [37–40], but terahertz spectroscopy has never been applied to it. Terahertz spectroscopy is well suited for studying cooperative motions of water due to the hydrogen bond network on time scales of picoseconds [41–44]. It is believed that the hydrogen bond network is a key to elucidate the anomalous properties of water. From terahertz spectroscopy of water above the melting point, Rønne and coworkers suggested that the relaxation time appears to diverge at ~228 K as the temperature is decreased [41, 42]. This implies that terahertz spectroscopy of water can provide information on the microscopic mechanism underlying the thermodynamic anomaly of water at low temperatures. The result of terahertz spectroscopy of water in reverse micellar solutions as a function of temperature is presented in Section 5 [45].

2. Terahertz waveform measurements using EODCP

2.1. EODCP

In THz-TDS using electro-optic (EO) detection, the detector module in **Figure 1(a)** consists of an EO crystal such as ZnTe, which is put between the two polarizers, and a photodiode for measurements of the probe laser intensity. The probe pulse is overlapped with the terahertz field pulse within the crystal, and the magnitude of the Pockels effect proportional to the amplitude of the terahertz field is obtained from measurements of the intensity of the probe pulse subject to the birefringence in the crossed-polarizers configuration.

The method using EODCP is schematically described in **Figure 3**. A femtosecond probe pulse is linearly chirped and temporally broadened to cover the terahertz waveform examined. The chirped probe pulse is overlapped with a terahertz field pulse within the EO crystal, modulated by the terahertz field, and dispersed onto a multichannel detector combined with a spectrometer. Since the wavelength axis can be converted to the time axis using the value of the chirp rate, the terahertz field waveform is derived from the two spectra of the probe pulses with and without terahertz field modulation. Thus, a single-shot measurement of the terahertz waveform is made using EODCP. Jiang and Zhang used a chirped pulse with a temporal width of ~30 ps and obtained a terahertz waveform three times broader than the original pulse width measured by THz-TDS with delay scanning using an unchirped probe pulse [9]. Further, Sun and coworkers analyzed the dependence of the EODCP-derived terahertz waveform on the chirp rate by assuming that the stationary phase method is applicable [10]. The EODCP-derived terahertz waveform is monocyclic for a monocyclic original terahertz field as long as the stationary phase method is used. On the other hand, the EODCP-derived terahertz waveform changes with the chirp rate and the spectral resolution of the spectrometer, and hence one requires an analytical expression of the EODCP-derived terahertz waveform without such an assumption for analysis of a variety of terahertz waveforms.

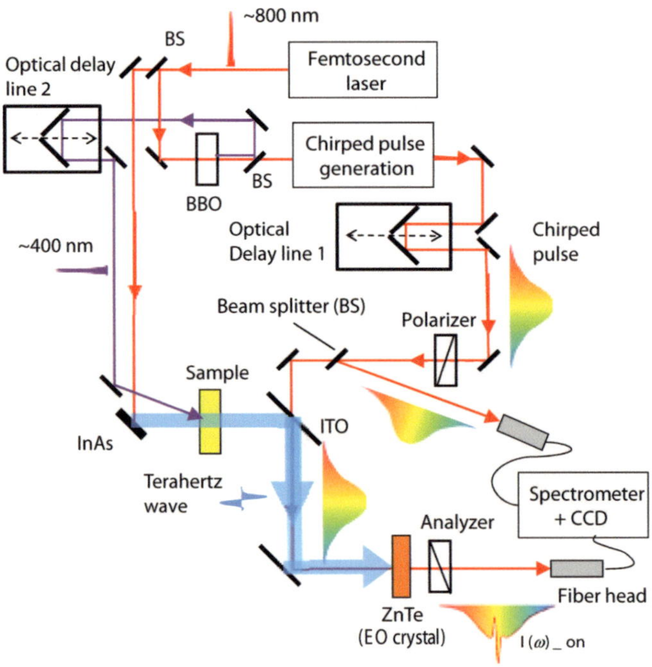

Figure 3. Schematic diagram of the experimental setup for time-resolved terahertz spectroscopy with electro-optic detection using a chirped probe pulse (EODCP) of the terahertz wave. A double-beam configuration was employed to obtain the chirped probe pulses with modulation due to the terahertz field and without it simultaneously. ITO is indium tin oxide.

http://dx.doi.org/10.5772/67195

2.2. Expression of the EODCP-derived terahertz waveform

We derive an expression of the EODCP-derived terahertz field waveform in this section. Denoting the electric field waveform of a chirped probe pulse by $C(t)$, we give the temporal shape of the chirped pulse modulated by a terahertz field $E_{THz}(t)$ as follows:

$$M(t) = C(t)\left[b + k\,E_{THz}(t-\tau)\right], \tag{1}$$

where k is a modulation coefficient, τ is the delay time between the terahertz field pulse and the chirped pulse, and b is the transmission coefficient because the probe pulse passes partly through the two crossed polarizers, e.g., owing to the inherent residual birefringence of the EO crystal or the finite extinction ratio of the polarizers. By comparing the EODCP-derived terahertz waveform with numerical results, Yellampalle and coworkers [46] showed that Eq. (1) is valid for an EO crystal with sufficient residual birefringence, as in the present study using a 1-mm-thick ZnTe crystal.

The waveform of the chirped pulse is defined as $C(t) = \exp(-t^2 T_c^{-2} - iat^2 - i\omega_0 t)$, where T_c, $2a$ and ω_0 are the width, chirp rate, and central frequency of the pulse, respectively. We give a monocyclic original terahertz field waveform by $E_{THz}(t) = \frac{t}{\Delta T}\exp(-t^2/\Delta T^2)$, which is a symmetrically bipolar function and represents a global feature of the terahertz field waveform generated by a femtosecond laser. If the modulated probe pulse is measured with a spectrometer with a multichannel detector, the spectral intensity $I(\omega)\_on$ is expressed by

$$I(\omega)\_\mathrm{on} \propto \int_{-\infty}^{\infty} g(\omega-\omega')\left|\int_{-\infty}^{\infty} M(t)\exp(i\omega' t)dt\right|^2 d\omega', \tag{2}$$

where $g(\omega-\omega')$ is the spectral response function of the spectrometer; we set $g(\omega-\omega') = \delta(\omega-\omega')$ in this derivation. One can solve Eq. (2) by using Gaussian integrals and dealing with complex numbers in the polar form. We follow the procedure of Jiang and Zhang to derive the terahertz waveform from EODCP [9], namely,

$$E_{mes}(\omega) = \frac{I(\omega)\_\mathrm{on} - I(\omega)\_\mathrm{off}}{I(\omega)\_\mathrm{off}}, \tag{3}$$

where $I(\omega)\_off$ is obtained by putting $M(t) = b\,C(t)$ in Eq. (2). Finally, we obtain an expression of the terahertz waveform as

$$\begin{aligned} E_{mes}(\omega) \propto & -\left\{\frac{2\tau}{T_c^2}\cos[\theta(\Delta\omega)] - (\Delta\omega + 2a\tau)\sin[\theta(\Delta\omega)]\right\} \\ & \times \exp\left[-\frac{\Delta\omega^2}{4}\left(\frac{1}{\beta + a^2\beta^{-1}} - \frac{1}{T_c^2\gamma}\right) - \frac{a\tau\,\Delta\omega}{\Delta T^2(\beta^2+a^2)}\right], \end{aligned} \tag{4}$$

with

$$\begin{aligned} \theta(\Delta\omega) = & \frac{a\,\Delta\omega^2}{4}\left(\frac{1}{\beta^2+a^2} - \frac{1}{\gamma}\right) - \frac{\Delta\omega\,\tau\beta}{\Delta T^2(\beta^2+a^2)} - \frac{a\tau^2}{\Delta T^4(\beta^2+a^2)} \\ & + \frac{1}{2}\arctan(a\,T_c^2) - \frac{3}{2}\arctan(a/\beta), \end{aligned} \tag{5}$$

where $\Delta\omega = \omega_0 - \omega$, $\beta = T_c^{-2} + \Delta T^{-2}$, and $\gamma = T_c^{-4} + a^2$. Here the quadratic term of k is neglected on the assumption that the modulation coefficient is very small and $k\,E_{THz} << b$. We confirmed that the waveform obtained from Eq. (4) agrees with that from Eq. (3) through the numerical integration of $I(\omega)\_on$ and $I(\omega)\_off$ with the same parameter values.

2.3. Experiment

A schematic diagram of the experimental setup for EODCP is illustrated in **Figure 3**. A regeneratively amplified femtosecond laser beam with a temporal width of ~150 fs (FWHM), a wavelength of 800 nm, a repetition rate of ~1 kHz, and a pulse energy of ~600 μJ was divided into two beams: one was irradiated onto an InAs (1 0 0) wafer to generate the terahertz field pulse and the other was chirped by passing it through a pair of gratings. We used two types of grating pairs, i.e., 1200 and 600 grooves/mm, to change the chirp rate. All the measurements in the present study were made in the atmosphere.

We employed a double-beam configuration to obtain $I(\omega)\_on$ and $I(\omega)\_off$ simultaneously. This configuration reduces the effect of the shot-to-shot fluctuation of the laser on the EODCP-derived terahertz waveform. The chirped beam was first passed through a polarizer and then divided into two beams. One beam, the reference $R(\omega)$ was led into a fiber that transferred the light to a spectrometer and was detected with a charge-coupled device (CCD) image sensor. The other, the signal $S(\omega)$ was aligned to travel collinearly with the terahertz field pulse, modulated by the terahertz field within a 1-mm-thick ZnTe crystal and was passed through an analyzer into the other fiber. The reference and signal beams were detected simultaneously on two different areas of the image sensor. $S(\omega)\_off$ and $R(\omega)\_off$ were measured simultaneously in the absence of the terahertz field, while $R(\omega)\_on$ and $S(\omega)\_on$ (= $I(\omega)\_on$) were measured in the presence of it. $I(\omega)\_off$ was obtained by $R(\omega)\_on \times S(\omega)\_off/R(\omega)\_off$. Thus, we could measure $I(\omega)\_off$ and $I(\omega)\_on$ simultaneously in the double-beam configuration and derive the terahertz field waveform using Eq. (3). Moreover, an optical delay line (delay line 1 in **Figure 3**) was employed in order to vary the time difference between the chirped probe pulse and the terahertz field pulse within the EO crystal. This was necessary to obtain the value of the chirp rate experimentally.

Figure 4 shows $I(\omega)\_on$ (black) and $I(\omega)\_off$ (red) obtained from a single-shot measurement at a chirp rate of −0.24 THz^2, together with the terahertz waveform derived from Eq. (3) (blue). The figure demonstrates that the modulation due to the terahertz field obviously appears in $I(\omega)\_on$ even for the single-shot measurement. However, the signal-to-noise ratio is roughly 10:1 and not so high. Further, the baseline is distorted and not zero. Thus, in the present study, the data were acquired by averaging over a few hundred shots to achieve a good signal-to-noise ratio.

2.4. Dependence of the terahertz field waveform on the chirp rate

Figure 5(a) displays the EODCP-derived terahertz waveform at a chirp rate of −0.24 THz^2. The waveform is shifted along the abscissa (wavelength) when the delay time between the chirped probe and terahertz field pulses is changed. **Figure 5(b)** was obtained by tracking corresponding peak positions (arrow in **Figure 5(a)**) in the terahertz waveforms at different delay times. We derived the value of the chirp rate from the slope of the line fitted to the data points. It was found that this value agrees with that obtained by a second-harmonic-generation frequency-resolved optical grating (SHG-FROG) measurement and the value estimated from the distance between the two gratings and the number of grooves ruled on the grating per unit length. The upper axis in **Figure 5(a)** is the time axis obtained from the wavelength

http://dx.doi.org/10.5772/67195

axis using the chirp rate value. The temporal width of the chirped pulse at this chirp rate was determined to be ~15 ps (FWHM) by cross-correlation measurement between the original femtosecond pulse and the chirped one.

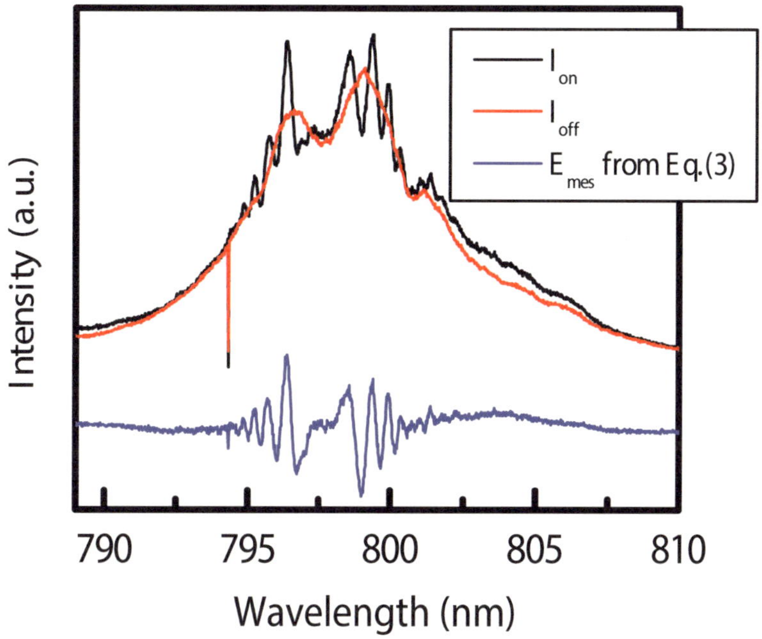

Figure 4. Chirped probe pulses with terahertz field modulation ($I(\omega)\_on$) and without it ($I(\omega)\_off$) obtained from a single-shot EODCP measurement, i.e., with no data accumulation; a negative spike at around 794 nm is due to the damage of the sensor element. The terahertz waveform derived from Eq. (3) is drawn by a blue line.

The EODCP-derived terahertz waveforms (blue line:−0.24 THz^2, red line:−1.9 THz^2) are compared with that obtained by THz-TDS with delay scanning (black line) in **Figure 5(c)**. We note that the temporal observation window becomes narrower with decreasing the chirp rate, and hence, it is a part of the full-time range in the figure for the chirp rate of −1.9 THz^2. The waveform due to the THz-TDS with delay scanning is monocyclic, although small long-lived oscillation due to absorption by water vapor in the terahertz frequency range is involved. Meanwhile, the waveforms measured by EODCP are multicyclic; the cycle number is considerably large at −0.24 THz^2, compared with at −1.9 THz^2. Such multicyclic behavior was not observed in the terahertz field waveform obtained using EODCP by Jiang and Zhang [9], although the temporal width of the chirped probe pulse in their measurement is comparable to that in the case of −0.24 THz^2. Further, it should be noted that the temporal period of the muticycle in the EODCP-derived terahertz waveform is roughly the same as that of the monocycle obtained by THz-TDS with delay scanning in **Figure 5(c)**, whereas the former was three times longer than the latter in the measurement by Jiang and Zhang [9]. This implies that the temporal resolution of EODCP is comparable to that of THz-TDS with delay scanning in our case.

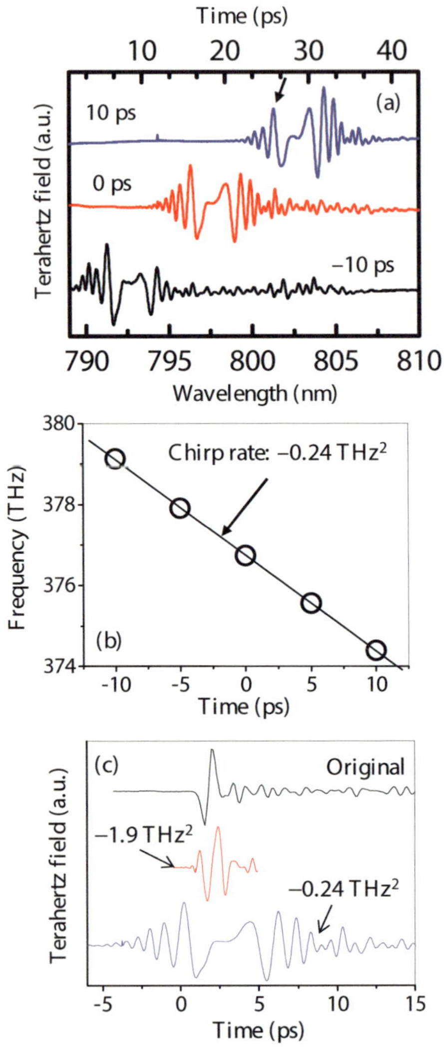

Figure 5. (a) The terahertz waveforms obtained by EODCP with a chirp rate of −0.24 THz^2 at three delay times between the input terahertz field and chirped probe pulses. The ordinate of each waveform is shifted for the sake of clarity. (b) Corresponding peak positions (arrow in (a)) in the terahertz waveform, represented by frequency, plotted as a function of the delay time (circles). The chirp rate is obtained from the slope of the solid line fitted to the data. The upper abscissa in Figure (a) is derived from the lower one using the chirp rate value. (c) The EODCP-derived terahertz waveforms (blue line:−0.24 THz^2, red line:−1.9 THz^2) are compared with that obtained by THz-TDS with delay scanning using a unchirped probe pulse (black line). The ordinate of each waveform is shifted for the sake of clarity. Reproduced from Murakami et al. J. Appl. Phys. **104**, 103111 (2008), with the permission of AIP Publishing.

The terahertz waveforms calculated from Eq. (4) at [a (THz^2), T_c (ps)] = (−0.12, 22), (−0.65, 4), and (−1.5, 1.8), with τ = 0 ps, ΔT = 0.45 ps, and ω_0 = 375 THz (=800 nm) are illustrated in **Figure 6**, together with the original terahertz field pulse (inset) used for the calculation; here we note that the time step corresponding to the wavelength one depends on the chirp rate and that the time step becomes smaller with decreasing the chirp rate. The waveform is found to change from monocyclic to multicyclic behavior with decreasing the magnitude of chirp rate. This tendency agrees with that seen in the experimental result of **Figure 5(c)**. It should be noted that the multicyclic behavior does not emerge in the waveform derived on the assumption that the stationary phase method is valid, as shown by Sun and coworkers [10]. Hence, the analysis based on the stationary phase method is not applicable to the case where the EODCP-derived terahertz waveform is multicyclic.

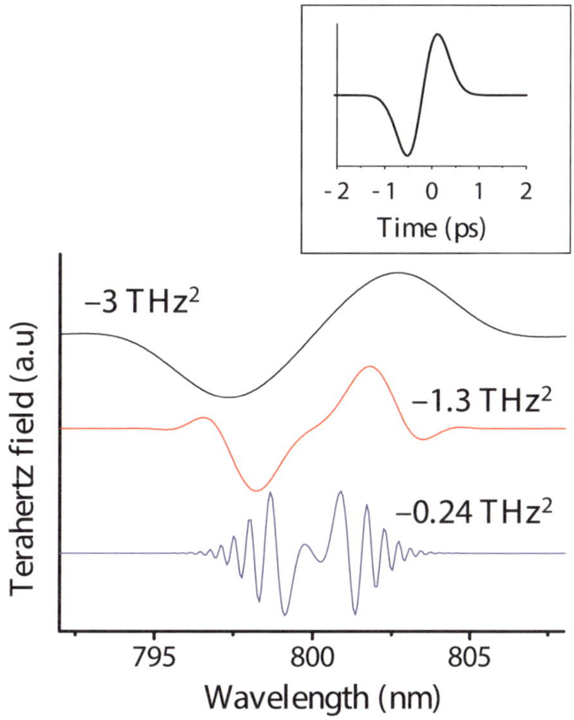

Figure 6. Chirp rate dependence of the terahertz waveform calculated from Eq. (4), where [a (THz^2), T_c (ps)] = (−0.12, 22), (−0.65, 4) and (−1.5, 1.8). The chirp rate value is given by $2a$. The ordinate of each waveform is shifted for the sake of clarity. We note that the time step corresponding to the wavelength one depends on the chirp rate, and that the time step becomes smaller with decreasing the chirp rate. The inset shows the original terahertz waveform used for the calculation. Reproduced from Murakami et al. J. Appl. Phys. **104**, 103111 (2008), with the permission of AIP Publishing.

2.5. Dependence of the EODCP-derived terahertz waveform on spectral resolution of the spectrometer

Figure 7 shows the dependence of the EODCP-derived terahertz waveform on the spectral resolution of the spectrometer at a chirp rate of −1.9 THz^2. The waveform does not exhibit monocyclic behavior at the high spectral resolution (0.2 nm, FWHM), whereas it becomes monocyclic and temporally broadened at the low resolution (3 nm). Therefore, it is found that the spectral resolution of the spectrometer plays an important role in determining the terahertz waveform obtained by EODCP.

The terahertz waveform measured by Jiang and Zhang does not show multicyclic features despite the fact that the chirp rate and temporal width of the chirped probe pulse are comparable to those in our case at −0.24 THz^2. This may be because the spectral resolution of the spectrometer used by them was not high enough to resolve the multicyclic waveform, although its value is not described in the literature [9].

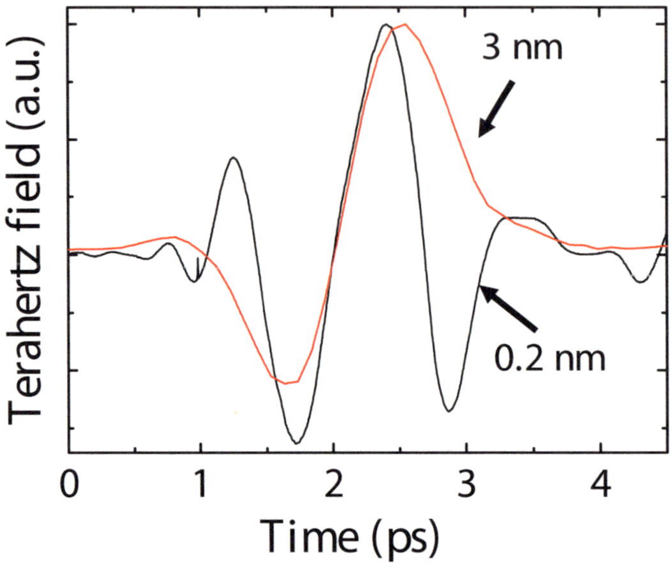

Figure 7. Terahertz waveforms measured by EODCP at two spectral resolutions of the spectrometer (0.2 nm (black line) and 3 nm (red line))at a chirp rate of −1.9 THz^2. Reproduced from Murakami et al. J. Appl. Phys. **104**, 103111 (2008), with the permission of AIP Publishing.

2.6. Terahertz spectroscopy using EODCP

In order to attain high spectral resolution in THz-TDS, the terahertz waveform needs to be measured in a given temporal observation window, according to time-frequency relation based on the Fourier transform. A time window of a few tens of picoseconds is required to obtain spectral resolution as small as 1 cm^{-1} (30 GHz). EODCP with a small chirp rate leads to a narrow temporal observation window under the limited spectral width of the probe pulse (~10 nm in the present study), as seen in **Figure 5(c)**, and so a high spectral resolution is not achieved. Therefore, EODCP with a small chirp rate has a crucial drawback for terahertz spectroscopy, even if the terahertz waveform approaches the original one. Thus, we have to use a chirped probe pulse with a chirp rate that assures an observation window of a few tens of picoseconds, although we will encounter multicyclic behavior in the terahertz waveform in such a case.

The multicyclic behavior in the terahertz waveform calculated from Eq. (4) at a chirp rate of −0.24 THz^2 leads to the oscillating behavior in the frequency spectrum, as shown in **Figure 8**; although the envelope connecting the peaks in the spectrum coincides with the spectrum obtained from the Fourier transform of the original terahertz field waveform (inset of **Figure 6**). Thus, we consider the effect of the oscillating behavior on the derivation of the

http://dx.doi.org/10.5772/67195

optical constant of materials by terahertz spectroscopy using EODCP. The quadratic term of the modulation coefficient is neglected in the derivation of Eq. (4), which implies that the EODCP instrument exhibits a linear response to the terahertz field. Under the condition of such a linear response, the EODCP-derived terahertz waveform is expressed by

$$E_{ref}(t) = \int_0^\infty \varphi(t - t_1)\, E_0(t_1) dt_1, \tag{6}$$

where $\varphi(t)$ and $E_0(t)$ are the time profile of the instrumental response and the original terahertz field, respectively. If the terahertz field is transmitted through a material under examination and measured by EODCP, its waveform is given by

$$E_{sig}(t) = \iint \varphi(t - t_1)\eta(t_1 - t_2)\, E_0(t_2) dt_1\, dt_2, \tag{7}$$

where the material is assumed to exhibit a linear response to the terahertz field and the response function is denoted by $\eta(t)$. Thus, the complex transmission coefficient $\tilde{\eta}(\omega)$ of the material is derived by

$$\tilde{\eta}(\omega) = \frac{\tilde{E}_{sig}(\omega)}{\tilde{E}_{ref}(\omega)} \tag{8}$$

with

$$\tilde{E}_{sig}(\omega) = \tilde{\varphi}(\omega)\tilde{\eta}(\omega)\, \tilde{E}_0(\omega) \quad \text{and} \quad \tilde{E}_{ref}(\omega) = \tilde{\phi}(\omega)\, \tilde{E}_0(\omega), \tag{9}$$

where the terms with a tilde, e.g.,$\tilde{E}_{sig}(\omega)$, are the Fourier transforms of the corresponding terms without a tilde, e.g., $E_{sig}(t)$. The convolution theorem is used in the derivation of Eq. (9). The Fourier-transformed spectra of the instrument response function and the original terahertz waveform are canceled out in Eq. (8), and so have no effect on derivation of the optical constant of the material. The complex refractive index is derived from the complex transmission coefficient [47].

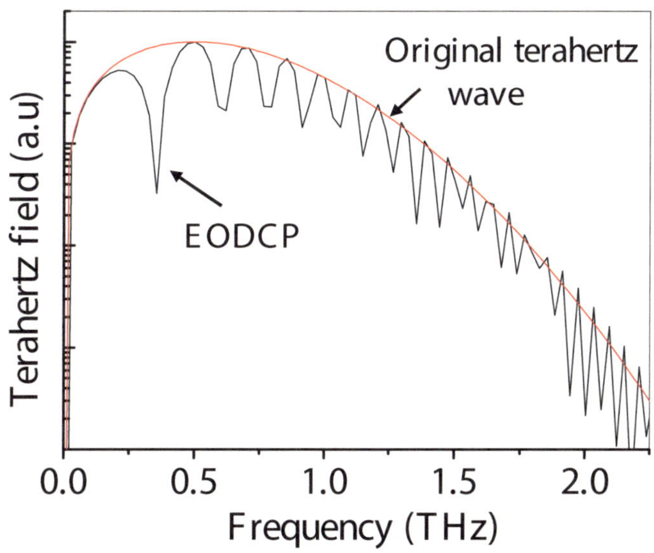

Figure 8. Frequency spectra obtained from the Fourier transform of the terahertz waveform calculated from Eq. (4) at a chirp rate of −0.24 THz² (black line) and the original waveform (red line). A logarithmic scale is used for the ordinate. Reproduced from Murakami et al. J. Appl. Phys. **104**, 103111 (2008), with the permission of AIP Publishing.

The EODCP-derived terahertz waveform depends on the delay time between the terahertz field and chirped probe pulses, as shown in **Figure 5(a)**. This implies that the instrument response of EODCP varies with the delay time. If the refractive index of a material examined is not unity in the terahertz spectral range, the terahertz field pulse transmitted through the material is delayed from that passing through free space by an amount determined by the refractive index. And thus, the two terahertz field pulses modulate temporally different portions of the chirped probe pulse. Therefore, the instrument response is different for the two terahertz fields. A method to remove the effect of variable instrument response is to make the two terahertz field pulses overlap with the same (temporal) portion of the chirped probe pulse using an optical delay line, although the time difference between the two terahertz field pulses must be measured in advance. The chirped probe beam can be modulated by the terahertz field at any temporal portion using the optical delay line (delay line 1 in **Figure 3**).

In order to examine the applicability of terahertz spectroscopy using EODCP, we measured the optical constant of a material by the method and compared the result with that by THz-TDS with delay scanning. A semi-insulating GaAs crystal (1 0 0) with a thickness of 500 μm was used as a sample. **Figure 9(a)** illustrates the terahertz waveforms observed using EODCP in the presence (red line) and absence (black line) of GaAs, where the optical delay line (delay line 1 in **Figure 3**) was not changed for the two measurements, i.e., we did not adjust the temporal overlapping between the chirped and terahertz field pulses. Both the waveforms exhibit multicycle behavior owing to the small magnitude of the chirp rate. The terahertz wave transmitted through the GaAs substrate changes its waveform and is delayed compared with the wave in the absence of the GaAs substrate. We derive the complex refractive index $\tilde{n}(\omega) = n(\omega) - i\kappa(\omega)$ of the material with a thickness of d from the following equation using the complex amplitude spectra obtained from the Fourier transform of the two terahertz waveforms,

$$\frac{E_{sig}(\omega)}{E_{ref}(\omega)} = \tilde{t}_{as}(\omega) \exp\left\{-i\frac{(\tilde{n}(\omega)-1)d\omega}{c}\right\}\left[\sum_{l=1}^{m}\left\{(\tilde{r}_{sa}(\omega))^2 \exp\left(-i\frac{2\tilde{n}(\omega)d\omega}{c}\right)\right\}^{l}\right]\tilde{t}_{sa}(\omega), \quad (10)$$

$$\tilde{t}_{as}(\omega) = \frac{2}{\tilde{n}(\omega)+1},\ \tilde{r}_{sa}(\omega) = \frac{\tilde{n}(\omega)-1}{\tilde{n}(\omega)+1},\ \tilde{t}_{sa}(\omega) = \frac{2\tilde{n}(\omega)}{\tilde{n}(\omega)+1},$$

where $\tilde{t}_{as}(\omega)$ and $\tilde{t}_{sa}(\omega)$ are a complex Fresnel coefficient for transmission at the sample surface, and a subscript of "as" or "sa" indicates that the terahertz wave proceeds into the sample or out of it. Further, $\tilde{r}_{sa}(\omega)$ is a complex Fresnel coefficient for reflection at the sample surface, where multireflection at the sample surface is taken into account. We use a successive approximation method for the numerical calculation. The complex refractive index thus derived is compared with that due to THz-TDS with delay scanning in **Figure 9(b)**. The agreement between the two results is good, and so the multicyclic profile of the terahertz waves obtained using EODCP does not affect derivation of the optical constant in this case.

A preliminary application of OPTP spectroscopy using EODCP was conducted for the same GaAs. The experimental system is schematically described in **Figure 3**. A second harmonic femtosecond pulse (~400 nm) generated with a BBO crystal was used as a pump light to produce charge carriers. The terahertz wave is reflected by the charge carriers, and so one can measure the lifetime of the charge carrier by time-resolved terahertz reflection or transmission spectroscopy. **Figure 10(a)** is the terahertz waveform after transmission through the GaAs substrate as a function of delay time between the optical pump and terahertz probe pulses using optical delay line 2 in **Figure 3**. The time-dependent change in the terahertz waveform is seen in **Figure 10(a)**. From the peak amplitudes of the frequency spectra, the time profile of the normalized reflectivity is derived, as plotted by circles in **Figure 10(b)**. On the other hand, OPTP spectroscopy using terahertz waveform measurements with delay scanning is applicable for GaAs, because the initial state is recovered in the period of the laser pulse repetition (~1 ms). From the peak amplitude of the terahertz waveform, the time-resolved normalized reflectivity was derived, as shown by a red line in **Figure 10(b)**. The results due to the two methods agree with each other and show a single exponential decay with a time constant of ~1 ns. Thus, it is demonstrated that terahertz waveform measurements using EODCP are a promising method for OPTP spectroscopy.

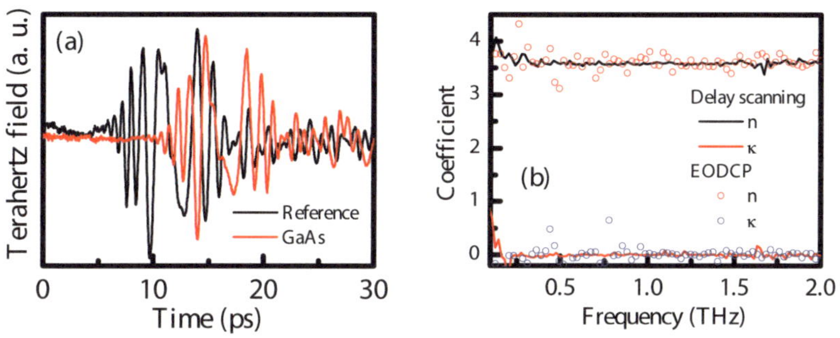

Figure 9. (a) Waveforms of the terahertz field pulses transmitted through a 0.5-mm-thick GaAs substrate (red line) and in the absence of the sample (black line), observed using EODCP. (b) The complex refractive index $\tilde{n}(\omega) = n(\omega) - i\kappa(\omega)$ of the sample derived from Eq. (10) using the complex amplitude spectrum obtained from the Fourier transform of the two terahertz waveforms in Figure (a) (circles), together with the result due to the terahertz waveforms measured by THz-TDS with delay scanning (lines).

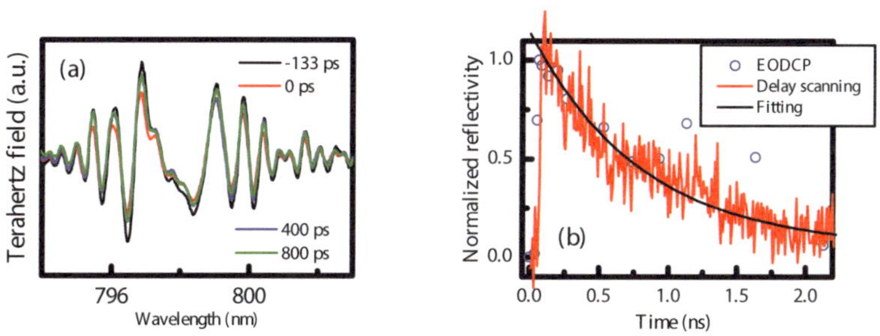

Figure 10. (a) Terahertz waveforms after transmission through the GaAs substrate measured by OPTP spectroscopy with EODCP at several pump-probe delay times. (b) The time-dependent normalized reflectivity (circles) derived from the peak amplitudes of the frequency spectra obtained from the Fourier transform of the terahertz waveforms, together with the result due to OPTP spectroscopy using THz-TDS with delay scanning (red line) and exponential decay curve with a time constant of ~1 ns (black line).

2.7. Future studies

Future work necessary for the development of the EODCP method is as follows. (1) We need to make the signal-to-noise ratio high for the single-shot measurement. One way for it is to use a detector with a wide dynamic range, for example, high-charge capacity to low readout noise in CCDs. The laser intensity is high enough to attain high signal-to-noise ratio. However, practically it is too strong for any sensors, and so the input intensity is usually much attenuated to prevent the sensor from saturation. (2) More pulse-to-pulse stability in the laser beam spot, polarization, and so on, in the femtosecond laser is required for the single-shot measurement. We used the double-beam configuration for the measurement, and so the pulse-to-pulse fluctuation in the laser frequency spectrum was corrected. However, the baseline of the terahertz waveform derived was distorted in the single-shot measurement (**Figure 4**). This implies that the laser fluctuation other than the spectral fluctuation affects the measurement. (3) It is interesting to perform the EODCP measurement using a chirped probe pulse with a wide spectral width and large magnitude of the chirp rate. In the present study, we used a chirped pulse with a small magnitude of it for the application of terahertz spectroscopy because of the narrow spectral width of the laser (~10 nm). There is a Ti:sapphire femtosecond laser whose spectral width is several tens of nanometers. It allows us to conduct EODCP measurements in a wide temporal observation window even by use of a chirped pulse with large magnitude of the chirp rate. As the magnitude of the chirp rate is raised, the EODCP-derived terahertz waveform approaches a monocyclic behavior, and correspondingly, the spectrum in the frequency domain shows less oscillating behavior. Thus, EODCP with such a chirped pulse will be advantageous for terahertz spectroscopy.

3. Experiment of THz-TDS of reverse micellar solution

Before describing the results of terahertz spectroscopy of the reverse micellar solutions, the experimental procedures are described in this section.

3.1. Sample

Myoglobin, AOT [=*bis*(2-ethyl-hexyl)sulfosuccinate], isooctane, and millipore water were used as a protein, surfactant, oil, and water, respectively, for preparation of the PCR micellar solutions. AOT was dried in a vacuum oven at 45°C. A Karl Fisher titrator was used to measure the water content in the samples. Two methods were applied for preparation of the PCR micellar solution. One is the so-called "injection method," i.e., a small amount of protein aqueous solution was injected into AOT/isooctane solution, and the other is a paste method, proposed by us; the two methods are described in detail elsewhere [48]. The reverse micellar solution (without a protein) was prepared by the injection method of a small amount of water.

3.2. Sample cell

A sample cell has an optical path length of 1 cm and two sample-chambers with the same dimension (dual cell); one is for a sample signal, and the other is for a reference signal, as seen in **Figure 1**. In the measurement of PCR micellar solution (Section 4), we used the PCR micellar solution for a sample signal and oil solvent for a reference signal. On the other hand, in the temperature-dependent THz-TDS of water in the reverse micellar solution (Section 5), we used reverse micellar solutions with water for a sample signal and without water for a reference signal. The dual cell is fixed on a computer-controlled stage, and the measurement of the two

http://dx.doi.org/10.5772/67195

samples was iteratively made by moving the stage for accumulation of the data. The measurement using the dual cell decreases the distortion of the absorption spectrum due to long-period fluctuation of the laser, in comparison with the measurement using a single cell, because the time interval between the measurements of the two samples is short, typically one minute.

3.3. Derivation of the optical constant in the measurement using a dual cell

We used THz-TDS with delay scanning for the reverse micellar solutions. The absorption coefficient $\alpha(\omega)$ and refractive index $n(\omega)$ of a sample were derived as follows: (1) the waveforms of the terahertz fields transmitted through the two solutions were converted to frequency-domain data (i.e., the intensity: $I_s(\omega)$ and $I_r(\omega)$, phase: $\phi_s(\omega)$ and $\phi_r(\omega)$; s : sample, and r : reference) using Fourier transform, (2) the absorption coefficient was derived from $\alpha(\omega) = d^{-1} ln [I_r(\omega)/I_s(\omega)]$, where d is the optical path length of the cell (1 cm in the present study), (3) the refractive index was derived with the equation $n_s(\omega) - n_r(\omega) = c(\omega d)^{-1}(\phi_s(\omega) - \phi_r(\omega))$, where $n_s(\omega)$ and $n_r(\omega)$ are the refractive indices of the sample and reference solutions, respectively, and c is the speed of light. A spherical reverse micelle is formed owing to the self-assembly of AOT molecules even if there is almost no water in the solution [49]. In the measurement of water in the reverse micellar solution (Section 5), the two solutions in the dual cell had the same concentration of AOT. Therefore, we assumed that the contribution of AOT molecules to the optical constant is canceled out in their derivation [50]. On the other hand, the absorption and phase shift of the terahertz wave by isooctane is very small because of its nonpolar nature, and so the effect of isooctane on the optical constant was considered negligible.

3.4. Absorption lineshape function

The absorption lineshape function $\tilde{C}(\omega)$ is given by

$$\tilde{C}(\omega) = \frac{A\alpha(\omega)n(\omega)}{\omega\left[1 - \exp\left(-\frac{\hbar\omega}{k_B T}\right)\right]} \tag{11}$$

$$= \frac{1}{2\pi}\int_{-\infty}^{+\infty} dt\, e^{-i\omega t}\, C(t), \tag{12}$$

$$C(t) = \langle M(0) \cdot M(t) \rangle \tag{13}$$

where $A = 3c\hbar(2\pi)^{-2}$, $\hbar$ is the reduced Planck constant, k_B is the Boltzmann constant, and T is the sample temperature [51, 52]. Eq. (11) indicates that the lineshape function is expressed by the Fourier transform of the time-correlation function $C(t)$ of the total dipole moment $M(t)$ of the system examined. The absorption lineshape function was calculated from Eq. (11) by use of the absorption coefficient and refractive index obtained from the measurements with THz-TDS, and then the relaxation process was examined using the correlation function.

3.5. Cryostat for temperature-dependent measurements

It is necessary that the reference and sample solutions are at the same temperature to obtain the correct optical constant of the sample at a given temperature. Therefore, the dual cell was fixed in a custom-made stainless-steel cryostat, and its temperature was controlled within approximately ±1 K using nitrogen gas flow from a liquid nitrogen vessel. The waveforms of the terahertz fields transmitted through the two solutions were alternately measured using a computer-controlled stage on which the cryostat was fixed.

We examined the baselines of the absorption coefficient and refractive index obtained from the THz-TDS using the temperature-controlled dual cell. To that end, the sample and reference cells were filled with the same AOT reverse micellar solution, and the temperature-dependent measurement of THz-TDS was made in the 273–296 K temperature range. The results are depicted in **Figure 11**, together with the optical constant of water in the reverse micellar solution with $w_0 = 35$ at 295 K. **Figure 11(a)** demonstrates that all the baselines of the absorption coefficient are flat in the frequency range between 0.1 and 1.2 THz; here, the standard deviation of the mean is around 6×10^{-3}. As for the refractive index, the baselines are flat and approximately unity, shifted only by ~3×10^{-3}, with the standard deviation of the mean of ~2×10^{-5} in **Figure 11(b)** in the same frequency range. The standard errors are small enough compared with the values of the optical constants, and hence, the experimental system used works for temperature-dependent measurements of the optical constant. Thus, we analyzed the optical constant in this terahertz frequency range.

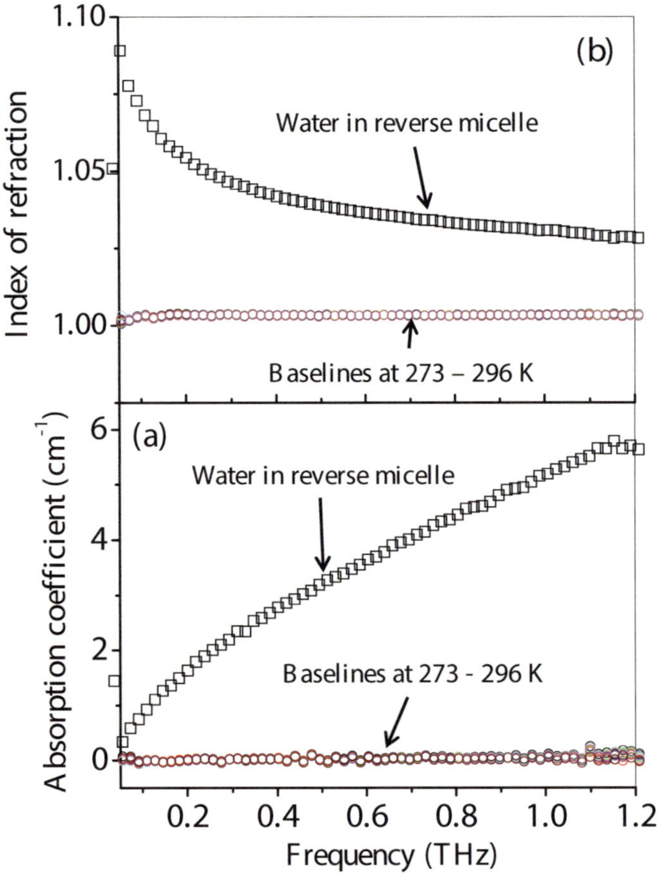

Figure 11. Temperature dependence of baselines for (a) the absorption coefficient and (b) the refractive index measured by THz-TDS with delay scanning using a temperature-controlled dual cell. The sample and reference cells were filled with the same AOT reverse micellar solution. The baselines at 273, 278, 283, 287, 293, and 296 K are shown by circles, and the optical constants of water in the AOT reverse micelle at $w_0 = 35$ and 295 K by squares. Adapted from Murakami et al. (2015), with permission of Elsevier.

4. Terahertz spectroscopy of PCR micellar solution

4.1. Protein concentration dependence of the absorption spectra

The protein concentration dependence of the absorption spectra of the PCR micellar solution at w_0 = 4.4 and 0.2 is displayed in **Figure 12(a)** and **(b)**, respectively. The amplitude of the spectrum becomes large with the raising of the protein concentration at w_0 = 0.2 in the spectral range examined, whereas it decreases at w_0 = 4.4. This is understood by the fact that the amplitude of the absorption spectrum of hydration water is smaller than that of bulk water in the terahertz spectral range [44, 53], as described below. Here we note that PCR micellar solution is a binary system of PCR micelles and protein-unfilled, i.e., water-containing reverse (WCR) micelles, because it is difficult that protein molecules are introduced into all the reverse micelles in the solution. The molar ratio of bulk-like to interfacial-hydrated water within the WCR micelle increases with increasing w_0 value [54–59]. The number of water molecules in the WCR micelle is estimated to be ~5 at w_0 = 0.2 and ~250 at w_0 = 4.4, whereas in the PCR micelle, it is estimated to be ~20 at w_0 = 0.2 and ~580 at w_0 = 4.4 [24, 48]. Moreover, the number of AOT molecules in the WCR micelle is estimated to be ~20 at w_0 = 0.2 and ~60 at w_0 = 4.4, whereas in the PCR micelle, it is estimated to be 110 at w_0 = 0.2 and ~130 at w_0 = 4.4 [24, 48]. Hence, almost all the water molecules within the WCR micelle will be bound to AOT molecules at w_0 = 0.2, whereas some of water molecules are bulk-like at w_0 = 4.4; we note that one AOT molecule has at least a few hydration sites [54]. In the PCR micelle, almost all the water molecules will be bound to the protein or AOT molecule at the two w_0 values; the first hydration shell of myoglobin includes several hundred water molecules [60]. Therefore, the bulk-like water becomes hydration one in the PCR micellar solution at w_0 = 4.4, when the WCR micelle disappears by formation of the PCR micelle. Thus, if the decrease in the amplitude of the absorption spectrum owing to hydration exceeds the increase due to the protein, the total spectral amplitude descends with increasing the protein concentration at w_0 = 4.4. On the other hand, the spectral amplitude increases by the contribution of protein with raising the protein concentration at w_0 = 0.2, because all the water molecules in both the WCR and PCR micelles are hydration water at w_0 = 0.2. Although there could be a difference between the spectra of hydration waters bound to AOT and protein, the effect of the difference is not observed in the abovementioned result of protein concentration dependence.

4.2. Derivation of terahertz absorption spectrum of a protein in a reverse micelle

We analyze the absorption spectra by

$$\alpha(\omega) = C_P\,\varepsilon_{\text{protein}}(\omega) + C_{\text{water}}\,\varepsilon_{\text{water}}(\omega) + C_S\,\varepsilon_S(\omega), \tag{14}$$

where $\varepsilon_{\text{protein}}(\omega)$, $\varepsilon_{\text{water}}(\omega)$, and $\varepsilon_S(\omega)$ are the molar extinction coefficients of protein, water, and surfactant in the PCR micellar solution, respectively, and C_P, C_{water} and C_S are the molar concentrations of protein, water, and surfactant, respectively. We do not take account of the effect of the interactions between the different constituents on the spectrum in Eq. (12). Moreover, the contribution of a low content of water will be negligible at w_0 = 0.2. On the other hand, the surfactant molecules are aggregated to form reverse micelles with and without a protein.

Therefore, we assume that the molar extinction coefficient of the surfactant in the PCR micellar solution is independent of the protein concentration. On these assumptions, we obtain the molar extinction coefficient of the protein from the difference spectrum between the absorption spectra at two protein concentrations, i.e.,

$$\begin{aligned} \Delta\alpha(\omega) &= \left(C_P' - C_P\right)\varepsilon_{\mathrm{protein}}(\omega) = \Delta C_P\,\varepsilon_{\mathrm{protein}}(\omega) \\ \varepsilon_{\mathrm{protein}}(\omega) &= \Delta\alpha(\omega)/\Delta C_P \end{aligned} \quad . \tag{15}$$

The molar extinction coefficient of the protein thus derived is depicted in **Figure 12(c)**, where combinations of the protein concentration for the difference spectrum are 1.5–0.72 mM (circles), and 1.5–0.34 mM (squares). The two spectra agree with each other in **Figure 12(c)**, and so the assumptions made above are considered to be valid. Thus, the terahertz absorption spectrum of the protein in the reverse micelle is derived from the protein concentration dependence of the absorption spectrum of the PCR micellar solution.

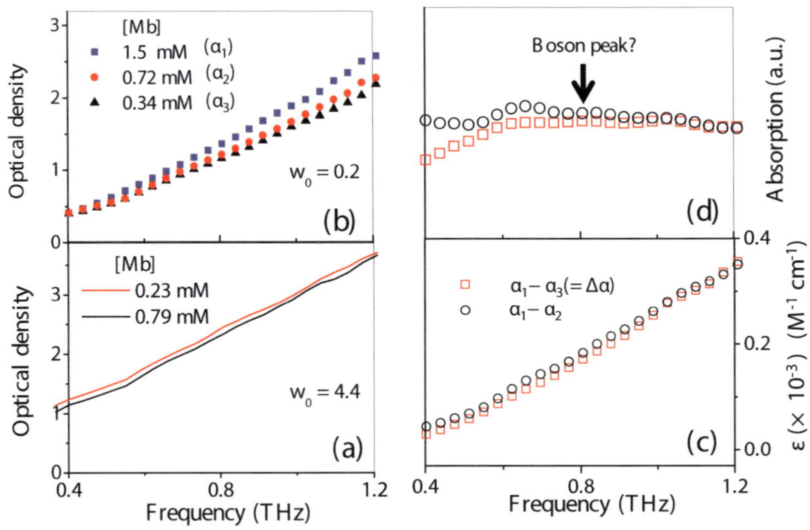

Figure 12. (a) Protein concentration dependence of the absorption spectra of the PCR micellar solution at $w_0 = 4.4$ and (b) $w_0 = 0.2$ at room temperature. (c) Molar extinction coefficient of myoglobin within the PCR micelle derived from the difference spectrum between the absorption spectra at different protein concentrations at $w_0 = 0.2$. (d) Absorption lineshape function of myoglobin obtained from Eq. (11). Adapted from Murakami et al. (2012), with permission of Elsevier.

4.3. Future studies

The terahertz spectrum in **Figure 12(c)** has no characteristic peaks, as seen for protein powder and liquid samples. This is considered to be so because a protein molecule has a large number of low frequency modes [61, 62] and shows structural fluctuations [63, 64]. The detailed interpretation of the terahertz spectrum of protein molecules is a future problem. Finally, this terahertz spectrum of myoglobin in relation to the boson peak of protein is discussed. It is well known from inelastic neutron and Raman scattering measurements of proteins that proteins show a broad spectrum with a peak position at around 1 THz, which is called the boson peak

http://dx.doi.org/10.5772/67195

[65–67]. The origin of the boson peak has not been completely understood. Further, the boson peak is observed not only for proteins but also for a wide range of glassy materials such as polymers and glass-forming liquids. The molar extinction coefficient is proportional to frequency and includes a Bose-Einstein factor. In order to compare the result of terahertz spectroscopy with those of inelastic neutron and Raman-scattering measurements, the absorption lineshape function must be derived by removing such frequency dependence, as described in Section 3.4, where we assume that the frequency dependence of the refractive index is negligible in this limited frequency range. **Figure 12(d)** is the absorption lineshape function thus derived. It is found that the spectrum shows a broad peak at around 0.8 THz, which is identical with that from inelastic neutron-scattering measurements of hydrated myoglobin [65]. An interesting thing is that this peak is observed at room temperature because the boson peak of proteins is observed below ~200 K in the inelastic neutron-scattering measurement. Thus, the broad peak observed in the PCR micellar solution could be a new candidate to clarify the mechanism of the boson peak, and further investigation will be performed and reported elsewhere.

5. Temperature-dependent terahertz spectroscopy of water in reverse micellar solutions

The Stokes radius of the reverse micelle examined ($w_0 = 35$) was ~12 nm at room temperature. It was found from temperature-dependent dynamic light-scattering measurements that this reverse micelle shows the water shedding at 286 K (**Figure 14(a)**) [45].

5.1. Spectral analysis

We analyzed the absorption lineshape function obtained from Eq. (11). The refractive index of water in the reverse micellar solution exhibits almost no frequency-dependent change in the frequency range examined and is close to unity, as seen in **Figure 11(b)**. Consequently, the refractive index does not affect the frequency dependence of the absorption lineshape function. Thus, we calculate $\tilde{C}(\omega)$ at $n(\omega) = 1$.

For curve fitting the absorption lineshape function, a sum of two exponential relaxations with time constants of τ_f for a fast component and τ_s for a slow one was employed for $C(t)$, i.e., $C(t) = A_f e^{-t/\tau_f} + A_s e^{-t/\tau_s}$, where A_f and A_s are the amplitudes for the fast and slow components, respectively. This leads to the sum of the two Lorentzian functions in the frequency domain by Eq. (12), i.e.,

$$F(\omega) = 2A_f\tau_f^{-1}\left(\omega^2+\tau_f^{-2}\right)^{-1} + 2A_s\tau_s^{-1}\left(\omega^2+\tau_s^{-2}\right)^{-1} + B. \quad (16)$$

This model can be regarded as the Debye model with two relaxation times [41–43, 52]. A term of B in Eq. (16) is a constant baseline necessary for a better fit. Water shows an ultrafast Debye relaxation of ~0.1 ps [41–43, 52], and its contribution to the absorption lineshape function is considered to be a constant baseline because the upper limit of the frequency range examined is too low to analyze the spectral shape due to the ultrafast relaxation. The fitting procedure using a nonlinear least squares method was identical for the absorption lineshape functions at all the temperatures examined.

The curve fitting using Eq. (16) does well for the absorption lineshape function of water in the reverse micelle solution at all the temperatures examined, as displayed in **Figure 13**. The fitting curve (black line) agrees well with the curve obtained from the experiment (circles), where the curves due to the fast component (blue line), the slow component (red line), and the baseline (green line) are also depicted. Thus, the time constants of the fast and slow components are acquired from the fitting. There is a question regarding the uniqueness of the time constants derived from the curve fitting, although multi-parameter fitting is a common method for spectral analysis [41–43, 68, 69]; in fact, as shown for liquid water in **Figure 14(c)**, there is a small difference between the time constants from the two literatures [41, 42, 68]. However, a systematic deviation due to the fitting procedure should not influence the conclusion on the basis of the temperature dependence of the values derived from the same procedure.

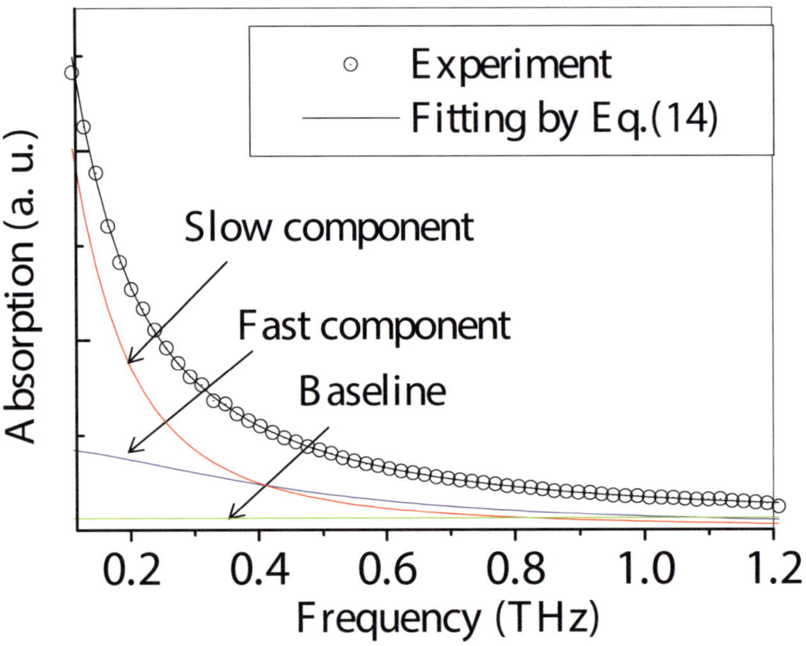

Figure 13. A result of the curve fitting using Eq. (16) of the absorption lineshape function obtained from Eq. (11) at 295 K. The contributions of the fast (blue) and slow (red) components and constant baseline (green) in Eq. (16) to the fitting curve are also illustrated. Reproduced from Murakami et al. (2015), with permission of Elsevier.

5.2. Temperature dependence of the relaxation time

Figure 14(b) shows the waveforms of the terahertz field pulses transmitted through the two solutions at 286 K (= T_c). Eight sets of data are superimposed in the figure. All the waveforms are reproduced for the reference solution, whereas the waveform for the sample solution abruptly changes during the repeated measurement. Since T_c is identical to the temperature at which the water shedding occurs in the reverse micellar solution, the change in the waveform for the sample solution is attributed to the water shedding.

http://dx.doi.org/10.5772/67195

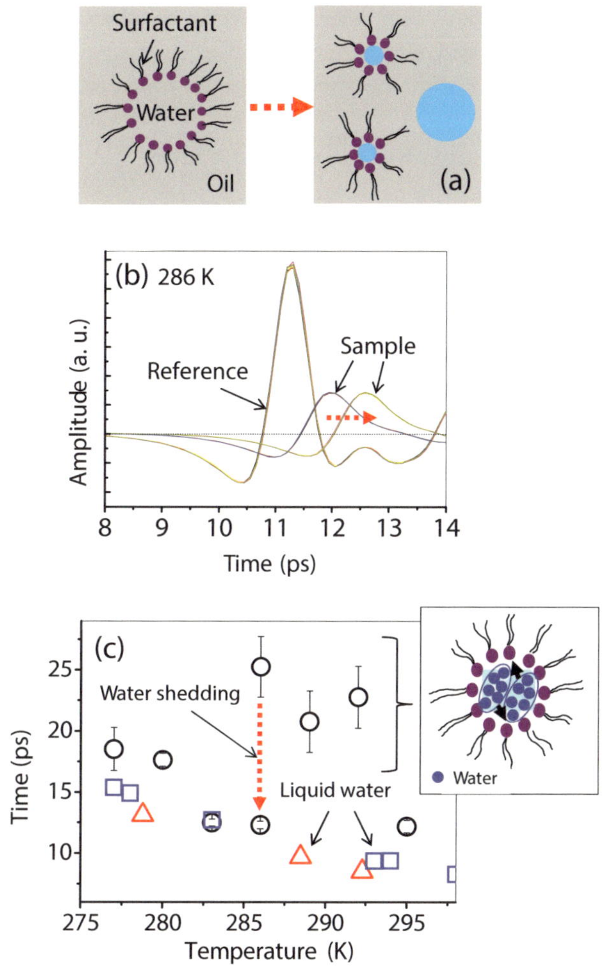

Figure 14. (a) Schematic cross-section of a reverse micelle before (left) and after the water shedding (right). (b) The terahertz waveforms after transmission through the sample and reference solutions at 286 K. Data obtained by THz-TDS (delay scanning) measurements repeated eight times are superimposed. The reference solution is AOT/isooctane solution without water, whereas the sample solution is that with water at $w_0 = 35$. The waveform for the sample solution changes (red arrowed dashed line) during the repeated measurement. (c) The time constant of the slow component obtained by the curve fitting using Eq. (16), together with the data of liquid water from literature [41, 42] (triangles) and [68] (squares). An abrupt drop (red arrowed dashed line) in the time constant occurs during the repeated measurement at 286 K. The inset shows that the confinement of the reverse micelles affects the collective motions of water with a cooperative length comparable with the aqueous cavity length. Adapted from Murakami et al. (2015), with permission of Elsevier.

The temperature dependence of the time constant of the slow component (circles) is depicted in **Figure 14(c)**, together with the results for liquid water (triangles [41, 42] and squares [68]) from literature. There are three characteristic findings as the temperature is reduced. First, considerable slowing down occurs above T_c compared with the case of liquid water. Second, the time constant dramatically decreases at T_c (red arrowed dashed line); this corresponds to a change in the terahertz waveform of the sample solution at T_c, as seen in **Figure 14(b)**.

Lastly, the temperature-dependent behavior of the time constant is similar to that of liquid water below T_c. On the other hand, the time constant of the fast component was found to be in the range from ~2 ps to 4 ps, and an abrupt drop in the time constant is observed at T_c [45].

5.3. Relaxation process before and after the water shedding

The slow component is due to the collective motions of water molecules observed in THz-TDS of liquid water. This is reasonable because almost all the water molecules in the reverse micelle are bulk-like at $w_0 = 35$. THz-TDS measurements of liquid water demonstrated that the relaxation time exhibits a power-law temperature dependence as $(T - T_s)^{-1})$ ($T_s = 228$ K) above the melting point of water [41, 42]. The power-law temperature dependence implies that the cooperative nature of the water motion becomes large with reducing temperature, i.e., a larger hydrogen-bond network is formed for the motion at lower temperatures; accordingly, the time scale of the water motion increases. Such a mechanism is valid for water confined in large reverse micelles. However, the confinement of the reverse micelles should affect the mechanism, as illustrated in the inset of **Figure 14(c)**, and hence, it is reasonable that the slowing down of the relaxation process with decreasing temperature is enhanced in the reverse micelle compared with that in liquid water.

A dramatic drop in the time constant at T_c in **Figure 14(c)** is due to water shedding from the reverse micelle. The expelled water is free from the confinement of the reverse micelle. Therefore, the relaxation process of the water will speed up compared with that of the water within the reverse micelle. Moreover, it is reasonable that the relaxation process of the expelled water is similar to that of liquid water below T_c, as seen in **Figure 14(c)**. We have recently shown for a reverse micelle containing a water-soluble probe molecule that the diffusion-like motion of the water molecules surrounding the probe molecule becomes active upon the water and dye molecule shedding from the reverse micelle [36]. This is a kind of speeding up of the water motion induced by the water shedding.

5.4. Future studies

The present study demonstrates that there is a distinct difference between the properties of water in and out of the cage, and that water extracted from the reverse micelle exhibits the relaxation process similar to that of liquid water. Therefore, the water droplet free from the cage is expected to be more appropriate for studies of the properties of supercooled water at low temperatures than water confined in the cage. The results on temperature-dependent THz-TDS of water in smaller reverse micelles that show the water-shedding below the melting point of water will be reported elsewhere.

6. Concluding remarks

Protein-containing reverse micelles are suitable systems for terahertz spectroscopy of protein molecules in liquid. Moreover, terahertz waveform measurements using EODCP is a promising method for OPTP spectroscopy. The next step is OPTP spectroscopy of PCR micellar solution. As seen in **Figure 12(b)**, the optical density of the protein molecule is very small in the terahertz range even for its high concentration, which is practically attained. Therefore, it will be necessary that the protein molecule exhibits large spectral change involved in the photoreaction induced by the pump optical pulse, where high quantum efficiency is needed for the photoreaction. If a protein exhibits a narrow band peak at the frequency characteristic of the collective motion coupled with the reaction, we expect to observe distinct change in the time-resolved terahertz spectrum. Further, it is noted that we used PCR micelles with a few tens of water molecules (w_0 = 0.2) to obtain the absorption spectrum of the protein molecule. On the other hand, at w_0 = 4.4, the spectrum of the protein molecule is buried in the large spectral contribution of ~600 water molecules around the protein one. A certain number of water molecules, with which the first hydration shell will be at least filled, will be required for the biological reaction of protein molecules. Hence, the water content in the reverse micelle has to be adjusted to the minimum under the condition that the biological reaction of protein occurs. Here, the hydration water is considered to contribute to the characteristic change in the time-resolved terahertz spectrum because it should move in concert with the protein's motion.

Acknowledgements

The author thanks M. Katsurada, K. Shimizu, Y. Toyota, T. Nishi, and S. Nashima for their collaboration. These studies were partly supported by Grants-in-Aid for Scientific Research (23654139 and 22310038).

Author details

Hiroshi Murakami

Address all correspondence to: murakami.hiroshi@qst.go.jp

National Institutes for Quantum and Radiological Science and Technology (QST), Kizuga City, Kyoto, Japan

References

[1] Y. -S. Lee, *Principles of terahertz science and technology* (New York, Springer, 2009).

[2] *Terahertz frequency detection and identification of materials and objects*, edited by R. E. Miles, X. -C. Zhang, H. Eisele, and A. Krotkus (Springer, Dordrecht, 2007).

[3] K. Y. Kim, B. Yellampalle, J. H. Glownia, A. J. Taylor, and G. Rodriguez, Phys. Rev. Lett. **100**, 135002 (2008).

[4] S. Kar, D. R. Mohapatra, E. Freysz, and A. K. Sood, Phys. Rev. B **90**, 165420 (2014).

[5] P. D. Cunningham and L. M. Hayden, J. Phys. Chem. C **112**, 7928 (2008).

[6] J. R. Danielson, Y. -S. Lee, J. P. Prineas, J. T. Steiner, M. Kira, and S. W. Koch, Phys. Rev. Lett. **99**, 237401 (2007).

[7] M. Jewariya, M. Nagai, and K. Tanaka, Phys. Rev. Lett. **105**, 203003 (2010).

[8] M. C. Hoffmann, J. Hebling, H. Y. Hwang, K. -L. Yeh, and K. A. Nelson, J. Opt. Soc. Am. B **26**, 29 (2009).

[9] Z. Jiang and X. -C. Zhang, Appl. Phys. Lett. **72**, 1945 (1998).

[10] F. G. Sun, Z. Jiang, and X. -C. Zhang, Appl. Phys. Lett. **73**, 2233 (1998).

[11] H. Murakami, K. Shimizu, M. Katsurada, and S. Nashima, J. Appl. Phys. **104**, 103111 (2008).

[12] D. A. Turton, H. M. Senn, T. Harwood, A. J. Lapthorn, E. M. Ellis, and K. Wynne, Nat. Commun. **5**, 3999 (2014).

[13] S. Hay and N. S. Scrutton, Nat. Chem. **4**:161 (2012).

[14] K. Itoh and M. Sasai, Proc. Natl. Acad. Sci. USA **101**, 14736 (2004).

[15] F. G. Parak, Rep. Prog. Phys. **66**, 103 (2003).

[16] E. Castro-Camus and M. B. Johnston, Chem. Phys. Lett. **455**, 289 (2008).

[17] S. E. Whitmire, D. Wolpert, A. G. Markelz, J. R. Hillebrecht, J. Galan, and R. R. Birge, Biophys. J. **85**, 1269 (2003).

[18] J. Xu, K. W. Plaxco, and S. J. Allen, J. Phys. Chem. B **110**, 24255 (2006).

[19] C. Zhang and S. M. Durbin, J. Phys. Chem. B **110**, 23607 (2006).

[20] S. Ebbinghaus, S. J. Kim, M. Heyden, X. Yu, U. Heugen, M. Gruebele, D. M. Leitner, M. Havenith, Proc. Natl. Acad. Sci. USA **104**, 20749 (2007).

[21] S. Ebbinghaus, S. J. Kim, M. Heyden, X. Yu, M. Gruebele, D. M. Leitner, and M. Havenith, J. Am. Chem. Soc. **130**, 2374 (2008).

[22] *Reverse micelles*, edited by P. L. Luisi, and B. E. Straub (Plenum Press, New York, 1984).

[23] *Structure and reactivity in reverse micelles*, edited by M. P. Pileni (Elsevier, Amsterdam, 1989).

[24] H. Murakami, Y. Toyota, T. Nishi, and S. Nashima, Chem. Phys. Lett. **519–520**, 105 (2012).

[25] B. van den Berg, R. Wain, C. M. Dobson, R. J. Ellis, EMBO J. **19**, 3870 (2000).

[26] C. Wichmann, P. T. Naumann, O. Spangenberg, M. Konrad, F. Mayer, and M. Hoppert, Biochem. Biophys. Res. Commun. **310**, 1104–1110 (2003).

[27] *Water: The forgotten biological molecule*, edited by D. L. Bihan and H. Fukuyama, (Pan Stanford, Singapore, 2011).

[28] S. K. Pal, J. Peon, B. Bagchi, and A. H. Zewail, J. Phys. Chem. B **106**, 12376 (2002).

[29] C. A. Angell, Ann. Rev. Phys. Chem. **34**, 593 (1983); in *Water : A Comprehensive treatise*, edited by F. Franks (Plenum, New York, 1982) Vol. 7, p. 1.

[30] O. Mishima and H. E. Stanley, Nature **396**, 329 (1998).

[31] H. E. Stanley, S. V. Buldyrev, P. Kumar, F. Mallamace, M. G. Mazza, K. Stokely, L. Xu, and G. Franzese, J. Non-cryst. Solids **357**, 629 (2011).

[32] L. Liu, S. -H. Chen, A. Faraone, C. -W. Yen, and C. -Y. Mou, Phys. Rev. Lett. **95**, 117802 (2005).

[33] P. Gallo, M. Rovere, and S. -H. Chen, J. Phys. Chem. Lett. **1,** 729 (2010).

[34] J. Hedström, J. Swenson, R. Bergman, H. Jansson, and S. Kittaka, Eur. Phys. J. Special Topics **141**, 53 (2007).

[35] K. Yoshida, T. Yamaguchi, S. Kittaka, M. -C. Bellissent-Funel, and P. Fouquet, J. Phys.: Condens. Matter **24,** 064101 (2012).

[36] H. Murakami, T. Sada, M. Yamada, and M. Harada, Phys. Rev. E **88**, 052304 (2013).

[37] C. Boned, J. Peyrelasse, and M. Moha-Ouchane, J. Phys. Chem. **90**, 634 (1986).

[38] A. K. Simorellis, W. D. V. Horn, and P. F. Flynn, J. Am. Chem. Soc. **128**, 5082 (2006).

[39] T. Spehr, B. Frick, I. Grillo, P. Falus, M. Müller, and B. Stühn, Phys. Rev. E **79**, 031404 (2009).

[40] C. A. Munson, G. A. Baker, S. N. Baker, and F. V. Bright, Langmuir, **20**, 1551 (2004).

[41] C. Rønne, P. -O. Åstrand, and S. R. Keiding, Phys. Rev. Lett. **82**, 2888 (1999).

[42] C. Rønne and S. R. Keiding, J. Mol. Liq. **101 (1–3)**, 199 (2002).

[43] J. T. Kindt and C. A. Schmuttenmaer, J. Phys. Chem. **100**, 10373 (1996).

[44] T. Arikawa, M. Nagai, and K. Tanaka, Chem. Phys. Lett. **457**, 12 (2008).

[45] H. Murakami, J. Mol. Liq. **210**, 37 (2015).

[46] B. Yellampalle, K. Y. Kim, G. Rodriguez, J. H. Glownia, and A. J. Taylor, Opt Exp. **15**, 1376 (2007).

[47] L. Duvillaret, F. Garet, and J. L. Coutaz, IEEE J. Sel. Top. Quantum Electron **2**, 739 (1996).

[48] H. Murakami, T. Nishi, and Y. Toyota, J. Phys. Chem. B **115**, 5877 (2011).

[49] M. Kotlarchyk, J. S. Huang, and S. -H. Chen, J. Phys. Chem. **89**, 4382 (1985).

[50] D. M. Mittleman, M. C. Nuss, and V. L. Colvin, Chem. Phys. Lett. **275**, 332 (1997).

[51] R. G. Gordon, Adv. Mag. Reson. **3**, 1 (1968).

[52] S. R. Keiding, J. Phys. Chem. A **101**, 5250 (1997).

[53] S. Mashimo, S. Kuwabara, S. Yagihara, K. Higasi, J. Phys. Chem. **91**, 6337 (1987).

[54] G. Onori, A. Santucci, J. Phys. Chem. **97**, 5430 (1993).

[55] D. E. Moilanen, E. E. Fenn, D. Wong, M. D. Fayer, J. Phys. Chem. B **113**, 8560 (2009).

[56] D. Cringus, A. Bakulin, J. Lindner, P. Vöhringer, M. S. Pshenichnikov, D. A. Wiersma, J. Phys. Chem. B **111**, 14193 (2007).

[57] A. M. Dokter, S. Woutersen, H. J. Bakker, Phys. Rev. Lett. **94**, 178301 (2005).

[58] P. Dutta, P. Sen, S. Mukherjee, A. Halder, K. Bhattacharyya, J. Phys. Chem. B **107**, 10815 (2003).

[59] R. E. Riter, D. M. Willard, N. E. Levinger, J. Phys. Chem. B **102**, 2705 (1998).

[60] I. D. Kuntz, J. Am. Chem. Soc. **93**, 514 (1971).

[61] B. K. Rai, E. W. Prohofsky, and S. M. Durbin, J. Phys. Chem. B **109**, 18983 (2005).

[62] A. Kitao, F. Hirata, N. Go, Chem. Phys. **158**, 447 (1991).

[63] A. Ansari, J. Berendzen, S. F. Bowne, H. Frauenfelder, I. E. T. Iben, T. B. Sauke, E. Shyamsunder, and R. D. Young, Proc. Natl. Acad. Sci. USA. **82**, 5000 (1985).

[64] K. Gunasekaran, B. Ma, and R. Nussinov, Proteins: Structure, Func.Bioinform. **57**, 433 (2004).

[65] S. Cusack and W. Doster, Biophys. J. **58**, 243 (1990).

[66] Y. Joti, A. Kitao, and N. Go, J. Am. Chem. Soc. **127**, 8705 (2005).

[67] P. Kumar, K. T. Wkfeldt, D. Schlesinger, L. G. M. Pettersson, and H. E. Stanley, Sci. Rep. **3**, 1980 (2013).

[68] U. Kaatze and V. Uhlendorf, Z. Phys. Chem. Neue Folge, **126**, 151 (1981).

[69] J. Barthel, K. Bachhuber, R. Buchner, and H. Hetzenauer, Chem. Phys. Lett. **165**, 369 (1990).

Section 3

Application of THz Spectroscopy

Chapter 9

Application of Terahertz Technology in Biomolecular Analysis and Medical Diagnosis

Xin Zhang and Zhuoyong Zhang

Additional information is available at the end of the chapter

http://dx.doi.org/10.5772/67090

Abstract

Terahertz technology is a nondestructive technique, which has progressed significantly in the scientific research and gains highly attention in the analysis of biological molecular, cellular, tissues and organs. In this decade, some studies were reported on the application of terahertz technology in medical testing and diagnosis. Here, we summarize the terahertz characters, terahertz spectroscopy, and terahertz imaging technology combined with chemometrics. This chapter focuses on introducing the research progress on analyzing the tissues of cancers using terahertz spectroscopy and terahertz imaging technology. Furthermore, the problems should be solved, and development directions of terahertz spectroscopy and terahertz imaging technology are discussed.

Keywords: medical diagnosis, medical testing, terahertz imaging, terahertz spectroscopy

1. Introduction

1.1. Development of terahertz spectroscopy in biomolecular detection and medical diagnosis

The terahertz (1 THz is equal to 10^{12} Hz) is an electromagnetic wave located between the infrared and microwave regions of the electromagnetic spectrum. Its frequency is defined from 0.1 up to 10 THz. In this region, the spectra can respond and display spectra absorption to low-frequency vibrational modes of molecules, such as torsional and collective vibrational modes and hydrogen-bond modes, and rotational modes [1, 2]. THz waves have a feature of low energy, non-ionizing which provide the advantage of harmless to analyze living tissues [3, 4]. THz spectroscopic methods have been used in the biological sciences for investigating DNA [5], proteins [6, 7], and tissues [8].

The characterization and quantification of DNA are often regarded as a complex laborious process in bioscience. A number of different techniques are therefore offering a variety of approaches, like spectrophotometry [9], UV-induced fluorescence [10, 11], chip-based nucleic acid analysis system [12], etc., for charactering DNA. The terahertz spectrum has been determined to be a promising candidate for the characterization of DNA. Several alternative methods, including fluorescent chromophore labeling and techniques that use terahertz radiation, have been proposed and are currently in use [13]. Terahertz spectroscopy can characteristic DNA samples pretreatment free, fast, and sensitively. Nagel et al. reported a promising approach for the label-free analysis of DNA molecules using direct probing of the binding state of DNA with terahertz spectroscopy [14]. In comparison with free-space detection scheme formerly used, this method provides an impressively promoted sensitivity enabling analysis down to femtomol levels. Debanjan Polley et al. reported a dielectric relaxation study using terahertz spectroscopy of extended hydration sheathe of dilute aqueous solution of salmon sperm (SS) and calf thymus (CT) DNA samples, which are always used as model organism [15]. They fitted the frequency-dependent complex dielectric response according to a Debye relaxation model, which assumes three relaxation modes in their work on SS DNA and CT DNA. The observed relaxation time constants have high relation with that of bulk water and vary from any particular trend indicating to the extended hydrogen-bonded network of DNA in marginal modification. Though a variety of methods were established for characterization of DNA, they have disadvantages like alteration to the nucleic acid sequence, requirement of a thick DNA testing layer, and conductor structure complexity. From that point, THz spectroscopy has its advantage to be applied on the area of DNA detection.

The THz frequency also corresponds to global correlated protein motions, molecular interaction between protein molecules, which were proposed to be essential to functional conformational changes. Niessen used THz microscopy to inhibitor binding sensitivity and test reproducibility of the narrow-band resonances for lysozyme protein crystals. To achieve the data analysis of THz spectra data, they applied a rapid data acquisition technique. The THz spectra were changed dramatically and can be reproducible with inhibitor binding [16]. Chen et al. proposed an approach for automatic identification of biomolecule terahertz (THz) spectra based on the most used chemometric methods, like principal component analysis (PCA) and fuzzy pattern recognition [17]. Chen investigated THz transmittance spectra of saccharide biomolecular samples, and some typical amino acid and their results demonstrate that THz spectroscopy can be utilized for identification of biomolecules efficiently.

In the application of medical testing and diagnosis, THz spectroscopy and THz imaging have been applied to complex analysis [18]. THz imaging has been used for detecting micrometastatic foci of early-stage cervical cancer in the lymph nodes [19]. The in vivo tissue spectroscopic response in vivo tissues is highly depending on the constituent materials and their physical arrangement for the heterogeneous of tissues. This means that the measurements of in vivo tissues will be different from spectroscopic measurements on homogeneous samples of DNA, saccharide, fat, or proteins. THz interactions with biological components of tissue were reviewed by Smye et al. [20]. Woodward et al. demonstrated the application of terahertz pulse imaging (TPI) on skin or related cancer tissues. Using this technique, they detected imaging in reflection approach for the study of skin tissue and corresponding cancer tissue both in vitro and in vivo. The sensitivity of terahertz radiation to polar molecules makes THz spectroscopy and imaging be used for analyzing the hydration levels in the skin. And, it also has potential to be applied on the preoperative determination of the lateral spread of skin cancer. The terahertz pulse shape in the time domain was studied, and the results show that they were able to differentiate diseased and normal tissues for the study of basal cell carcinoma [21].

The study of Nazarov et al. showed that in the terahertz frequency range, small organic molecules had characteristic absorption lines. Large molecules and tissues had crucial absorption linearly increases with frequency. THz refection spectroscopy provides possibility to study strongly absorbing substances [22]. Using differences in THz absorbance, pathologic diagnosis between normal and cancerous tissues has been reported in the medical literatures [23, 24]. Knobloch's results show that different kinds of tissues can be clearly distinguished from both the larynx of a pig and cancerous human liver using THz spectroscopy. Cherkasova et al. studied human and rat skin reflection spectra in vivo and the effect of glucose and glycerol on these spectra by THz spectroscopy [25]. Variations in skin optical properties were found in the frequency of 0.1 THz.

Hyperspectral imaging can provide information in both space and spectral dimensions. THz imaging has potential for tissue analysis and medical diagnosis as a very promising harmless approach for future imaging applications. It is hopefully to develop the THz imaging system in the future, which requires high-frequency resolution and a cost-effective and much more compact setup that does not necessarily require a laboratory environment based on the development of THz techniques.

1.2. Application of chemometrics on terahertz spectra analysis

THz spectroscopy should take advantage of chemometrics, which has applied in other fields of spectroscopy like infrared, near-infrared, Raman, or fluorescence. Chemometrics [26] provides multivariate tools for exploring the relationships among the objects and tested variables in collected dataset as well as classifiers. Chemometrics has been applied for qualitative and quantitative analysis of the THz spectra of not very complex mixture systems. Absorption intensity of THz spectra is proportional to the concentration of analytes in commonly used dynamic ranges; thus, normal linear modeling methods in chemometrics can be used in their calibration and prediction [27].

A review of terahertz pulsed spectroscopy summarized the most common chemometric methods applied for processing the THz spectra including the way of quantitative univariate and multivariate methods, and it can be found from this chapter [28]. Two quantitative analysis approaches are mostly used: the first is the method applied to a single spectrum without any calibration but using the intensity of the spectra for modeling, and the other is calibration based on a series of THz spectra of reference samples in order to predict quantitative information from unknown samples.

THz spectroscopy combine with chemometrics was reported for quantitative or qualitative analysis of mixture systems in environment, food, agriculture, material, biology, and medicine. Otsuka et al. did quantitative analysis of mefenamic acid polymorphs by terahertz spectroscopy with chemometric methods [29]. They studied the effect of spectra data preprocessing on the chemometric parameters of the calibration models. Hua et al. mostly used regression models like partial least squares (PLS) and principle component regression (PCR) methods for quantitative evaluation of cyfluthrin in n-hexane by THz-TDS [30]. Partial least squares (PLS) is one of the most effective and reliable methods normally being applied for quantitative analysis of various spectra. El Haddad et al. applied principal component analysis (PCA), PLS, and artificial neural networks (ANN) to quantitative analysis of ternary mixtures by THz-TDS [31], and they obtained good results. Ellrich et al. presented a postscanner by THz spectroscopy using chemometric methods for the evaluation of detected THz fingerprints [32].

Assessment of THz spectroscopy with chemometrics is still under studying including preprocessing, data selection and calibration methods. We strongly believe that THz spectroscopy should take advantage of multivariate analysis for advanced data processing, classification and calibration methods of chemometrics.

2. Instrumentation

For all of the work introduced in this chapter, a transmission THz-TDS cell configuration was used, as depicted in **Figure 1**. **Figure 2** gives schematic of a terahertz time-domain transmission spectrometer system used in this chapter. The THz spectroscopy system was equipped with a commercially available femtosecond laser (SPECIM, MaiTai) for generating the THz pulse. The femtosecond laser light is separated into two beams using a prism. One beam is the probe, and it travels across a free space to focus on the detecting antenna, so that the probe beam provides a relative time delay periodically. The other beam goes through a GaAs-based semiconductor antenna to generate the THz pulse. Then, a parabolic mirror with a hemispherical silicon lens is applied for improving the coupling efficiency of the THz radiation. The beam that passed through the sample placed at the focus of the parabolic mirror, and it is collected by another parabolic mirror. Finally, a photoconductive detector is used for signature collection [33]. In the experiments, the volume of the THz spectra system was filled with dry nitrogen (N_2) to reduce absorption caused by vapor in air.

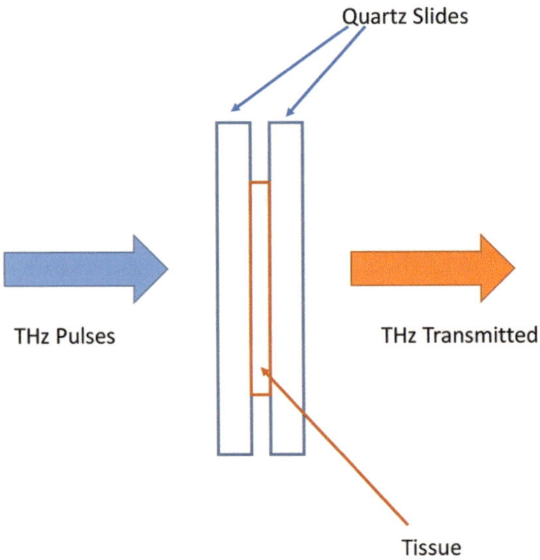

Figure 1. Schematic representation of tissue samples for TDS measurement.

http://dx.doi.org/10.5772/67090

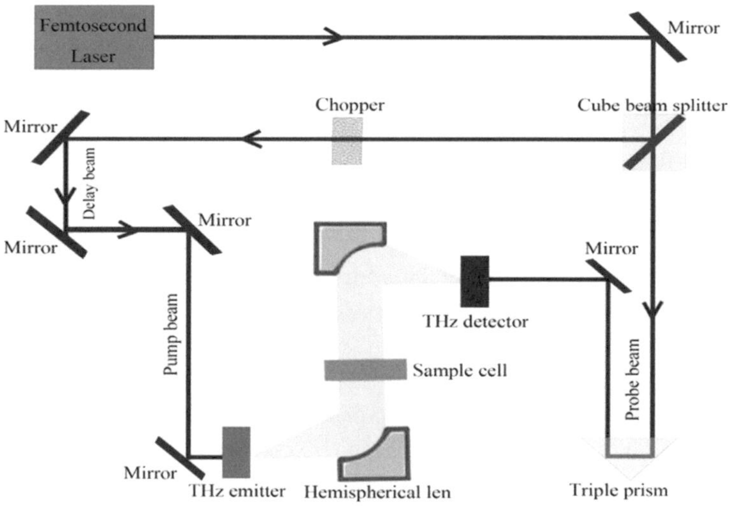

Figure 2. Schematic of a terahertz time-domain transmission spectrometer system used in this chapter.

3. Theory and practical applications

3.1. Parameter extraction from THz spectra

The measurement of the reference pulse and the sample pulse are necessary for calculating the THz absorption coefficient of a sample. When the tissues are analyzed by THz spectra for medical testing or diagnosis, the sample pulse collected is transmitted through the tissue slides, and the reference signal is the THz signal transmitted without the tissue slides. The THz electric field pulses can be calculated as a function of time and the frequency for both signal crossed the tissue sample and reference passed nothing. The frequency domain spectra are obtained by the fast Fourier transform (FFT) in this work. The refractive index $n(\omega)$ describing the dispersion and absorption coefficient $\alpha(\omega)$, describing the absorption characteristics, can be calculated through the following equations [33]:

$$n(\omega) = \frac{\varphi(\omega)c}{\omega d} + 1 \tag{1}$$

$$\alpha(\omega) = \frac{2\kappa(\omega)\omega}{c} = \frac{2}{d}\ln\frac{4n(\omega)}{A(\omega)\,(n(\omega)+1)^2} \tag{2}$$

for which ω is the frequency, and $\rho(\omega)$,$k(\omega)$ and $\phi(\omega)$ are functions for the amplitude ratio, extinction coefficient, and phase difference of the sample and reference signals, respectively. d is the thickness of sample, and c is the velocity of light in vacuum.

3.2. Chemometric methods applied in this chapter

3.2.1. Data preprocessing methods

The Savitzky-Golay (SG) method is a polynomial filter that performs numerical differentiation and smoothing [34]. This filter removes the noise in the dataset analyzed and simplifies the computation during the model building. SG method has the ability to process the signals with little delay and with no shifts of the peaks, and it can be performed in a computationally efficient procedure by applying least squares on subsets of the data. In the dataset, a window moves forward within 2m + 1 points fitting by a polynomial of degree p (in which $p \leq 2$ m). The dth ($0 \leq d \leq p$) differentiation of the original data at the midpoint is obtained by performing the fitness polynomial. Finally, the convolution of the entire input data with a digital filter of length 2m + 1 is performed by running least squares polynomial fitting [35, 36].

The multiplicative scatter correction (MSC) is a transformation method used to cope with scaling and offset effects, which is mostly applied in spectral data analysis [37]. It is used to counterbalance for additive and/or multiplicative effects in spectral data. MSC assumed that each spectrum collected from the samples is determined on one hand by the actual sample characteristics and on the other by the particle size. It can also decrease or remove physical effects like particle size and surface blaze, and it corrects differences in the baseline and in the trend. The sample preparation for THz spectra collection always has problems about the difference of particle size during grind and compactness during tableting. So, it has an advantage that the transformed spectra are similar to the original spectra without the effect of baseline effect and the trend of a standard spectrum and that an optical interpretation is therefore more easily accessible [38].

Orthogonal signal correction (OSC) can be applied to remove systematic noise such as baseline variation and multiplicative scatter effects, which is a data processing technique introduced by Fearn [39]. The basic idea of the OSC method is to remove the systematic variations in the collected data that are orthogonal or not related to the properties of the dependent variables. The removed information can be structured noise, such as baseline, instrument variation and measurement conditions. Some reports show that the use of OSC may not result in calibration models with lower prediction errors than models based on raw data. The advantage of using OSC lies in the analysis and interpretation of the corrected data but not in decreased prediction errors. By removing orthogonal information, the important calibration information will be concentrated in fewer principal components instead of being distributed among many linearly dependent variables.

The PC-OSC method used the constraints based on OSC and also applied the theory of principal component analysis (PCA). The detailed procedure of PC-OSC can be referred in Ref. [40]. The emphatic orthogonal signal correction (EOSC) method can be used for the baseline correction of Raman spectra or near- infrared spectra, and it is a method that can be extended to apply on THz spectra. The theory and procedures of EOSC can be found in Refs. [4, 41].

Asymmetric least squares (AsLS) method calculates complex baseline shapes by adjusting the asymmetry parameter and the smoothness parameter. The asymmetry parameter is related to the position of the baseline, and the smoothness parameter related to the flexibility in the shape of the baseline. The AsLS method has been proposed by Zhang et al. [42]. By minimizing the penalized least squares function based on the Whittaker smoother, AsLS method estimates a background contribution and removes or decreases the baseline [43]. The application of this method can be found in Ref. [44]. The initial range of lambda is 10^2–10^5, and p is 0.099, which are experienced parameters based on literatures and our laboratory work experience. We adapted the results to PLS models, respectively, and obtained the best result when lambda is 10^3 [27].

Wavelet transform not only can compress the data to extract feature information but also can remove the noise in the spectral data. In brief, wavelet analysis is based on the wavelet transform of analysis signal. The signal of different spatial scales (frequency) is divided into high-frequency and low-frequency part, and the position of each component on the time axis remains the same. A more detailed description of wavelet analysis algorithm can be found elsewhere [45].

Figure 3a displays raw THz absorption spectra of the prepared samples of binary amino acid mixtures of L-glutamic acid and L-glutamine. **Figure 3b**, **3c** and **3d** displays the THz spectra of the binary amino acid mixtures with preprocessing of SG smoothing, MSC, and AsLS, respectively. From **Figure 3**, we can observe that the absorption spectra with processing of SG smoothing have eliminated the effect of noise and display spectral characteristics more clearly. By using MSC, the scaling and offset effects were removed. Normally, using OSC can get very similar results to that applied in MSC. AsLS can eliminate or decrease the baselines in the THz spectra of the samples. The absorption bands are more easily attributed to certain wavelength in this work. The binary mixtures under different concentration ratios display certain quantitative relation on THz absorption coefficients, which would be the basis for quantitative analysis by THz-TDS transmission spectroscopy. We did not apply wavelet transform on the sample analysis in this chapter, for the other preprocessing could obtain good result already. Anyway, wavelet transform still should be considered in the future work, since it is a powerful method for de-noising, compressing the dataset, removing the background information, etc.

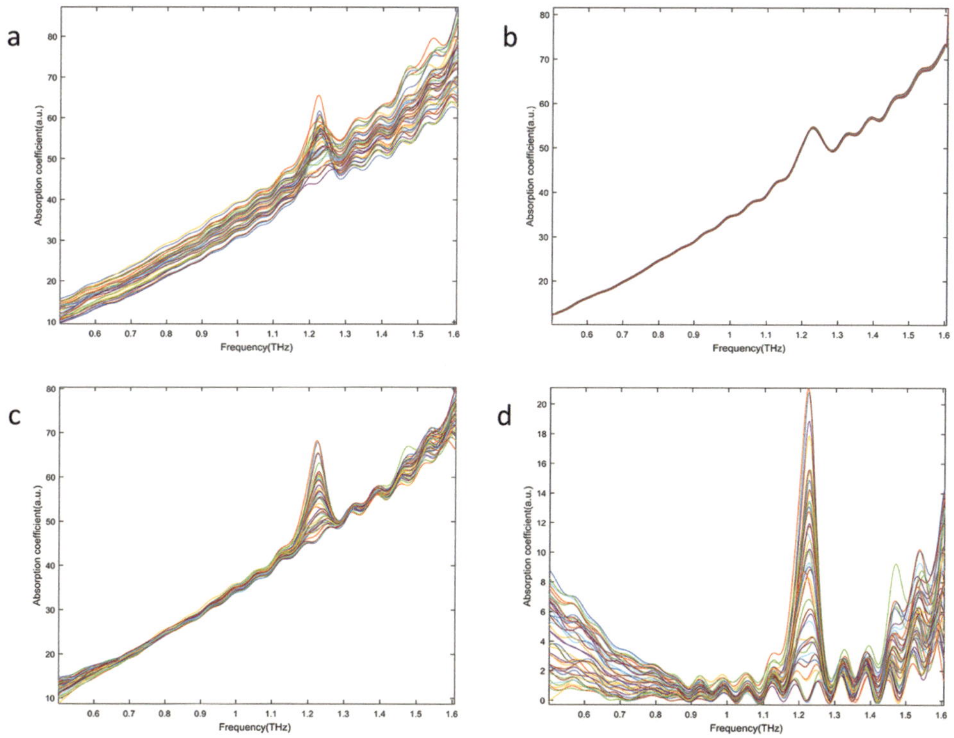

Figure 3. The spectra of the L-glutamic acid and L-glutamine mixture samples after different preprocessing are (a) raw THz spectra, (b) the spectra after smoothing, (c) the spectra after MSC, and (d) the spectra after AsLS.

3.2.2. Examples of chemometric methods for regression model building of THz spectra

3.2.2.1. Principal component analysis

Principal component analysis (PCA) is commonly used to reduce the number of predictive variables. PCA condenses all the spectral information into a few linear combinations of the latent variables instead of the original variables. The linear combinations of the variables can be used to summarize the data without losing too much information but remove noise in the process [46]. It can be used to identify the underlying structure of large datasets and can be used to identify groups within the data from complex mixtures. It can also be used for removing any contribution from noise.

3.2.2.2. Partial least squares

Partial least squares (PLS) is one of the most used chemometric techniques to build quantitative model based on principal component analysis and principal component regression. PLS extracts the orthogonal features from the spectrum and then constructs the correlation between the spectra matrix (independent variable, X) and concentration matrix (dependent variable, Y). The detailed procedure of PLS can be referred to in Ref. [47].

We have investigated PLS for quantitative analysis of L-glutamic acid and L-glutamine using THz spectroscopy. Also, we compared the difference between iPLS and PLS. iPLS divided the whole spectrum into several intervals and builds PLS model for each subset to evaluate the most suitable sub-dataset for a stable model. The subset or several subsets with the lowest root mean square error of cross validation (RMSECV) are chosen for the PLS model building [27]. The iPLS yielded better results with low RMSEP (0.39 ± 0.02%, 0.39 ± 0.02%), and higher R^2 values (0.9904, 0.9906) for glutamine and glutamic acid comparing to the conventional PLS models.

We also analyzed binary isomer of saccharide mixtures, D-(−)fructose and D-(+)galactose anhydrous quantitatively using THz-TDS combined with PLS. The result showed that correlation coefficient (R^2) between true and predicted values is higher than 0.9773. The mean value of root mean square error of prediction (RMSEP) in cross validation set was less than 1.26%.

Therefore, THz-TDS combined with chemometrics is feasible for quantitative analysis of the biomolecular mixtures and may also be extended to analysis of more component mixtures.

3.2.3. Examples of chemometric for classification model building of THz spectra

3.2.3.1. Partial least squares-discriminant analysis

Partial least squares-discriminant analysis (PLS-DA) is a supervised classification method based upon partial least squares regression [47]. The PLS-DA algorithm models the relationship between the measured variable of the dataset and the target variables corresponding to the class label [48]. PLS-DA extracts the latent variables by reducing the dimension of the dataset like principal component analysis and finds the maximum separation among the classes. The latent variables explain both the variance of the THz spectral data and the high correlation with the response matrix that encodes the class membership [33]. Component number, which has high relative with the accuracy rate and the percentage of the explained variable, is a very important parameter for prediction accuracy and explanation of the model that needs to be estimated for PLS-DA.

3.2.3.2. Support vector machine

Support vector machine (SVM) is a powerful machine learning method with associated learning algorithms that analyze data used for classification and regression analysis [49]. For classification tasks, based on the structural risk minimization principle, this method attempts to find the separating hyperplane which has the largest distance from the nearest training data points. LIBSVM was one of the mostly used toolboxes, and the SVM calculations in the example in this chapter used the linear kernel function [50].

The two essential factors that affected SVM classification performance are (1) error penalty parameter C, which is the compromise between the proportion of error classification samples and algorithm complexity, and (2) form of kernel function and its parameters. Different kernel functions have influence on the classification performance, while different parameters of same kernel function may also affect the results.

3.2.3.3. Fuzzy rule-building expert system (FuRES)

FuRES is a classification tree model using fuzzy entropy of classification which each rule is a temperature-constrained sigmoid logistic function [51]. The rules of FuRES are similar to the processing units in most feed-forward artificial neural networks, while FuRES processing units differ in that the weight vector is constrained to an optical length. The inductive dichotomizer 3 is used for classification by minimizing $H(C|A)$, the classification entropy. The weight vector w should be normalized before modeling. A temperature parameter t is used to control the fuzziness of each rule. When optimizing the computational temperature, the maximized extend of the entropy for classification can be found [4]. The equations for this method are given below:

$$\chi_A(x_k) = (1 + e^{-(\mathbf{x}_k\mathbf{w}-a)/t})^{-1} \tag{3}$$

for which a is the bias value and $x_A(x_k)$ is the degree of fuzzy membership of object x_k. [4]. The conditional probability $p(c_i|a_j)$ is obtained by summing the membership functions with the attribute a_i and the class of c_i.The equation is given as follows:

$$p\left(c_i \middle| a_j\right) = \sum_{k=1}^{n_i} \chi_A(x_k) \Big/ \sum_{k=1}^{n} \chi_A(x_k) \tag{4}$$

where n_i is the number of objects in class c_i. The classification entropy $H(C|a_i)$ of the attribute a_i is given by

$$H\left(C \middle| a_j\right) = -\sum_{i=1}^{n} p\left(c_i \middle| a_j\right) \ln\left[p\left(c_i \middle| a_j\right)\right] \tag{5}$$

The classification entropy $H(C|A)$ of the system is the weighted sum of the entropy for each attribute:

$$H(C|A) = \sum_{j=1}^{2} p(a_j)H\left(C\middle| a_j\right) \quad (6)$$

The FuRES model provides inductive logic in the tree structure of the classifier. In this way, it can accommodate overlapping the data and avoid overfitting the data by the temperature constraint. The advantage of fuzzy classification trees comparing to network classifiers is that they furnish a simple inductive structure that is amenable to interpretation.

3.2.3.4. *Fuzzy optimal associative memory (FOAM)*

An optimal associative memory (OAM) is using a one-way data to replace binary image of encoded multivariate data as a two-way binary image for fuzzy method [52]. Bipolar matrix with similar size grid unit is built first. A vector of ν variables is converted to ν × *h* bipolar matrix. After removing *u* unused grid, the number of grid is *k* ((ν × *h*) − *u*) [4]. The FOAM stores pattern in a weight matrix **W,** which expressed by

$$\mathbf{W} = \sum_{i=1}^{n} \mathbf{y}_i \mathbf{y}_i^T \quad (7)$$

for which $\mathbf{y}_i$ is the ith bipolarly encoded approach [4]. The stored grid-encoded spectra are orthogonalized to form a basis using singular value decomposition. The encoded predicted background scan $\mathbf{z}_f$ can be obtained by

$$\mathbf{z}_f = \mathbf{V}(\mathbf{V}^T \mathbf{z}_i) \quad (8)$$

for which **V** is a matrix following the orthogonalized pattern of the collected data variables. Then, z_f is decoded to a spectrum vector by changing the gridding procedure. The data object can be assigned to the corresponding classes to make the building fit with the minimum error after reconstructing the raw data. In this chapter, FOAM used its standard configuration of 100 intensity grids and a 19-point triangular fuzzy membership function [4].

3.2.3.5. *Validation*

Bootstrapped Latin partitions (BPLs) are generalized for evaluation of calibration methods based on cross validation and random sampling verification [53]. Unbiased evaluation of classification or calibration methods is important, especially as these methods are applied to increasingly complex datasets that are under-determined like THz spectra dataset. Precision bounds, such as confidence intervals, are required for interpreting any experimental result. By using BPLs unbiased and reliable evaluation can be gotten by systematically model with samples drew from an arbitrary discrete distribution.

3.2.3.6. *Example of diagnosis of cervical cancer*

In this work, THz-TDS system was applied to detect the normal and malignant tissue sections as an example for chemometric applications on THz spectral analysis. The classification models combined with different pretreatment methods were established to build a new diagnosis technique for cervical cancer diagnosis based on terahertz spectroscopy. The effects of different preprocessing methods on THz spectra data to de-noise, remove baseline and optimize model were investigated.

The normal and cancerous cervical tissues were collected and provided by Beijing Haidian Maternal & Child Health Hospital. To keep the tissues, all the cervical tissues were put into 4% formaldehyde solution. The tissues were washed with ethanol solutions for dehydration when we analyzed the samples. The tissues were put into xylene for hyalinization and then embedded by paraffin wax before sliced into 8 μm thick sections. The water-flatted sections were spread upon quartz plates and then put in a regulated heating oven and dried. Two replicate slides were prepared for each of the tissue sections [33].

To establish a model for diagnosis of cervical cancer, PLS-DA, FuRES, FOAM, and SVM were used to build classification models. The parameter for FuRES and FOAM was determined using a self-optimizing PLS-DA from the training datasets. A method to verify the accuracy of classification and calibration models, bootstrapped Latin partition, was used for cross validation of the calibration dataset. When we built PLS-DA and SVM models for this study, the matrix of category variables was used, one for the normal samples and two for the cancer tissue samples. When the predicted value of an external test tissue sample in PLS-DA model was smaller than 1.5, the sample was assigned to normal class or assigned to cancer class otherwise [33].

The classification results of these methods were compared when the data was processed with different preprocessing. MSC, SG smoothing, SG first derivative, EOSC, and PC-OSC were used for pretreatment of the THz spectra, respectively, and the data were normalized before modeling. The performance of preprocessing methods applied in this work was compared. The results of the modeling approaches after pretreatment are evaluated by the pooled prediction rates. The raw data were divided as training datasets and prediction sets based on KS method. The pretreatments were constructed from the training datasets and applied to the prediction sets. For cross validation, five Latin partitions bootstrapped 50 times were applied to evaluate the prediction accuracy of the classification models with different parameters and different pretreatments. For each bootstrap, the data were separate as training set and test set, and each spectrum was used only once in the test set. Each time, four Latin partitions were used for calibration during model building, and the fifth was used for prediction. The predicted results of the five test sets from each partition were pooled back after each validation was finished. This approach was used for evaluations of all the classification models in this work. The average prediction results were calculated with 50 bootstraps to give 95% confidence intervals [33]. The number of components of the OSC model was selected by finding the maximum average classification rate across internal 100×5 bootstrap Latin partitions. All model optimization and construction were performed in MATLAB (MATLAB 7.14.0.334. The MathWorks Inc.).

The obtained results showed that the FuRES and FOAM model using Savitzky-Golay smooth by the first derivative and PC-OSC as pretreatment methods had provided a good predictive results and their classification rates are 92.9 ± 0.4 and 92.5 ± 0.4 %, respectively [4]. The results of the proposed methods show that terahertz spectroscopy combined with fuzzy classifiers could supply a technology which has potential for diagnosis of cancerous tissue. Combining SG first derivative with PC-OSC as signal pretreatment procedure, the prediction accuracies of the optimal SVM and PLS-DA were 94.0 ± 0.4 and 94.0 ± 0.5%, respectively. Therefore, SVM and PLS-DA with the combination of SG first derivative and PC-OSC based on terahertz spectroscopy of tissue can also provide a good application for diagnosis of cervical carcinoma.

Comparing the classification accuracies pretreated by different preprocessing methods, it indicated that the classification models applied based on terahertz spectroscopy of tissue could provide a better application for early diagnosis of cervical carcinoma, with high classification accuracies. Coupled with terahertz technology, the proposed procedure could provide a convenient, solvent-free, and environmentally friendly application that had a potential development as cancer diagnosis method.

3.2.4. Examples of chemometric for resolution of THz spectra data

3.2.4.1. Multivariate curve resolution

Multivariate curve resolution (MCR) is designed to solve the analysis problem of mixture systems following bilinear model. The MCR methods decompose the raw mixed measurement datasets into matrix corresponding to pure concentration profiles and pure spectra. Constrains following physical and chemical property can be flexibly used during iterations when alternating least squares is applied and the maximum variance of the raw measurement data is explained. The profiles in the bilinear model resolved by MCR are physically and chemically meaningful and correspond to interpretable patterns of variation of principal components [27, 54, 55].

In our study, MCR-ALS was applied to resolve binary amino acid mixtures of L-glutamic acid and L-glutamine analyzed by THz-TDS. Non-negativity constraint was applied on both spectra and concentration directions during the iterations. The spectra of the pure analytes obtained from the MCR displayed are corresponding to glutamine and glutamic acid. MCR results provided fitting error in % (exp) is 6.731, and percent of variance explained (R^2) at the optimum is 99.55, and this results show that the MCR model can fit the raw data well in this case. The correlation coefficients (r^2) between the reference THz spectra of pure analytes and those resolved pure spectra for each principal component by MCR are 0.9990 and 0.9979 for glutamine and glutamic acid, respectively. The spectrum of glutamine resolved by MCR is in good agreement with that measured by THz in the laboratory. The fitting constant is 0.9999.

MCR-ALS was also applied to resolve the binary isomer mixtures of D-(−)fructose and D-(+) galactose analyzed by THz-TDS. The absorption spectra of the two components obtained by MCR-ALS were well fitting to spectra of pure D-(−)fructose and D-(+)galactose anhydrous, which were obtained from experiment data in the same condition, respectively. The results of correlation coefficient between THz spectra obtained by MCR-ALS and pure D-(−)fructose and D-(+)galactose samples are 0.9974 and 0.9933, respectively. Relative concentrations of the two components were resolved by MCR-ALS, and they can fit the true concentrations well. MCR-ALS successfully resolved pure THz spectra of the components in the binary isomer mixtures and their corresponding concentrations.

http://dx.doi.org/10.5772/67090

Hyperspectral images based on infrared, near-infrared, Raman, and fluorescence are an active area of research that has grown quickly since a decade ago [56]. THz time-domain imaging is an emerging modality and has attracted a lot of interest since THz spectra got successes [57]. MCR-ALS can easily resolve the pure spectra and their corresponding concentration distributions from hyperspectral imaging, as well as external spectral can be used for local rank constraints. In the future, MCR-ALS can be proposed to resolve the biomedical images based on THz spectroscopy based on the development of THz hyperspectral imaging [56]. We believe that THz time-domain imaging will provide potential for medical diagnosis.

4. Does terahertz radiation lead to DNA or tissue damage?

Investigations of the interaction between nonionizing electromagnetic radiation and biological systems are necessary before the application of electromagnetic radiation in medical diagnosis. Recent emergence and growing use of terahertz radiation for medical imaging and public security screening raise questions on reasonable levels of exposure and health consequences of it. In particular, picosecond-duration THz pulses have shown promise for novel diagnostic imaging techniques. From different studies, the researcher got different conclusions on the effects of THz pulses on human cells and tissues. Titova et al. studied the biological effects of THz radiation on artificial human skin tissues [58], and their work shows that THz pulse irradiation may cause DNA damage in exposed skin tissue when intense THz pulses are applied. They consider that DNA damage repair mechanisms are quickly activated after THz radiation. But they found that the cellular response to pulsed THz radiation is significantly different from that induced by exposure to UVA (400 nm). However, Hintzsche et al. investigated power intensities ranged from 0.03 to 0.9 mW/cm and the cells were exposed for 2 and 8 h. Chromosomal damage and DNA damage were not detected in the disdained condition. Cell proliferation was also found to be unaffected by the exposure [59]. Bogomazova studied DNA damage and transcriptome responses in human embryonic stem cells (hESCs). They did not observe any effect on the mitotic index or morphology of the hESCs following THz exposure [60]. Anyway, THz is still one of the most harmless techniques for biomolecular analysis and medical diagnosis, and it will be powerful to detect the information from DNA, protein, or tissues.

5. Conclusion

Terahertz technology is progressing in biological and medical diagnosis in recent years. It has potential to be applied on analysis of biological molecular, cellular, tissues and organs, since terahertz is a nondestructive technique. The study displayed in this chapter indicates that terahertz spectroscopy and terahertz imaging technology combined with chemometrics can give accurate classification for normal and cancer tissues, predict the concentration of different compounds, and resolve the pure THz spectra of biomolecules in mixture systems. Comparing the classification accuracies pretreated by different preprocessing methods, it indicated that the models applied based on terahertz spectroscopy could provide a better application for medical diagnosis. Terahertz is still one of the most harmless techniques for medical diagnosis, and it is hopefully to be developed as a powerful technique to detect the information from biomolecules or tissues in the future.

Acknowledgements

This work was supported by the National Instrumentation Program (2012YQ140005) and the National Science Foundation of China (21275101).

Author details

Xin Zhang and Zhuoyong Zhang*

*Address all correspondence to: zhangzhuoyong@cnu.edu.cn

Department of Chemistry, Capital Normal University, Beijing, China

References

[1] B. Yu, F. Zeng, Y. Yang, Q. Xing, A. Chechin, X. Xin, I. Zeylikovich, and R. R. Alfano, "Torsional vibrational modes of tryptophan studied by terahertz time-domain spectroscopy," *Biophysical Journal*, vol. 86, pp. 1649–1654, Mar 2004.

[2] M. Walther, P. Plochocka, B. Fischer, H. Helm, and P. U. Jepsen, "Collective vibrational modes in biological molecules investigated by terahertz time-domain spectroscopy," *Biopolymers*, vol. 67, pp. 310–313, 2002.

[3] A. J. Fitzgerald, E. Berry, N. N. Zinovev, G. C. Walker, M. A. Smith, and J. M. Chamberlain, "An introduction to medical imaging with coherent terahertz frequency radiation," *Physics in Medicine and Biology*, vol. 47, pp. R67–R84, Apr 7 2002.

[4] N. Qi, Z. Zhang, Y. Xiang, Y. Yang, and P. d. B. Harrington, "Terahertz time-domain spectroscopy combined with fuzzy rule-building expert system and fuzzy optimal associative memory applied to diagnosis of cervical carcinoma," *Medical Oncology*, vol. 32, pp. 1–6, 2015.

[5] M. Brucherseifer, M. Nagel, P. H. Bolivar, H. Kurz, A. Bosserhoff, and R. Büttner, "Label-free probing of the binding state of DNA by time-domain terahertz sensing," *Applied Physics Letters*, vol. 77, pp. 4049–4051, 2000.

[6] J. Xu, K. W. Plaxco, and S. J. Allen, "Probing the collective vibrational dynamics of a protein in liquid water by terahertz absorption spectroscopy," *Protein Science*, vol. 15, pp. 1175–1181, 2006.

[7] H. Yoshida, Y. Ogawa, Y. Kawai, S. Hayashi, A. Hayashi, C. Otani, E. Kato, F. Miyamaru, and K. Kawase, "Terahertz sensing method for protein detection using a thin metallic mesh," *Applied Physics Letters*, vol. 91, p. 253901, 2007.

[8] S. Nakajima, H. Hoshina, M. Yamashita, C. Otani, and N. Miyoshi, "Terahertz imaging diagnostics of cancer tissues with a chemometrics technique," *Applied Physics Letters*, vol. 90, p. 041102, 2007.

[9] G. Sozzi, D. Conte, L. Mariani, S. L. Vullo, L. Roz, C. Lombardo, M. A. Pierotti, and L. Tavecchio, "Analysis of circulating tumor DNA in plasma at diagnosis and during follow-up of lung cancer patients," *Cancer Research*, vol. 61, pp. 4675–4678, 2001.

[10] M. Sirajuddin, S. Ali, and A. Badshah, "Drug–DNA interactions and their study by UV–Visible, fluorescence spectroscopies and cyclic voltametry," *Journal of Photochemistry and Photobiology, B: Biology*, vol. 124, pp. 1–19, 2013.

[11] D. L. Sokol, X. Zhang, P. Lu, and A. M. Gewirtz, "Real time detection of DNA RNA hybridization in living cells," *Proceedings of the National Academy of Sciences*, vol. 95, pp. 11538–11543, 1998.

[12] J. Khandurina, T. E. McKnight, S. C. Jacobson, L. C. Waters, R. S. Foote, and J. M. Ramsey, "Integrated system for rapid PCR-based DNA analysis in microfluidic devices," *Analytical Chemistry*, vol. 72, pp. 2995–3000, 2000.

[13] S. Stranieri, M. Karna, and D. Shreiber, "Terahertz Characterization of DNA: Enabling a Novel Approach". No. ARL-CR-0788. ILLINOIS UNIV AT URBANA-CHAMPAIGN, p.20, 2015.

[14] M. Nagel, P. H. Bolivar, M. Brucherseifer, H. Kurz, A. Bosserhoff, and R. Büttner, "Integrated THz technology for label-free genetic diagnostics," *Applied Physics Letters*, vol. 80, pp. 154–156, 2002.

[15] D. Polley, A. Patra, and R. K. Mitra, "Dielectric relaxation of the extended hydration sheathe of DNA in the THz frequency region," *Chemical Physics Letters*, vol. 586, pp. 143–147, 2013.

[16] K. A. Niessen, M. Xu, E. Snell, and A. Markelz, "THz microscopy measurements on inhibitor dependence of protein intramolecular modes," in *2013 38th International Conference on Infrared, Millimeter, and Terahertz Waves (IRMMW-THz)*, Mainz, Germany, 1–6 September 2013.

[17] T. Chen, Z. Li, and M. Wei, "Identification of biomolecules by terahertz spectroscopy and fuzzy pattern recognition," *Spectrochimica Acta Part A Molecular & Biomolecular Spectroscopy*, vol. 106, pp. 48–53, 2013.

[18] N. Qi, Z. Zhang, and Y. Xiang, "Application of terahertz technology in medical testing and diagnosis," *Spectroscopy and Spectral Analysis*, vol. 33, pp. 2064–2070, Aug 2013.

[19] E. Jung, M. Lim, K. Moon, Y. Do, S. Lee, H. Han, H. J. Choi, K. S. Cho, and K. R. Kim, "Terahertz pulse imaging of micro-metastatic lymph nodes in early-stage cervical cancer patients," *Journal of the Optical Society of Korea*, vol. 15, pp. 155–160, Jun 2011.

[20] S. Smye, J. Chamberlain, A. Fitzgerald, and E. Berry, "The interaction between terahertz radiation and biological tissue," *Physics in Medicine and Biology*, vol. 46, p. R101, 2001.

[21] R. M. Woodward, B. E. Cole, V. P. Wallace, R. J. Pye, D. D. Arnone, E. H. Linfield, and M. Pepper, "Terahertz pulse imaging in reflection geometry of human skin cancer and skin tissue," *Physics in Medicine and Biology*, vol. 47, p. 3853, 2002.

[22] M. M. Nazarov, A. P. Shkurinov, E. A. Kuleshov, and V. V. Tuchin, "Terahertz time-domain spectroscopy of biological tissues," *Quantum Electronics*, vol. 38, p. 647, 2008.

[23] R. M. Woodward, B. E. Cole, V. P. Wallace, R. J. Pye, D. D. Arnone, E. H. Linfield, and M. Pepper, "Terahertz pulse imaging in reflection geometry of human skin cancer and skin tissue," *Physics in Medicine and Biology*, vol. 47, pp. 3853–3863, Nov 7 2002.

[24] P. Knobloch, C. Schildknecht, T. Kleine-Ostmann, M. Koch, S. Hoffmann, M. Hofmann, E. Rehberg, M. Sperling, K. Donhuijsen, G. Hein, and K. Pierz, "Medical THz imaging: an investigation of histo-pathological samples," *Physics in Medicine and Biology*, vol. 47, pp. 3875–3884, Nov 7 2002.

[25] O. P. Cherkasova, M. M. Nazarov, E. E. Berlovskaya, A. A. Angeluts, A. M. Makurenkov, and A. P. Shkurinov, "Studying human and animal skin optical properties by terahertz time-domain spectroscopy," *Bulletin of the Russian Academy of Sciences: Physics*, vol. 80, pp. 479–483, 2016.

[26] M. A. Sharaf, D. L. Illman, and B. R. Kowalski, *Chemometrics* vol. 82: John Wiley & Sons: New York, 1986.

[27] S. Lu, X. Zhang, Z. Zhang, Y. Yang, and Y. Xiang, "Quantitative measurements of binary amino acids mixtures in yellow foxtail millet by terahertz time domain spectroscopy," *Food Chemistry*, vol. 211, pp. 494–501, 2016.

[28] J. El Haddad, B. Bousquet, L. Canioni, and P. Mounaix, "Review in terahertz spectral analysis," *TrAC, Trends in Analytical Chemistry*, vol. 44, pp. 98–105, 2013.

[29] M. Otsuka, J. i. Nishizawa, J. Shibata, and M. Ito, "Quantitative evaluation of mefenamic acid polymorphs by terahertz–chemometrics," *Journal of Pharmaceutical Sciences*, vol. 99, pp. 4048–4053, 2010.

[30] Y. Hua, H. Zhang, and H. Zhou, "Quantitative determination of cyfluthrin in n-hexane by terahertz time-domain spectroscopy with chemometrics methods," in *Instrumentation and Measurement Technology Conference, 2009. I2MTC'09. IEEE*, 2009, pp. 670–674.

[31] J. El Haddad, F. de Miollis, J. Bou Sleiman, L. Canioni, P. Mounaix, and B. Bousquet, "Chemometrics applied to quantitative analysis of ternary mixtures by terahertz spectroscopy," *Analytical Chemistry*, vol. 86, pp. 4927–4933, 2014.

[32] F. Ellrich, G. Torosyan, S. Wohnsiedler, S. Bachtler, A. Hachimi, J. Jonuscheit, R. Beigang, F. Platte, K. Nalpantidis, and T. Sprenger, "Chemometric tools for analysing terahertz fingerprints in a postscanner," in *2012 37th International Conference on Infrared, Millimeter, and Terahertz Waves*, Wollongong, NSW, Australia, 23–28 September 2012.

[33] N. Qi, Z. Zhang, Y. Xiang, Y. Yang, X. Liang, and P. d. B. Harrington, "Terahertz time-domain spectroscopy combined with support vector machines and partial least squares-discriminant analysis applied for the diagnosis of cervical carcinoma," *Analytical Methods*, vol. 7, pp. 2333–2338, 2015.

[34] A. Savitzky and M. J. Golay, "Smoothing and differentiation of data by simplified least squares procedures," *Analytical Chemistry*, vol. 36, pp. 1627–1639, 1964.

[35] H. H. Madden, "Comments on the Savitzky-Golay convolution method for least-squares-fit smoothing and differentiation of digital data," *Analytical Chemistry*, vol. 50, pp. 1383–1386, 1978.

[36] A. X. Zhao, X. J. Tang, Z. H. Zhang, and J. H. Liu, "The parameters optimization selection of Savitzky-Golay filter and its application in smoothing pretreatment for FTIR spectra," in *2014 9th IEEE Conference on Industrial Electronics and Applications*, Hangzhou, China, 9–11 June 2014, pp. 516–521.

[37] T. Isaksson and T. Næs, "The effect of multiplicative scatter correction (MSC) and linearity improvement in NIR spectroscopy," *Applied Spectroscopy*, vol. 42, pp. 1273–1284, 1988.

[38] M. Maleki, A. Mouazen, H. Ramon, and J. De Baerdemaeker, "Multiplicative scatter correction during on-line measurement with near infrared spectroscopy," *Biosystems Engineering*, vol. 96, pp. 427–433, 2007.

[39] T. Fearn, "On orthogonal signal correction," *Chemometrics and Intelligent Laboratory Systems*, vol. 50, pp. 47–52, 2000.

[40] J. Zhang, Z. Zhang, Y. Xiang, Y. Dai, and P. d. B. Harrington, "An emphatic orthogonal signal correction-support vector machine method for the classification of tissue sections of endometrial carcinoma by near infrared spectroscopy," *Talanta*, vol. 83, pp. 1401–1409, 2011.

[41] Y. Roggo, P. Chalus, L. Maurer, C. Lema-Martinez, A. Edmond, and N. Jent, "A review of near infrared spectroscopy and chemometrics in pharmaceutical technologies," *Journal of Pharmaceutical and Biomedical Analysis*, vol. 44, pp. 683–700, 2007.

[42] Z. M. Zhang, S. Chen, and Y. Z. Liang, "Baseline correction using adaptive iteratively reweighted penalized least squares," *Analyst*, vol. 135, pp. 1138–1146, 2010.

[43] P. H. Eilers, "A perfect smoother," *Analytical Chemistry*, vol. 75, pp. 3631–3636, 2003.

[44] X. Zhang, A. de Juan, and R. Tauler, "Multivariate curve resolution applied to hyperspectral imaging analysis of chocolate samples," *Applied Spectroscopy*, vol. 69, pp. 993–1003, 2015.

[45] C. Torrence and G. P. Compo, "A practical guide to wavelet analysis," *Bulletin of the American Meteorological Society*, vol. 79, pp. 61–78, 1998.

[46] S. Wold, K. Esbensen, and P. Geladi, "Principal component analysis," *Chemometrics and Intelligent Laboratory Systems*, vol. 2, pp. 37–52, 1987.

[47] P. Geladi and B. R. Kowalski, "Partial least-squares regression: a tutorial," *Analytica Chimica Acta*, vol. 185, pp. 1–17, 1986.

[48] M. Barker and W. Rayens, "Partial least squares for discrimination," *Journal of Chemometrics*, vol. 17, pp. 166–173, 2003.

[49] B. E. Boser, I. M. Guyon, and V. N. Vapnik, "A training algorithm for optimal margin classifiers," in *Proceedings of the Fifth Annual Workshop on Computational Learning theory*, Pittsburgh, Pennsylvania, USA. 1992, pp. 144–152.

[50] C. C. Chang and C. J. Lin, "LIBSVM: a library for support vector machines," *ACM Transactions on Intelligent Systems and Technology (TIST)*, vol. 2, p. 27, 2011.

[51] P. d. B. Harrington, "Fuzzy multivariate rule-building expert systems: minimal neural networks," *Journal of Chemometrics*, vol. 5, pp. 467–486, 1991.

[52] B. W. Wabuyele and P. d. B. Harrington, "Fuzzy optimal associative memory for background prediction of near-infrared spectra," *Applied Spectroscopy*, vol. 50, pp. 35–42, 1996.

[53] P. d. B. Harrington, "Statistical validation of classification and calibration models using bootstrapped Latin partitions," *TrAC Trends in Analytical Chemistry*, vol. 25, pp. 1112–1124, 2006.

[54] S. Navea, A. d. Juan, and R. Tauler, "Detection and resolution of intermediate species in protein folding processes using fluorescence and circular dichroism spectroscopies and multivariate curve resolution," *Analytical Chemistry*, vol. 74, pp. 6031–6039, 2002.

[55] E. Pere-Trepat, M. Petrovic, D. Barcelo, and R. Tauler, "Application of chemometric methods to the investigation of main microcontaminant sources of endocrine disruptors in coastal and harbour waters and sediments," *Analytical and Bioanalytical Chemistry*, vol. 378, pp. 642–654, Feb 2004.

[56] S. Piqueras, L. Duponchel, R. Tauler, and A. De Juan, "Resolution and segmentation of hyperspectral biomedical images by multivariate curve resolution-alternating least squares," *Analytica Chimica Acta*, vol. 705, pp. 182–192, 2011.

[57] Z. Xu and E. Y. Lam, "Image reconstruction using spectroscopic and hyperspectral information for compressive terahertz imaging," *Journal of the Optical Society of America A: Optics and Image Science, and Vision*, vol. 27, pp. 1638–1646, 2010.

[58] L. V. Titova, A. K. Ayesheshim, A. Golubov, D. Fogen, R. Rodriguez-Juarez, F. A. Hegmann, and O. Kovalchuk, "Intense THz pulses cause H2AX phosphorylation and activate DNA damage response in human skin tissue," *Biomedical Optics Express*, vol. 4, pp. 559–568, 2013/04/01 2013.

[59] H. Hintzsche, C. Jastrow, B. Heinen, K. Baaske, T. Kleine-Ostmann, M. Schwerdtfeger, M. K. Shakfa, U. Kärst, M. Koch, and T. Schrader, "Terahertz radiation at 0.380 THz and 2.520 THz does not lead to DNA damage in skin cells in vitro," *Radiation Research*, vol. 179, pp. 38–45, 2012.

[60] A. Bogomazova, E. Vassina, T. Goryachkovskaya, V. Popik, A. Sokolov, N. Kolchanov, M. Lagarkova, S. Kiselev, and S. Peltek, "No DNA damage response and negligible genome-wide transcriptional changes in human embryonic stem cells exposed to terahertz radiation," *Scientific Reports*, vol. 5, p. 7749, 2015.

Chapter 10

Broadly Tunable CW Terahertz Sources Using Intrinsic Josephson Junction Stacks in High-Temperature Superconductors

Manabu Tsujimoto, Takanari Kashiwagi,
Hidetoshi Minami and Kazuo Kadowaki

Additional information is available at the end of the chapter

http://dx.doi.org/Chapter DOI: 10.5772/67087

Abstract

Electromagnetic waves in the 0.3–3.0 THz frequency range are considered to have great potential in research and industry; thus, compact, solid-state and continuous-wave (CW) terahertz sources have been developed throughout the vast field of science and technology. Since the first demonstration of terahertz emission from intrinsic Josephson junctions (IJJs) in the high-temperature (high-T_c) superconductor $Bi_2Sr_2CaCu_2O_{8+\delta}$, terahertz generation utilizing stacks of IJJs has become a major topic of research, both experimentally and theoretically. In this chapter, we describe recent progress on the development of high-T_c superconducting terahertz sources. We demonstrate that these superconducting terahertz sources emit continuous terahertz radiation and generate power in the microwatt range at broadly tunable frequencies in the range of 0.5–2.4 THz. The solid-state source is extremely small in size and its output power is sufficiently stable during operation. In addition, we also established a transmission imaging system using high-T_c sources to promote effective use in various applications.

Keywords: CW terahertz source, high-temperature superconductor, intrinsic Josephson effect

1. Introduction

The intrinsic Josephson effect as a tunneling of cooper pairs between adjacent CuO_2 planes inside highly anisotropic high-temperature (high-T_c) cuprate superconductors has attracted much experimental and theoretical attention since its experimental observation in 1992 by Kleiner et al. [1]. They observed the characteristic quasiparticle current-voltage branches corresponding to the individual switching of intrinsic Josephson junctions (IJJs) in single crystalline $Bi_2Sr_2CaCu_2O_{8+\delta}$ (Bi-2212), which consists of an atomic-scale stack of superconducting CuO_2 layers in between insulating Bi_2O_2 layers. **Figure 1(a)** shows the crystal structure of Bi-2212 and scheme of the IJJ. This experimental verification had a considerable influence on the concept of high-T_c superconductivity because it essentially ensures inhomogeneity of the superconducting order parameters. This rather surprising consequence naturally leads one to the new material concept of high-T_c superconductors as stacks of superconductor-insulator-superconductor devices. Some reviews of the phenomenon and the present status of research regarding the theoretical understanding have already been published [2]. In 1962, Josephson predicted that the tunneling current of cooper pairs through weakly coupled superconductors can be expressed as $J = J_c \sin(\varphi_2 - \varphi_1)$, where J_c is the critical current density and $(\varphi_2 - \varphi_1)$ is the phase difference of the wave functions in two superconductors [3]. When the current reaches J_c, a chemical potential difference $\mu_2 - \mu_1 = 2\mathrm{eV}$ appears between two superconductors. The phase difference alters with time by $\varphi_2 - \varphi_1 = -(2\mathrm{eV})t/\hbar$. Hence, an AC Josephson current, $J = J_c \sin(\omega_J t + \varphi_0)$, can be produced at the Josephson frequency

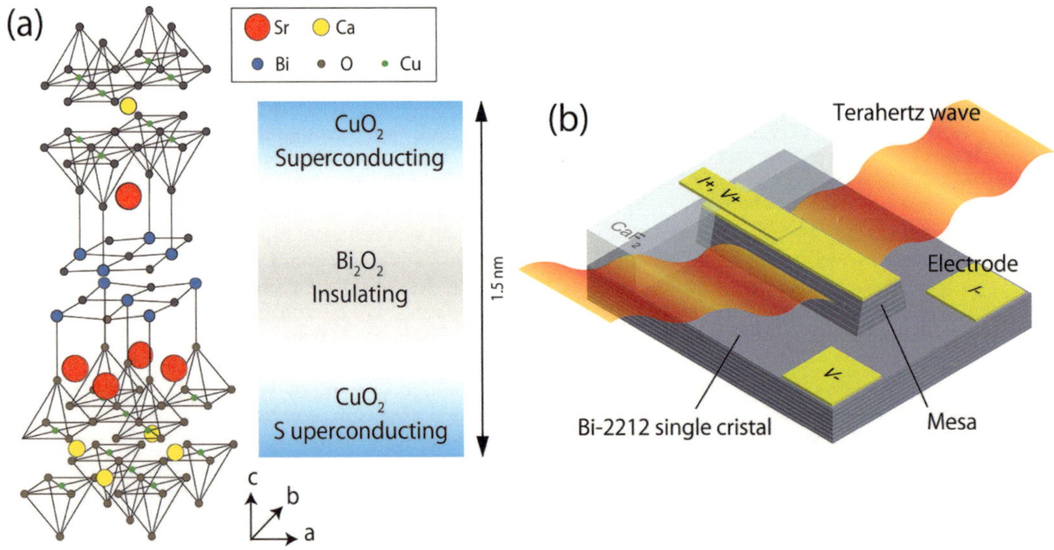

Figure 1. (a) Crystal structure of $Bi_2Sr_2CaCu_2O_{8+\delta}$ (Bi-2212) and scheme of the intrinsic Josephson junction. (b) Schematic view of the mesa-type high-T_c superconducting terahertz source.

$$f_J = \frac{\omega_J}{2\pi} = \frac{2e}{h}\mathrm{V}. \tag{1}$$

As this universal relation is determined only by fundamental physical constants, the Josephson junction can act as a perfect DC/AC convertor: for example, application of 1 mV corresponds exactly to the generation of electromagnetic oscillation at 0.483 THz. A recent highlight was the development of flux-flow oscillators based on high-quality Josephson junctions [4]. The maximum electromagnetic radiation power generated typically ranges from several nanowatts to a few microwatts [5, 6], whereas the frequency tunability is strictly limited by the superconducting gap energy, which is typically 750 GHz for a niobium-based junction. The use of discrete arrays of Josephson junctions enables the emitted power, the linewidth and the impedance to be optimized by adjusting the topology of the circuit. Nevertheless, the maximum power is insufficient to permit Josephson junctions to be used as practical high-frequency sources. The reasons for this include poor impedance matching, microwave losses and significant power generation at higher harmonics. The voltage fluctuation causes a broadening of the spectral linewidth; for example, a typical linewidth of 160 MHz has been obtained [7]. However, a linewidth below 1 MHz is required to use Josephson junctions as local oscillators of a subTHz frequency band receiver.

In 1986, high-T_c superconductivity was discovered by Bednorz and Müller in the Ba-La-Cu-O system [8]. Subsequently, Maeda et al. found a new phase of Bi-2212, which is the representative compound in a series of Bi-based cuprate superconductors $Bi_2Sr_2Ca_{n-1}Cu_nO_y$ $(n = 1, 2, 3, \cdots)$, exhibiting relatively high-T_c [9]; i.e., the T_c of $Bi_2Sr_2Ca_2Cu_3O_{10+\delta}$ with $n = 3$ exceeded 100 K. In a subsequent study, T_c records were achieved with Tl- and Hg-based cuprates of 125 K [10] and 135 K [11]. The reason why so many studies have been concentrating on the Bi-2212 phase is that large and high-quality single crystals of this material are easily grown by the floating zone method [12]. The cleavable crystal enables us to obtain a clear crystal surface, which is indispensable for studying the material properties. Accordingly, a number of interesting phenomena have been observed and discussed in the Bi-2212 system. For example, clear experimental evidence for the melting of the classic Abrikosov vortex lattice in a wide temperature range below T_c [13] was obtained as a consequence of a number of experimental and theoretical studies.

In 2007, Ozyuzer et al. [14] published the first report on the intense, continuous and coherent terahertz electromagnetic wave radiation from the IJJs in Bi-2212. **Figure 1(b)** shows a schematic view of the mesa-type IJJ terahertz source. Since this discovery, a great deal of interest has been drawn not only to the physical mechanism of the radiation but also to the possible variety of applications in the vast field of science and technology. In this chapter, the authors describe a study on the fundamental mechanism of terahertz radiation from the high-T_c superconductor Bi-2212 and recent developments in potential applications. Although a considerable number of theoretical studies have been conducted to understand the mechanism of terahertz radiation, our focus remains on the experimental work because of space limitations.

2. Experimental

2.1. Sample preparation

Single crystals of Bi-2212 were prepared by a traveling solvent floating zone (TS-FZ) technique [12]. This technique is appropriate for growing incongruent melting materials including high-T_c cuprate superconductors. The TS-FZ technique has some remarkable advantages in comparison with other conventional techniques such as the flux method. For instance, the TS-FZ technique enables us to obtain high-purity single crystals because there is no need to use a container or crucible. In addition, the TS-FZ technique enables us to overcome the problem associated with the uncontrollable doping process of cationic elements, which is a key ingredient required for high-T_c superconductivity. Single crystals of Bi-2212 were grown in air at a growth rate of 0.5–1.0 mm/h using an infrared convergence-type floating zone furnace [15]. A molten zone forming between the crystal and the feed rod was placed at the center of the ellipsoidal mirrors. Counter-rotation of both the feed rod and crystal was used to achieve homogeneous heating of the zone and especially to promote mixing. The single crystal then grows as the feed rod and the crystal moves down simultaneously. An as-grown Bi-2212 single crystal rod can be easily cleaved into thin slabs using a knife and Scotch tape. In this chapter, as-grown crystals were annealed at 650°C for 24 hours in argon gas mixed with 0.1% oxygen in order to obtain slightly underdoped crystals. The temperature dependence of the *c*-axis resistance shows behavior typical of slightly underdoped Bi-2212, as shown in the inset of **Figure 3(a)**.

Bi-2212 mesa samples were fabricated either by the focused ion beam (FIB) milling technique or photolithography technique. A piece of cleaved Bi-2212 crystal with the dimensions 110.01 mm^3 was glued onto a sapphire substrate and silver and gold thin layers were evaporated onto it. In this study, the author attempted to fabricate three different types of mesa: *terrace-type*, *groove-type* and *stand-alone-type* mesas, as presented in **Figure 2(a)–(c)**, respectively. The *terrace-type* mesa was milled by using conventional photolithography and the argon ion milling technique. The *groove-type* mesa was fabricated by patterning a 10-μm wide groove around the mesa by FIB milling. The *stand-alone-type* mesa was made by depositing a narrow gold strip on a single crystal, followed by argon ion milling and cleavage from the milled substrate. Then, the mesa was attached to a thin gold layer covering the sapphire substrate. Thin gold wire electrodes were fixed onto the top surface of these mesas and the substrate by silver paste. In this study, one rectangular *terrace-type* mesa, one rectangular *stand-alone-type* mesa and *groove-type* mesas with three different geometrical shapes (rectangle, square and disk) were fabricated. The sample dimensions and profile curves were measured using atomic force microscopy (AFM). The obtained cross-sectional profile is considerably slanted and rounded at the edges, resulting in an approximately trapezoidal shape. Typically, the cross section of the upper section is approximately 10% smaller than that of the lower section. The total number of embedded junctions is estimated from the mesa height *d*.

2.2. Measurement setup

The current-voltage characteristics (IVCs) were measured with the conventional electrical circuit for a resistance measurement, as shown in **Figure 2(e)**. The load resistance R_{LO}, the standard resistor $R_S = 10\ \Omega$ and the mesa sample are connected in series to the function generator. The current I is obtained from the voltage drop at R_S. The mesa voltage V is measured directly using a DC differential amplifier. To measure the multiple IVC branches, we use the constant voltage source instead of the function generator. The load resistance R_{LO} is typically set to 100 Ω.

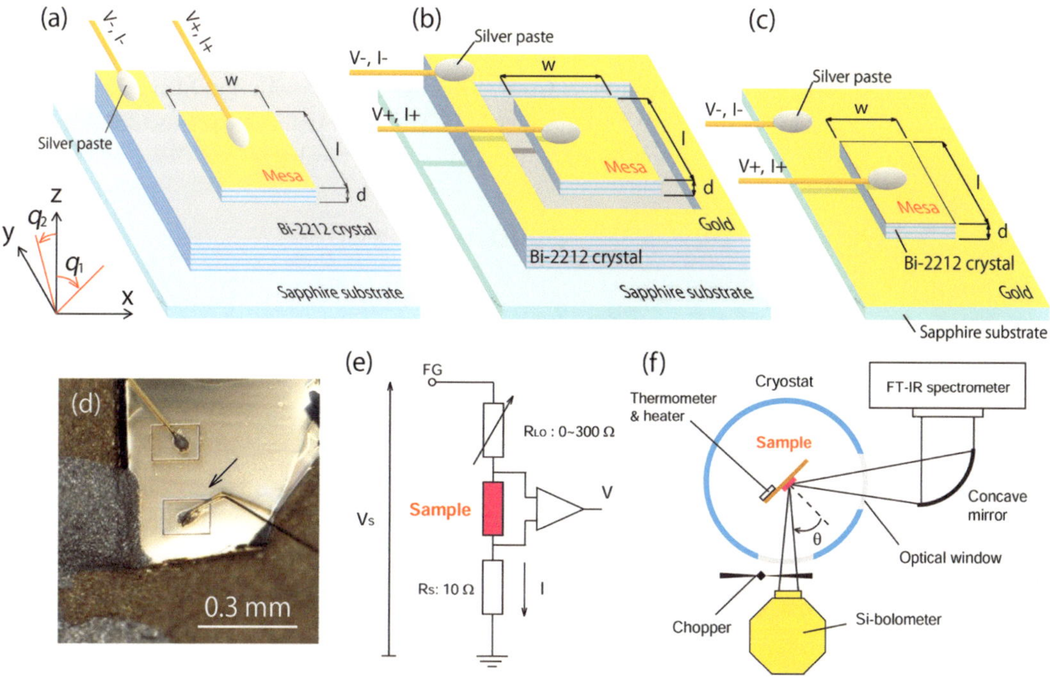

Figure 2. Schematic views of (a) terrace-type, (b) groove-type and (c) stand-alone-type mesa samples. (d) Optical microscopy image of the groove-type mesa sample. (e) Electrical circuit diagram for measuring IVCs. (f) Optical system for detecting the terahertz emission and measuring the emission frequencies.

Figure 2(f) shows an optical system for detecting the terahertz waves emitted from the sample. The output intensity is monitored during the bias scan using a Si bolometer. A He-flow type cryostat with two optical polyethylene windows was used for cooling the sample. The sample was mounted on a cold finger. The bath temperature T_b was monitored using a RhFe thermometer attached close to the cold finger. The detection angle θ can be adjusted by rotating the sample holder. The optical chopper was used to modulate the emitted terahertz wave at 80 Hz for the lock-in detection. We used a Fourier transform infrared (FT-IR) spectrometer to measure the emission frequencies.

3. Geometrical cavity resonance

The IVCs and terahertz-emission intensity for disk mesa D3 (*groove-type*, radius $a = 61.5–65.0$ μm and $d = 1.6$ μm) at $T_b = 25.0$ K are displayed in **Figure 3(a)** in black and red, respectively. In **Figure 3(b)**, the emission region is presented at an expanded scale. The most intense emission occurs on the return branch of the outermost IVC curve, where all stacking IJJs are in the resistive state. In most cases, the radiation is observed in a narrow $I–V$ range; for example, $V = 0.96–0.98$ V and $I = 10.8$ mA for sample D3. As the bias current is decreased, emission suddenly ceases due to a jump to an internal IVC branch. It is interesting to note that emission usually occurs within a sample-dependent range of the base temperature of the mesa.

For example, mesa D3 emits between 10 and 50 K. During the emission, since a DC of I~11 mA is supplied to mesa D3, it is inevitably heated at 11 mW power. This huge heating power density cannot be removed quickly enough from the mesa to maintain equilibrium, resulting in a considerable increase of the mesa temperature [16]. The local heating may induce a chaotic nonequilibrium state and may adversely affect the emission of terahertz radiation.

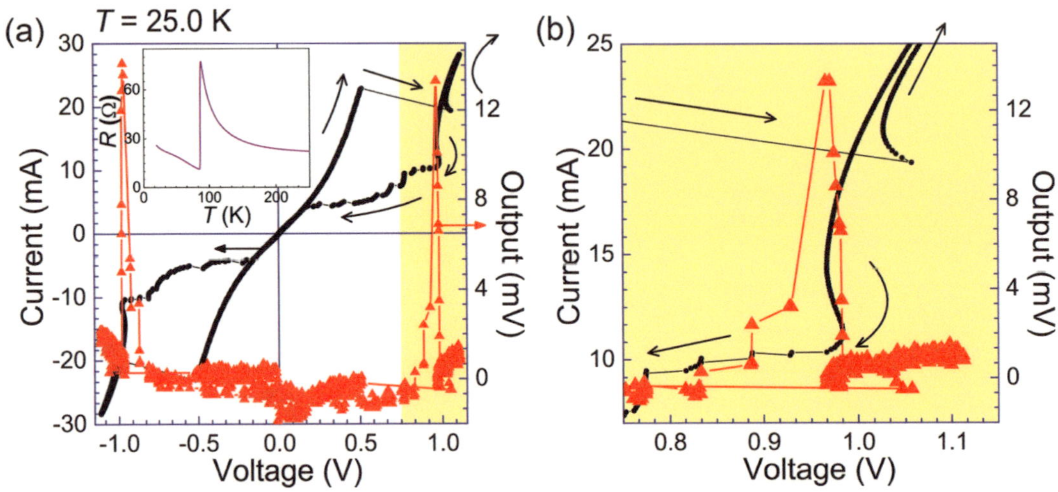

Figure 3. (a) IVCs (black, scale on the left) and output intensity detected using the Si bolometer (red, scale on the right) for disk mesa D3 at $T_b = 25.0$ K. (b) Details of the shaded high-bias regime in **Figure 3(a)** where the emissions are observed.

Figure 4(a) shows the far-infrared spectra of the terahertz radiation from three disk mesas with different radii: D1 ($a = 33.9$–38.9 μm), D1 ($a = 48.9$–51.5 μm) and D1 ($a = 61.5$–65.0 μm). The emission frequencies increase with decreasing radius, thereby indicating the geometrical cavity resonance effect. In **Figure 4(c)**, the emission frequency f is plotted as a function of $1/2a$. The dashed line represents the calculated cavity resonance frequency

$$f_{11}^{c} = \frac{\chi_{11} c_0}{2\pi n a}, \tag{2}$$

where $\chi_{11} = 1.841$ for the TM(1, 1) mode [17]. The emission frequency f is clearly proportional to $1/2a$. The data were fitted with $n = 4.2$, which is in good agreement with previous results [18]. Note that this n value is about 50% larger than that obtained from infrared spectroscopy [19]. In addition, for each disk mesa, the second harmonic at $2f$ is clearly visible and these frequencies are distinguishable from those of the nearest higher cavity modes [20].

The resonance frequencies of the TM(m, p) modes for a rectangular cavity of width w and length ℓ are in the form [21]:

$$f_{mp}^{c} = \frac{c_0}{2n}\sqrt{\left(\frac{m}{w}\right)^2 + \left(\frac{p}{\ell}\right)^2}. \tag{3}$$

It is very curious that the TM(0.1) mode for rectangular mesas has never been observed in previous experiments. Since the mesa is considerably heated by the DC current, the http://dx.doi.org/Chapter DOI: 10.5772/67087

inhomogeneous temperature distribution inside the mesa prevents the formation of standing electromagnetic waves, particularly along the longer rectangular dimension. However, such hot spots with T even exceeding T_c, as observed by Wang et al. [22], do not seem to be a problem in our experiments, since the fundamental frequencies of disk mesas excellently obey the linear relationship of the cavity resonance frequency with the inverse of a. The discrepancies in relatively long rectangular mesas with $w \ll \ell = 300$–400 μm may have a different origin. This was verified by the fabrication of rectangular mesas with different ℓ/w ratios. **Figure 4(b)** shows the observed frequencies for rectangular mesas as a function of $1/w$. The dashed lines represent the cavity resonance frequencies for the TM(1,0) modes. It is obvious that the frequency obeys the TM(1,0) cavity resonance mode very well, definitely not the TM(0,1) resonance mode.

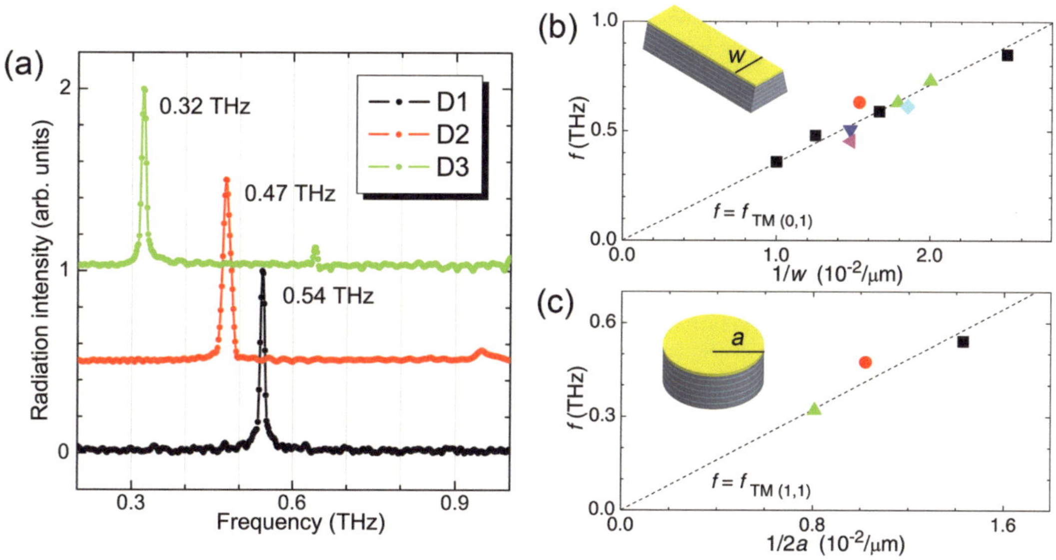

Figure 4. (a) Radiation spectra measured using a FT-IR spectrometer for three disk mesas. Observed frequencies f vs. $1/w$ for rectangular mesas (b) and $1/2a$ for disk mesas (c). The dashed lines represent the cavity resonance frequencies for TM (0,1) for the rectangular mesa and TM(1.1) for the disk mesa.

4. Anisotropic spatial emission patterns

The basic antenna theory excellently explains the spatial emission patterns of electromagnetic waves from the conventional antenna element, which consists of a normal metal, the dielectric material, etc., by solving Maxwell's equations for the electromagnetic fields. These electromagnetic waves originate from either an electric current flowing through the conducting surface or an electron charge orderly distributed within the element. Hence, measuring spatial emission patterns from the emitting device enables us to investigate both the internal electromagnetic fields and the surface current distribution. In the case of the IJJ terahertz source, the situation must be substantially different from the conventional antenna element due to its superconducting characteristics. The excitation mode inside the disk mesa was investigated by measuring the angular dependence of the radiation intensity $I(\theta)$, where θ is the variable detection angle defined in **Figure 2(f)**.

Figure 5(a) shows $I(\theta)$ for disk mesa D3 in an arbitrarily fixed plane. The shadowing effect of the radiation from the superconducting Bi-2212 crystal wall outside the groove is expected to be negligibly small in the first approximation. In **Figure 5(a)**, the following characteristic features are noted: first, $I(\theta)$ is strongly anisotropic, having a maximum around $\theta = \theta_{\max} = 20–35°$ from the top ($\theta = 0°$), where a local minimum occurs with the intensity ratio $I(20)/I(0)$~1.5–2.0—. This shallow minimum feature is almost the same as for rectangular mesas as discussed later, although $\theta_{\max}$ is somewhat less than the corresponding rectangular mesa value. Second, $I(\theta)$ rapidly diminishes as θ approaches –90°. $I_{\text{cav}}(\theta)$ is calculated by assuming the TM(1,1) cavity mode alone which is shown as the black dashed curve in **Figure 5(a)** [20]. Clearly, the calculated $I_{\text{cav}}(\theta)$ does not fit the experimental data, especially near to $\theta = 0$, where $I_{\text{cav}}(\theta)$ is a maximum [20]. This disagreement can be removed by introducing the superposition of the radiation from the uniform AC Josephson current source with the same Josephson frequency, as described for rectangular mesas [18, 20]. Subsequent to similar calculations that were conducted by introducing the mixing parameter $\alpha = 1.44$ into the dual source model, where the mixing parameter corresponds to 41% of the radiation arising from the uniform AC Josephson current source, a more optimal fit is obtained for the experimental data as shown by the solid orange curve in **Figure 5(a)**. It is significant that the intensity from the uniform mode is comparable to that of the fundamental cavity mode, as in rectangular mesas [18]. Neither higher disk cavity excitation frequencies of the Bessel type nor subharmonics were observed in **Figure 4(a)**. This experimental evidence clarifies unambiguously that the terahertz radiation is mainly generated by the uniform mode of the AC Josephson current. However, it does not necessarily mean that the geometrical cavity resonance is not entirely unimportant for the determination of the actual radiation frequency values. An extended study of the spatial radiation patterns of the higher harmonics could provide additional supporting information for these conclusions.

The θ dependence of the radiation intensity for the rectangular mesa R3 (*terrace-type*, $w = 77.4–4.5$ μm, $\ell = 400$ μm and $d = 1.2$ μm) in the yz-plane $I(\theta_1)$ and in the xz-plane $I(\theta_2)$ are shown in **Figure 5(b)** and **(c)**, respectively, where $I(\theta_1)$ and $I(\theta_2)$ and the coordinates are defined in **Figure 2(a)**. As clearly seen that the observed emission has a considerably characteristic anisotropic pattern in comparison with the usual antenna elements. Typically in our results with other samples, the emission is usually stronger in the yz-plane than in the xz-plane, suggesting that the standing electromagnetic wave is certainly excited along the shorter side. The radiation intensities $I(\theta_1)$ in these figures are normalized at the maximum value obtained at $\theta_1 = \theta_2 = 0°$. As a result, the following characteristic features of spatial patterns are eventually obtained: first, $I(\theta_1)$ in the yz-plane has a maximum value, $I(\theta_1 = –30)/I(\theta_1 = 0)$~2, with a nearly symmetrical pattern. Second, $I(\theta_1)$ has a local minimum at the top, although it differs from sample to sample [18]. These commonly observed sample dependencies may arise from various differences in the properties among mesa samples, which are, for example, spatial inhomogeneities in the AC Josephson current due to a heating effect. Third, small but clear minor lobes are observed at $\theta_1 = \theta_2$~75° as shown in **Figure 5(b)** and **(c)**, whereas their integrated intensity appears to be only a few percentage less than the total. These lobes might be caused by the higher harmonics. The higher harmonics seems to hardly contribute to $I(\theta)$ because of the weakness as mentioned before. The standard antenna theory predicts that the radiation intensity $I(\theta)$ should reach a certain maximum value at the top of the mesa [23], where $\theta_1 = \theta_2 = 0°$ in the present system. However, our experimental results clearly show a different behavior from this prediction, in which the electromagnetic wave excitation is simply induced by the fundamental rectangular cavity mode. Here, if we assume a oscillating superconducting current spatially uniform in surface direction as well as in the case of disk mesa, $I(\theta)$ would rapidly diminish as θ approaches to 0° and would reach a maximum value at 90° resulting from the dipole radiation. Furthermore, in the actual case, the superconducting substrate of Bi-2212 and other obstructions in the optical path may also cause some complicated problems.

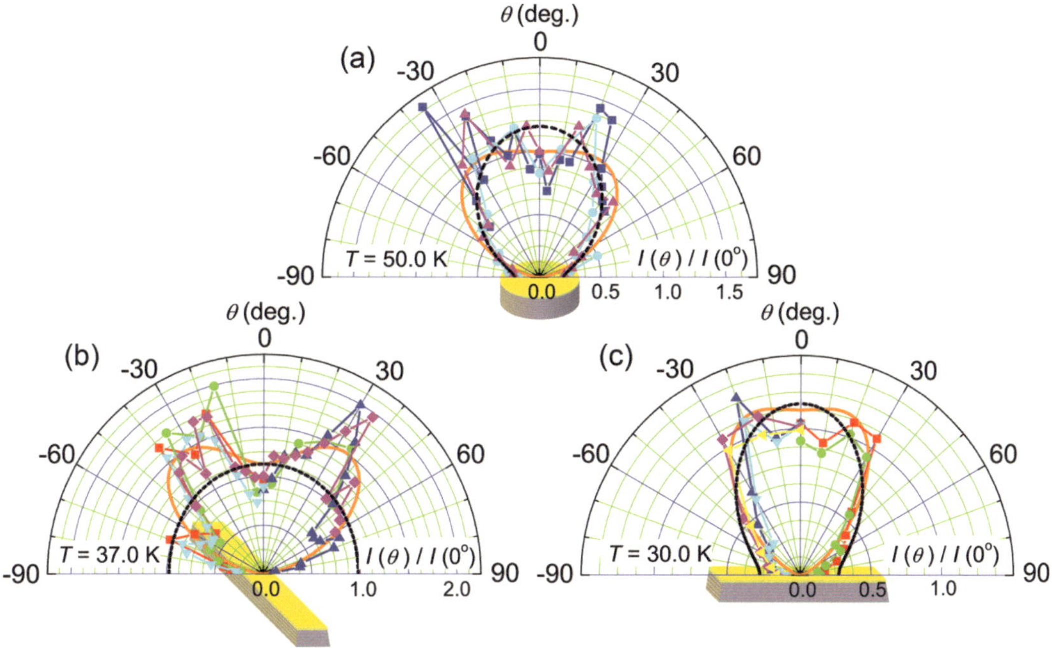

Figure 5. Polar plots of the radiation intensity $I(\theta)$ for (a) disk mesa and rectangular mesa in (b) *yz*-plane and (c) *xz*-plane. The solid and dashed lines are the best fits to the dual-source model and its cavity component, respectively.

5. Tunable radiation from internal current-voltage branches

Most workers considered the enhancement of the power of the output radiation by the excitation of an internal cavity mode to be such that the radiation from the AC Josephson current source alone was too weak to observe. However, recently, the contributions to the output power from the AC Josephson current source alone and that enhanced by resonance with an internal electromagnetic cavity source were found to be comparable in magnitude [18, 24]. Here, the authors show clear evidence that the mesas can emit radiation at many frequencies, without strong interaction with an internal cavity mode. More importantly, the resulting radiation is tunable over a broad range of frequencies, allowing the possible construction of a powerful device that could fill the terahertz gap [25].

The authors found that electromagnetic emission also occurs at many points in the inner region of the multiply branched IVCs, where the number of the resistive IJJs $N = 1, 2, \cdots, N_{\max}$ is fixed but different for each branch. $N_{\max}$ is the number of IJJs in the stack estimated from the mesa height d. In **Figure 6(a)** and **(b)**, the radiation frequencies f are plotted as color-scaled symbols on the high-bias regions of the multiply branched IVC structures for R3 (*groove-type*, $w = 99.2$–102 μm, $\ell = 137$–140 μm and $d = 1.3$ μm) at $T_b = 35.0$ K and R2 (*stand-alone-type*, $w = 61.1$–63.7 μm, $\ell = 331$–333 μm and $d = 1.5$ μm) at $T_b = 52.5$ K, respectively. The insets show the full IVCs. All of the R1 curves in **Figure 6(a)** bend backwards with increasing current, indicative of Joule heating. However, R2 is less susceptible to heating effects and its IVCs in **Figure 6(b)** are monotonic. The radiation spectra were obtained at as many I–V bias points as possible. At some bias points, denoted by open diamonds, no emission was detected.

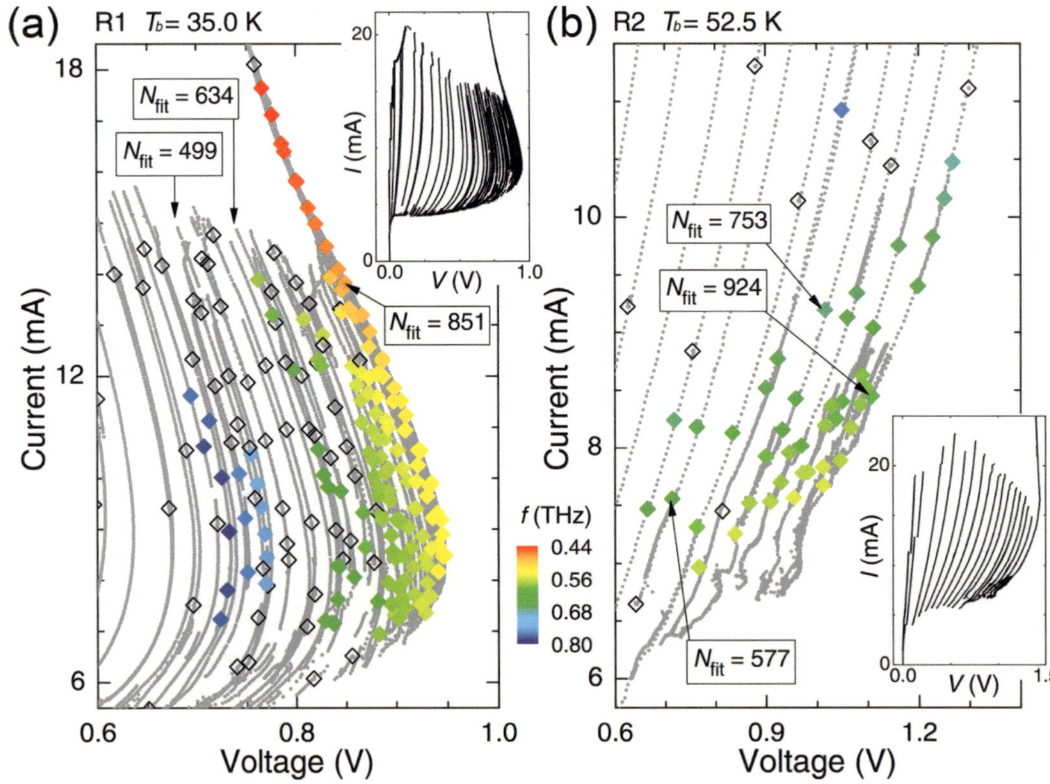

Figure 6. Emission at the color-coded frequencies was observed at the filled diamonds for R1 (a) and for R2 (b). No radiation was detected at the open diamonds. Arrows indicate the numbers of resistive junctions N_{fit} obtained from fits to the AC Josephson relation. The two insets show the full IVCs for each sample.

By repeated measurements of the emission from a particular IVC branch with constant N, the author confirmed that f satisfies the AC Josephson relation $f = f_J = (2e/h)V/N$. In **Figure 7(a)**, the emission data from the IVCs of R1 shown in **Figure 6(a)** was replotted in terms of $f(V)$, representing the data from each branch in terms of unique symbols and colors. By http://dx.doi.org/Chapter DOI: 10.5772/67087

fitting the data for a particular branch to the AC Josephson relation with $N = N_{\mathrm{fit}}$, the experimental best-fit value N_{fit} for each emitting branch was determined. Three examples each for R1, respectively, are indicated by the dashed lines in **Figure 7(a)** and the arrows in **Figure 6(a)**. In the inset in **Figure 7(a)**, the entire emission data from the respective inner branches are plotted as $f(V/N_{\mathrm{fit}})$. For both R1 and R2, the AC Josephson relation $f = f_J = (2e/h)V/N_{\mathrm{fit}}$ is very well obeyed. As **Figure 7(a)** clearly demonstrates that f is slightly tunable on each branch (indicated by a fixed symbol and color), in **Figure 7(b)** and **(c)**, we replotted the data from each respective R1 and R2 branch as $f(N_{\mathrm{fit}})$. Although for both samples f is tunable as indicated by the vertical bars for each fixed branch number N_{fit}, R1 and R2 display rather different aspects of tunability. The overall tunability of the two samples is nearly the same, with R1 and R2, respectively, being tunable from 0.44 to 0.78 THz and from 0.43 to 0.76 THz, respectively. However, for R1, this range is primarily due to the f-dependence on the IVC branch number, whereas for R2, the tunability is greatest on a single branch.

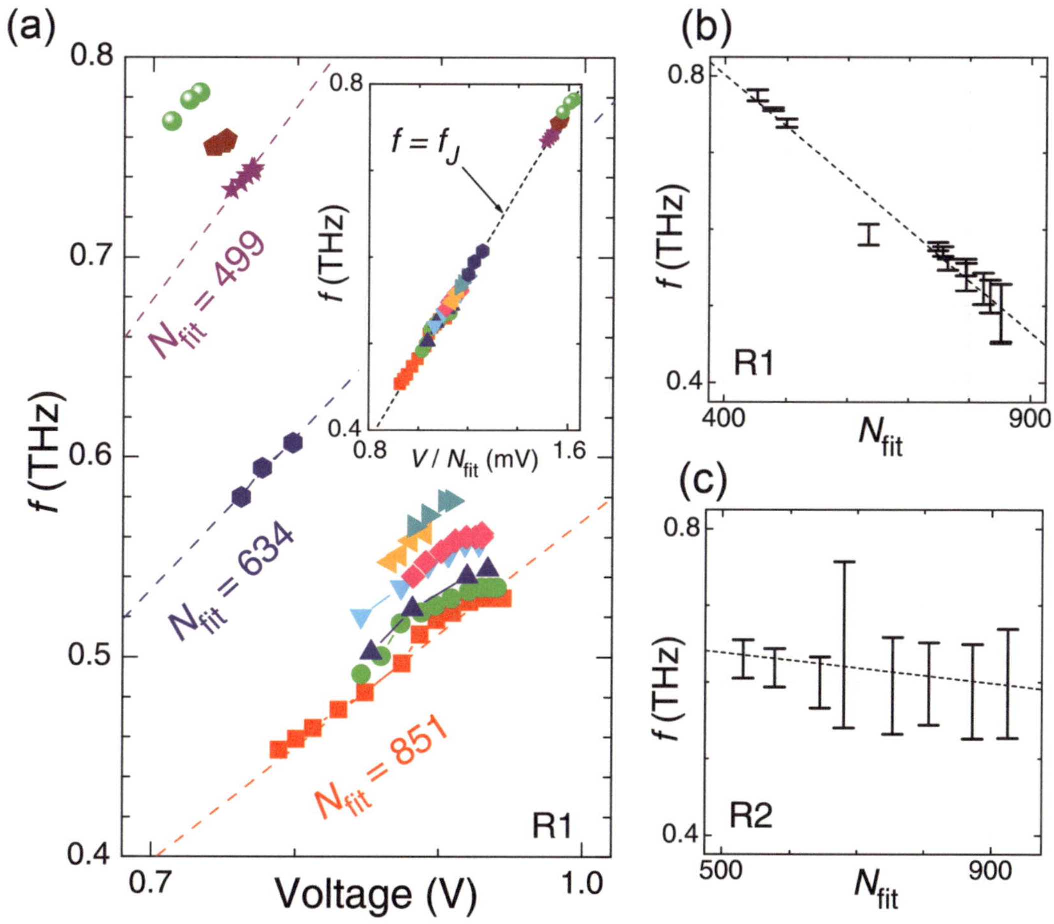

Figure 7. (a) Plots of $f(V)$ for 11 IVC branches for R1. The three dashed lines represent the theoretical fits to $f = f_J = (2e/h)V/N_{\mathrm{fit}}$. The inset shows the plot of $f(V/N_{\mathrm{fit}})$ for all R1 data with a $f = f_J$ line. Plots of $f(N_{\mathrm{fit}})$ ranges for R1 (b) and R2 (c). The dashed curves serve to guide the eye.

6. Applications: terahertz imaging system

Radiation in the frequency range of 0.1–10 THz, known as terahertz waves, is known to be the most unique across the entire range of electromagnetic wavelengths, because of the unavailability of compact and convenient sources [25] of radiation of these wavelengths. In recent years, much effort has been made to overcome the above difficulty, particularly in the field of semiconductors and laser technologies. Resonant tunneling diodes [26], unitraveling-carrier photodiodes [27] and a quantum cascade laser [28] have been developed as useful terahertz sources. Recently, graphene and carbon nanotubes have also been reported as good candidates for terahertz wave generators [29, 30].

Because terahertz waves have the properties of being able to pass through various objects, such as plastics, paper, ceramics, semiconductors, liquids and proteins, similar to radio waves, they enable us to obtain images of various cm-by-cm wide substances [31], where the spatial resolution is actually determined by the wavelength of the terahertz wave, typically 1 mm. Since the radiation damage to these soft objects is known to be negligibly small in contrast to that inflicted by X-rays, in fact there is much demand for the nondestructive and noninvasive imaging technique in the field of the material inspection, medical diagnostics, security checks and environmental monitoring. Moreover, measuring the absorption spectrum in the terahertz region, where various molecular vibration modes of organic and high-molecular compounds dominate the electromagnetic wave absorption, one can identify the chemical substances both qualitatively and quantitatively. To realize these applications, compact, solid state and reliable source devices are strongly desired in spite of their technical difficulties.

Previous studies found that the present IJJ system can be used as a terahertz source device emitting continuous and microwatt power terahertz waves at tunable frequencies between 0.5 and 2.4 THz [32]. These device characteristics enable us to accumulate the data for an individual measurement point at high resolution in a few milliseconds or less. A solid-state IJJ device is extremely small in size and its output power is sufficiently stable during operation. We have attempted to demonstrate a terahertz imaging experiment using IJJ sources in order to promote the effective use of the present IJJ device for various practical applications.

The imaging system consists of a He-flow cryostat, two off-axis parabolic mirrors, a scan stage and a fast hot-electron (HE) bolometer as shown in **Figure 8(a)**. Two off-axis parabolic mirrors 1 and 2 with focal lengths of 152.4 and 220 mm and diameters of 75 mm are set to focus the terahertz waves on the sample. The imaged object is fixed on a 1-mm thick quartz plate and scanned in both the *X*-(horizontal) and *Z*-(vertical) directions using a biaxial scanner at a variable speed below 130 mm/s. The practical maximum speed should be 80 mm/s, which corresponds to a 5-ms time constant per data point when the measurement step is set to submillimeter levels. Although the IJJ emitting device can be modulated much faster above 500 kHz, in the present setup the maximum scan speed only depends on the minimum time constant of the lock-in amplifier. Nevertheless, the actual imaging has been performed at the speed of 32 mm/s in order to obtain sufficient data points. Before starting every scan, the optical path presented in **Figure 8(a)** is precisely adjusted by using the visible light from a photodiode attached to the source position as reference. A high-speed InSb HE bolometer placed directly behind the scan stage is used for measuring the output intensity of the terahertz waves passing through the sample. The IVCs of the IJJ device and the terahertz output intensity are simultaneously monitored using two oscilloscopes (OSCs). The photograph in the lower left part of **Figure 8(a)** shows the imaging system viewed from the top.

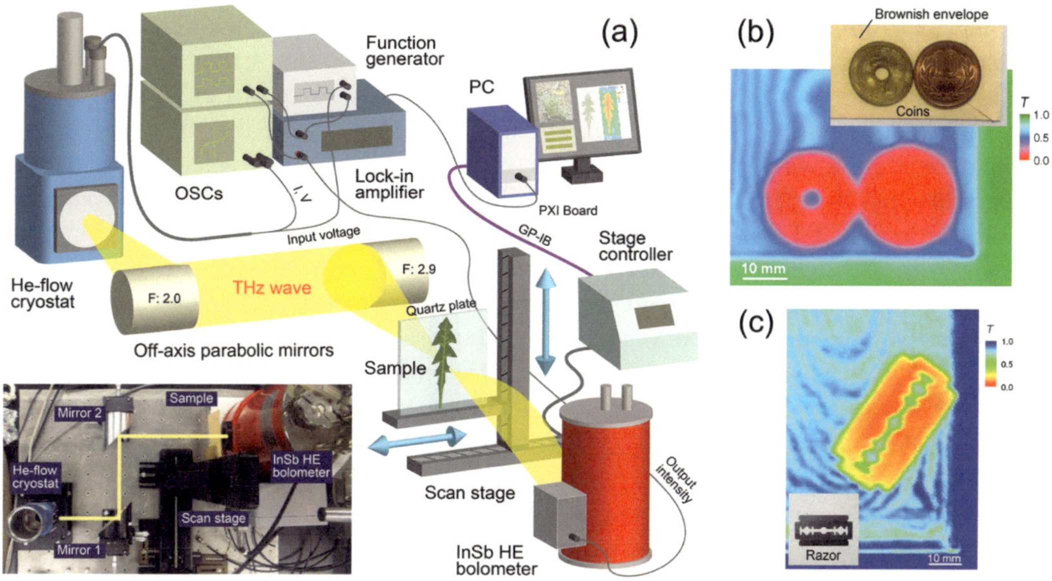

Figure 8. (a) Schematic diagram of the terahertz imaging system. The left lower photograph presents the actual system viewed from above. Terahertz images of 5-yen (left) and 10-yen (right) coins (b) and a razor blade (c) placed inside a brownish paper envelope. In these figures, the transmittance T is plotted in a color-coded scale. The photograph of each object is presented beside their terahertz image.

An imaging experiment was performed with the following procedure: first, pulse-modulated terahertz waves were generated from the IJJ source as follows: the DC source voltage is modulated by a low-amplitude square wave below 100 mV at 10 kHz and then applied to the device using the function generator. Maximum intensity is obtained by adjusting the offset level of the DC voltage and stabilized well to minimize power fluctuations by adjusting the amplitude of the square wave. The input current and the voltage were simultaneously monitored using an OSC (see **Figure 8(a)**). The imaging data are accumulated by simultaneous measurement of the sample position in the X- and Z-directions and the detected intensity. The intensity is acquired by an AD converter installed in a PXI bus system (National Instruments Co.) through the analog output channel of the lock-in amplifier. Some parameters such as scanning speed, measurement area and measurement step could be set by a homemade LabVIEW program. The obtained data are directly stored on the hard-disk drive and are simultaneously displayed on the PC screen in real time.

As test examples, the terahertz transmission images of two Japanese coins and a thin razor blade placed inside a brownish envelope presented in **Figure 8(b)** and **(c)**, respectively, were recorded. The photographs of these objects are presented beside their terahertz images. The emission frequency was set to 0.54 THz. Note that the envelopes were attached to the scan stage without using a quartz plate. The screen size of the image in **Figure 8(b)** is 330250 pixels, which corresponds to a measurement step of 0.20.2 mm^2, whereas those of **Figure 8(c)** are 6080 pixels and 0.40.2 mm^2, which were appropriately set in terms of the required spatial resolution. It took at least 20 minutes to acquire each image. The observed image clearly reflects the transmittance T of pixels corresponding to the measurement step of 0.20.2 mm^2 of the terahertz wave through objects. The hole in the center of the 5-yen coin, which has a diameter of exactly 5 mm, is vividly shown in **Figure 8(b)** (left). The estimated spatial resolution of ~1 mm is comparable with the Rayleigh limit. Based on the results shown in **Figure 8(b)**, the calculated transmittance of one sheet of paper is 79% and it is clear that the coins are not transparent at all. It is interesting to note that several interference fringe-like patterns in the deeper blue area in **Figure 8(b)** and **(c)** are clearly observed. Assuming that terahertz waves with a wavelength of $\lambda = c_0/f = 0.56$ mm are reflected at the inner walls of the envelope and interfere with each other, the interference condition would be expressed as $2t_{\mathrm{m}} = (\lambda/2)2m$, where t_m is the interspace distance between the inner walls of the envelope and m is an integer. Hence, the distance between two neighboring fringes should be expressed as $t_{\mathrm{m+1}}–t_{\mathrm{m}} = \lambda/2 = 0.28$ mm. Since the thickness of the coin is 1.5 mm, at least $1.5/0.28\tilde{5}$–6 concentric fringes are expected to appear around the coins. This simple estimation agrees with the experimental result shown in **Figure 8(b)**. The interference effect is a unique property of monochromatic terahertz waves, which may be useful for versatile purposes in applications of interferometry and a wide variety of communications.

7. Summary

This chapter presents studies of terahertz electromagnetic wave emission from intrinsic Josephson junctions in the high-temperature superconductor $Bi_2Sr_2CaCu_2O_{8+\delta}$ (Bi-2212) and its device applications. This chapter started with the introduction of the Josephson effect and other basic formulations. Some historical developments concerning electromagnetic wave emission from conventional junctions and arrays were mentioned as a pioneering contribution to the present study. Sample preparation techniques and experimental setups were introduced. Photolithography, argon ion milling and focused ion beam milling were discussed as techniques that were used for the fabrication of *terrace-type*, *groove-type* and *stand-alone-type* mesa structures. The geometrical resonance conditions were investigated using several mesas with different geometrical shapes and sizes. The experimental results clearly demonstrate the validity of the cavity resonance model. The spatial radiation patterns were measured and analyzed for disk and rectangular mesas. As a result, significant anisotropic emission patterns depending on the mesa geometry were observed. These results can naturally be understood by introducing the mechanism that combines two radiation sources: the AC Josephson current and cavity mode excitation. The current-voltage characteristics of the internal branches, which were found to emit radiation across a broad frequency range, were examined to determine whether the excitation of an internal cavity mode was an essential feature of the coherent radiation. Hence, it was concluded that the primary source of the intense, coherent subTHz radiation is the AC Josephson current and that the internal electromagnetic cavity produced by the geometrical shape of the emitting mesa is at the best of minor importance. The final part of this chapter presents a demonstration of the terahertz imaging system using high-T_c superconducting terahertz sources. As text examples, terahertz images of coins and a razor blade placed inside a brownish paper envelope were presented. These images enable the interference fringes caused by the monochromatic terahertz waves to be clearly observed. These demonstrations strongly confirm that this imaging system would be useful in many practical applications, such as medical tissue diagnosis, food inspection, biological and biomedical imaging and environmental monitoring.

Several phenomena relating to terahertz radiation from high-T_c superconductors remain poorly understood. For instance, in some cases, two kinds of emission were found to be taking place in two current-voltage regions: stable emission near the fully resistive region and unstable emission in the retrapping region. Understanding a complicated situation, it requires a thorough investigation of the thermal properties of the system, such as the temperature distribution inside the mesa. In addition, it is not obvious whether there is an upper limit to the radiation intensity and even to the radiation frequency. The effect of a magnetic field on terahertz emission continues to remain unclear, although selected theoretical predictions regarding the enhancement of the radiation intensity have been published. Hence, further studies are undoubtedly needed in order to elucidate these fascinating phenomena.

Author details

Manabu Tsujimoto*, Takanari Kashiwagi, Hidetoshi Minami and Kazuo Kadowaki

*Address all correspondence to: tsujimoto@ims.tsukuba.ac.jp

Division of Materials Science, Faculty of Pure and Applied Sciences, University of Tsukuba, Tsukuba, Japan

References

[1] R. Kleiner, F. Steinmeyer, G. Kunkel and P. Müller, Phys. Rev. Lett. **68**, 2394 (1992).

[2] A. Yurgens, Supercond. Sci. Technol. **13**, R85 (2000).

[3] B. D. Josephson, Phys. Lett. **1**, 251 (1962).

[4] V. P. Koshelets, S. V. Shitov, L. V. Filippenko, A. M. Baryshev, W. Luinge, H. Golstein, H. van de Stadt, J.R. Gao and T. de Graauw, IEEE Trans. Appl. Supercond. **7**, 3589 (1997).

[5] A. V. Ustinov, S. V. Shitov, N. Iosad and H. Kohlstedt, IEEE Trans. Appl. Supercond. **7**, 3601 (1997).

[6] J. Mygind, V. P. Koshelets, A. V. Shchukin, S. V. Shitov and I. L. Lapytskaya, IEEE Trans. Appl. Supercond. **5**, 2951 (1995).

[7] A. K. Jain, K. K. Likharev, J. E. Lukens and J. E. Sauvageau, Phys. Rep. **109**, 309 (1984).

[8] J. G. Bednorz and K. A. Müller, Zeitschrift Für Phys. B Condens. Matter **64**, 189 (1986).

[9] H. Maeda, Y. Tanaka, M. Fukutomi and T. Asano, Jpn. J. Appl. Phys. **27**, L209 (1988).

[10] Z. Z. Sheng and A. M. Hermann, Nature **332**, 138 (1988).

[11] A. Schilling, M. Cantoni, J. D. Guo and H. R. Ott, Nature **363**, 56 (1993).

[12] M. J. V. Menken, A. J. M. Winkelman and A. A. Menovsky, J. Cryst. Growth **113**, 9 (1991).

[13] E. Zeldov, D. Majer, M. Konczykowski, V. B. Geshkenbein, V. M. Vinokur and H. Shtrikman, Nature **375**, 373 (1995).

[14] L. Ozyuzer, A. E. Koshelev, C. Kurter, N. Gopalsami, Q. Li, M. Tachiki, K. Kadowaki, T. Yamamoto, H. Minami, H. Yamaguchi, T. Tachiki, K. E. Gray, W.K. Kwok and U. Welp, Science **318**(80), 1291 (2007).

[15] T. Mochiku and K. Kadowaki, Phys. C Supercond. **235–240**, 523 (1994).

[16] A. Yurgens, Phys. Rev. B **83**, 184501 (2011).

[17] A. Derneryd, IEEE Trans. Antennas Propag. **27**, 660 (1979).

[18] K. Kadowaki, M. Tsujimoto, K. Yamaki, T. Yamamoto, T. Kashiwagi, H. Minami, M. Tachiki and R. A. Klemm, J. Phys. Soc. Japan **79**, 23703 (2010).

[19] S. Tajima, G. D. Gu, S. Miyamoto, A. Odagawa and N. Koshizuka, Phys. Rev. B **48**, 16164 (1993).

[20] R. A. Klemm and K. Kadowaki, J. Phys. Condens. Matter **22**, 375701 (2010).

[21] M. Tsujimoto, I. Kakeya, T. Kashiwagi, H. Minami and K. Kadowaki, Opt. Express **24**, 4591 (2016).

[22] H. B. Wang, S. Guénon, B. Gross, J. Yuan, Z. G. Jiang, Y. Y. Zhong, M. Grünzweig, A. Iishi, P. H. Wu, T. Hatano, D. Koelle and R. Kleiner, Phys. Rev. Lett. **105**, 57002 (2010).

[23] M. Leone, IEEE Trans. Electromagn. Compat. **45**, 486 (2003).

[24] M. Tsujimoto, K. Yamaki, K. Deguchi, T. Yamamoto, T. Kashiwagi, H. Minami, M. Tachiki, K. Kadowaki and R. A. Klemm, Phys. Rev. Lett. **105**, 37005 (2010).

[25] M. Tonouchi, Nat. Photonics **1**, 97 (2007).

[26] E. R. Brown, J. R. Söderström, C. D. Parker, L. J. Mahoney, K. M. Molvar and T. C. McGill, Appl. Phys. Lett. **58**, 2291 (1991).

[27] H. Ito, T. Furuta, F. Nakajima, K. Yoshino and T. Ishibashi, J. Light. Technol. **23**, 4016 (2005).

[28] J. Faist, F. Capasso, D. L. Sivco, C. Sirtori, A. L. Hutchinson and A. Y. Cho, Science. **264** (80), 553 (1994).

[29] N. L. Rangel and J. M. Seminario, J. Phys. Chem. A **112**, 13699 (2008).

[30] O. V Kibis, M. Rosenau da Costa and M. E. Portnoi, Nano Lett. **7**, 3414 (2007).

[31] K. Kawase, Opt. Photonics News **15**, 34 (2004).

[32] T. Kashiwagi, K. Sakamoto, H. Kubo, Y. Shibano, T. Enomoto, T. Kitamura, K. Asanuma, T. Yasui, C. Watanabe, K. Nakade, Y. Saiwai, T. Katsuragawa, M. Tsujimoto, R. Yoshizaki, T. Yamamoto, H. Minami, R. A. Klemm and K. Kadowaki, Appl. Phys. Lett. **107**, 82601 (2015).

Chapter 11

Terahertz Pulse Detection Techniques and Imaging Applications

Sung-Liang Chen and L. Jay Guo

Additional information is available at the end of the chapter

http://dx.doi.org/10.5772/67089

Abstract

Recent years have witnessed successful developments of detection techniques of terahertz (THz) pulse radiation and its imaging applications such as security, medicine and environmental sensing, to name an important few. Progress of detection techniques has been made in many aspects, including detection sensitivity, real-time detection, room-temperature operation, detection bandwidth and dynamic range, spatial (wavefront) and temporal profiles and so on. New detection techniques utilizing cutting-edge materials, sensors, systems and even novel detection mechanisms contribute to advances in terahertz pulse detection. While detection techniques continuously improve, terahertz pulsed imaging (TPI) also finds broad and intriguing applications. For instance, TPI has shown applications in nondestructive evaluation in pharmaceutics, biomedical characterization of tissues, medical diagnosis of cancers, identification of explosive hazards and examination of art and archeology. The chapter highlights recent progress of terahertz pulse detection techniques and imaging applications.

Keywords: Terahertz, Detection, Imaging, pulsed terahertz detection, terahertz pulsed imaging

1. Introduction

Terahertz (THz) electromagnetic spectrum from 0.1 to 10 THz (3 mm–30 μm wavelength) is a scientifically rich frequency band that involves research in physics, chemistry, material science, biology and medicine. In the past five years, there has been a prosperous rise of research activity related to THz pulses, partly due to the advancement of various new technologies. There is a range of applications based on the study of the interaction between matter (solid, liquid, or gaseous) and THz pulses. To achieve further understanding and utilization of THz pulses, it is essential to advance THz pulse detection techniques. Furthermore, there is a need to develop novel and reliable THz pulse detectors that can facilitate diverse THz pulse applications. Another motivation for ameliorating detection techniques is the applications of THz pulsed imaging (TPI). TPI has exceptional potential for applications in security, nondestructive evaluation, biological sciences and medicine.

The objective of this chapter is to review the state-of-the-art technology for THz pulse detection and imaging developed in recent five years. A search in the Web of Knowledge (Thomson Reuters) with "THz pulse" in "Title" has returned 589 articles during 2011–2015, while the number of articles returned is 379 during 2006–2010 (the search was conducted on June 24, 2016). Another search on Google Scholar with the same keyword "THz pulse" generates 4020 and 2540 articles for the five years 2011–2015 and 2006–2010, respectively. The two results show an increase of 55% and 58%, respectively, in the total number of publications between the two five-year periods, which indicates the increasing interest in the area of THz pulses. As stated earlier, the research field of THz pulse detection and TPI is of unique significance. Therefore, the chapter is limited to, what the authors consider, the most interesting recent research findings in the area of THz pulse detection techniques and TPI applications including THz spectroscopy. The chapter is presented in the following arrangement: Section 2 will elaborate on the significance of pulsed THz technology, Section 3 will present recent THz pulse detection techniques, Section 4 will focus on recent TPI applications and Section 5 will provide a summary and future outlook in this field.

2. Significance of THz pulses

The THz pulse radiation is of great interest by several unique features. First, the use of a short THz pulse enables the study of THz fields and the collection of information in time domain, which is the underlying principle of THz time-domain spectroscopy (THz-TDS). A variety of physical phenomena and material characteristics can be studied utilizing short THz pulses [1]. For example, by studying the absorption of THz photons in doped semiconductors, carrier dynamics can be studied [2]. For intrinsic semiconductors, the complex permittivity or THz absorption coefficient and refractive index can be determined [3]. Second, since a single-cycle THz pulse can be intense and short, it is experimentally possible to tap into the regime of extreme nonlinear optics [4] where the usual approximation (e.g., the conditions for complete state transfer) no longer holds. THz pulse radiation with high energies has found many applications such as nonlinear spectroscopy [5], high harmonic generation [6], molecular alignment [7] and charged particle acceleration [8, 9], to name an important few. Third, the spectral bandwidth of pulsed THz waves can be of the order of several hundred percent of the center frequency and therefore, pulsed THz waves show promise in short-distance data transmission at high bit rates.

Conventionally, optical sampling is used to perform time-resolved measurement of THz responses. Free-space electro-optic (EO) sampling is one of the most common sampling techniques. Another common technique for detecting THz pulses is photoconductive antennas [10, 11]. The single-cycle THz pulse can cover a wide spectral range (0.1–50 THz). The two detection schemes are ideal for THz-TDS to investigate the spectral response of materials. On the other hand, commercially available thermal detectors such as bolometers, pyroelectric detectors and Golay-cell detectors are generally used for measuring THz pulse energy. The three devices are maturely developed and can provide stable performance. However, bolometer provides high sensitivity, but cryogenic cooling is necessary, while the other two detectors show low sensitivity at room temperature, restricting detailed measurement. The rapid development of THz pulse detection techniques such as time-domain profile detection and energy measurement offers many advantages and will be discussed in Section 3.

One primary application of THz pulse detection is TPI, which can provide a three-dimensional (3D) map of the object by using the time of flight of THz pulses [12]. TPI can be regarded as an extension of the THz-TDS. TPI can acquire not only valuable spectral information but also 3D images. In image acquisition, a THz pulse is launched to the sample and the reflected echo is measured in amplitude and/or phase. The time-of-flight information of the echo pulse indicates the presence of the boundaries or inner structures along the propagation direction of the THz, which extracts the one-dimensional depth profile. By performing a two-dimensional (2D) scan from pixel to pixel, a 3D image of the target can be visualized. Thus, TPI is possible to provide 3D views into a layered structure. Unlike THz CW imaging, TPI can attain distinctive knowledge of the target, such as the spectral and depth information, by using the acquired amplitude and phase information of THz waves in the time domain. The critical benefit renders TPI valuable for diverse applications, such as detection of breast cancer [13] and inspection of pharmaceutical tablets [14–17], to name but a few.

Note that in the subject of THz pulses, there are issues relating to generation and detection, devices and systems and applications. We focus our discussion on recent development, basically in five years, in THz pulse detection techniques and TPI applications.

3. Terahertz pulse detection techniques

In order to attain the potential offered by pulsed THz technology, the generation and characterization of THz pulses play an important role in THz pulse applications. There are many methods for THz pulse emission, such as photoconductive antenna [18], optical rectification [19] and laser-induced plasma [20–24]. On the other hand, several approaches for THz pulse measurement are well established because of historically long-term needs in detection technologies such as radio astronomy. The newly developed technology for THz pulse detection in recent years will be described in this section. The progress has been made due to the advance of novel detection mechanism and material science.

3.1. Time-domain THz sampling

In recent years major efforts have been made to improve the performance of photoconductive and EO sampling of pulsed THz wave. Photoconductive and EO methods are able to provide large detection bandwidth for THz pulses, which is useful for spectroscopy. As for photoconductive sampling, there are many studies and promising results [25–32]. (i) The sensitivity, THz bandwidth and dynamic range of THz pulse detection are improved by fabricating photoconductive antennas on InGaAs/InAlAs multilayer heterostructures [25, 26]. (ii) A nanowire-based detector can be well suited for near-field THz sensing [27]. (iii) The detection of highly confined THz fields is demonstrated by employing nanostructure of optical materials [28]. (iv) Plasmonic contact electrodes are used to enhance THz detection sensitivity [29]. (v) Novel optical gating technique is used to realize subpicosecond temporal resolution in pulse detection [30]. (vi) Some work is related to adapting photoconductive THz detectors to THz-TDS systems [31, 32]. As for EO sampling, similarly, there are various progress such as improvement of detection efficiency [33, 34] and dynamic range [35], polarization sensing of THz pulses [36] and a new detection scheme based on the amplitude variation of optical pulse [37].

3.2. Energy or power measurement

One important need in THz pulse applications is to measure the energy or power of THz pulses. Recently there have been many investigations on new technologies.

We demonstrated a novel scheme based on photoacoustic conversion of carbon nanotube (CNT) nanocomposite to realize efficient and real-time measurement of THz pulse energy [38]. Conventionally used thermal detectors utilize continuous heat integration to measure the power of pulsed THz radiation. The power can be converted to energy with the pulse repetition frequency (PRF). Due to the mechanism of heat integration, most thermal detectors have slow response times, which limit the characterization of energy of each THz pulse at high values of PRF. Unlike conventional thermal detectors, we utilize photoacoustic effect to realize real-time detection of THz pulse energy. Specifically, the transient and localized heating in an absorber by the absorption of THz pulse energy produces ultrasound, which is subsequently detected by a sensitive acoustic sensor. Moreover, our method responds only to the pulse excitation while rejecting other continuous radiations. In other words, in THz pulse detection, the ultimate sensitivity will not be restricted by the background continuous radiation, thus showing the potential for the efficient detection of THz pulse energy. In order to achieve efficient detection, it is essential to optimize the efficiency for photoacoustic conversion and subsequent acoustic sensing. We choose a CNT-polydimethylsiloxane (PDMS) nanocomposite to achieve efficient THz-to-ultrasound conversion. This is because CNTs can efficiently absorb THz radiation and then convert it into heat via THz absorption capability and low specific heat of CNTs, while PDMS has a high thermal coefficient of volume expansion. On the other hand, we employ a photonic device, a polymer microring resonator, as a highly sensitive acoustic sensor. The resonator has a high optical quality (Q) factor of 1.3×10^5. The acoustic pressure impinges on the microring resonator, thus changing its effective refractive index and the resonance wavelength. To enable sensitive conversion of the ultrasound pulse to the modulated optical intensity, we identify a wavelength where the local slope in the transmission spectrum is high and then probe the microring using this wavelength. The modulated optical intensity is further recorded by a low-noise high-speed photodetector. A high Q factor of the resonator correlates with a high slope in the transmission spectrum and therefore high sensitivity of ultrasound detection. The setup of the PA detection of THz pulses is shown in **Figure 1**. A two-color air ionization scheme is used to generate broadband THz pulses. The noise-equivalent detectable energy of the technique is calibrated as ~220 pJ. We expect that three orders-of-magnitude improvement in sensitivity are possible by configuring the nanocomposite in the form of an acoustic lens as well as employing a microring resonator with a higher Q factor. The fast response time less than 0.1 μs is achieved, which is several orders faster than that of a commercial pyroelectric detector (~0.1 s). In addition, the novel method possesses other advantages such as room-temperature operation, compact detector size in mm scale and wide spectral response for THz spectroscopy.

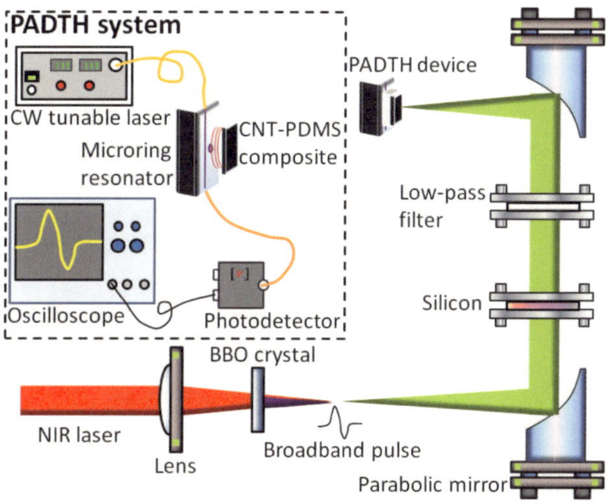

Figure 1. Experimental setup for the photoacoustic detection of THz (PADTH) pulse radiation. Using a β-barium borate (BBO) crystal, we can obtain a second-harmonic laser field from its fundamental one. Then, mixing the fundamental and the second-harmonic laser fields at the focus produces a broadband THz pulse radiation. We use a parabolic mirror to collect and also collimate the THz radiation from the focus, and then we use a silicon wafer and a low-pass filter in order to choose THz radiation with frequencies less than 6 THz. Finally, the THz radiation is refocused using another parabolic mirror for detection test of the PADTH device, consisting of a CNT-PDMS composite for high-efficiency THz-to-ultrasound conversion, ultrasound gel for sound coupling, and an optical microring resonator for sensitive ultrasound detection. In the PADTH system, a CW tunable laser is used to probe the microring resonator at a high-slope region in its transmission spectrum, and a high-speed photodetector is used to record the temporal optical intensity, which duplicates the temporal acoustic pressure. Reprinted with permission from Ref. [38].

Sensitive room-temperature detection of THz radiation is highly difficult. A detection mechanism based on the hot-electron photothermoelectric effect in graphene can be a promising approach [39]. First, photo-excited carriers thermalize rapidly due to strong electron-electron interactions but lose energy to the lattice slowly. Next, the electron diffusion due to the electron temperature gradient, as well as asymmetry or dissimilar contact metals, produces a net current by the thermoelectric effect. **Figure 2** shows optical and atomic-force micrographs of the graphene THz detector made by microfabrication technologies. Each of the two metal electrodes consists of partially overlapping Cr and Au regions and contacts the monolayer graphene flake. Electron temperature gradient ΔT is produced across the device, then resulting in potential gradient ΔV. The photoresponse is obtained by integrating ΔV over the length of the device. The graphene thermoelectric THz detector shows sensitivity exceeding 10 V/W (700 V/W) at room temperature and noise-equivalent power less than 1100 pW/√Hz (20 pW/√Hz), referenced to the incident (absorbed) power. The sensitivity is comparable with that of the best room-temperature THz detectors. Further improvements on orders-of-magnitude sensitivity is possible indicated by studying a model of the response including contact asymmetries. A fast intrinsic response time of 10.5 ps due to electron-phonon relaxation is estimated for THz time-domain measurements.

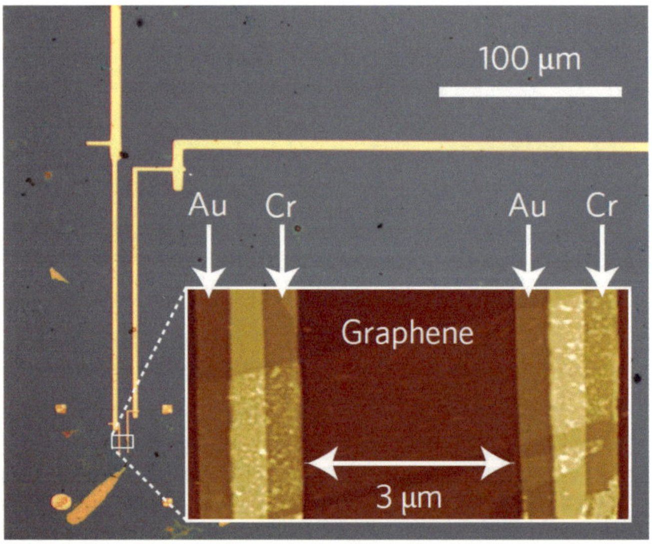

Figure 2. Graphene photothermoelectric detector device. The optical micrograph shows electrical contacts; the atomic-force micrograph in inset shows bimetallic contacts connected to an exfoliated graphene layer. Reprinted with permission from Ref. [39].

There are more examples besides the above two highlights. (i) Sensitive THz pulse energy detection is demonstrated by wavelength conversion from THz waves to near-IR light using $LiNbO_3$ crystals [40]. (ii) Sensitive THz detection is possible through THz light amplification in optically pumped graphene [41]. (iii) A THz line array detector with 20 elements is demonstrated with an average noise equivalent power of 106.6 pW/vHz and the -3-dB bandwidth of 0.18 THz [42]. (iv) A fast response time of 45 ps allowing the measurement of ultrashort

THz pulses is achieved [43]. (v) Electroluminescence effect is utilized to develop a low-cost, probe-beam-free THz detection system [44].

3.3. Other sensing works

There are some intriguing research works for THz pulse sensing. We incorporate some examples here. (i) THz detection capable of acquiring the entire spatiotemporal profile of THz radiation in a single laser shot is demonstrated, which is based on space-to-time grading by using a converging probe intensity front [45]. The approach does not require any specially designed optics or precise alignment. The scheme has several merits such as a simple setup, high temporal resolution and fast acquisition. Compared to the conventional EO sampling, the technique reduces the time taken for data acquisition and thus may offer a decent option for real-time detection of THz radiation. (ii) The wavefront characterization of THz pulses is presented [46], which is realized by using a Hartmann sensor associated with a 2D EO imaging system composed of a ZnTe crystal and a CMOS camera. The wavefront sensing is crucial to applications such as optimization of far-field intensity distribution of time-domain imaging or enhancement of the peak power of intense THz sources. (iii) Measurement of spectra of THz pulses is realized by using a system consisting of channels for measuring amplitudes of pulses and an algorithm based on the iteration method or the amplitude-frequency method [47]. The spectrum measurement is essential to THz-TDS applications.

4. Terahertz pulsed imaging applications

THz imaging systems have been improved in recent years. The systems are intrinsically safe, nondestructive and noninvasive and can answer many of the questions left unresolved by complementary techniques, such as optical, Raman and infrared imaging [48]. Two types of imaging, TPI and THz CW imaging, are used with their respective strengths and weaknesses [49]. TPI renders data richer in information. Specifically, depth information can be retrieved in TPI, while THz CW imaging acquires intensity image data without depth information. In this section, recent progress on the development of TPI systems will be introduced. Besides, we will give an overview over a broad range of TPI applications such as medical imaging and diagnosis, evaluation of tablets in pharmaceutics, painting investigation for art and archeology, material characterization and detection of concealed weapons.

4.1. THz pulsed imaging system

TPI system has shown progress in different aspects such as development of a compact system design suited to specific applications and enhancement of system performance including resolution, imaging time and dynamic range. For example, the performance of THz computed tomography (CT) is improved [50, 51]. THz CT can acquire and render 3D images in the THz frequency range, as in the optical, infrared, or X-ray regions of the electromagnetic spectrum [50]. A THz CT system using an injection-seeded parametric source for frequency-tunable, Fourier transform-limited and high-power THz emission and a heterodyne detector for sensitive THz detection is demonstrated, as shown in **Figure 3** [51]. This system covers a frequency range of 0.95–2.7 THz and achieves a dynamic range greater than 90 dB, enabling high-resolution 3D THz CT images of samples with strong THz absorption. For illustration, 3D imaging of a pencil and a plastic product is obtained. The hidden lead as the internal structure of the pencil and the internal defect of the plastic product can be successfully revealed, demonstrating the system's capability and potential in nondestructive testing and evaluation of a variety of industrial products, such as semiconductors, pharmaceuticals, plastics and ceramics.

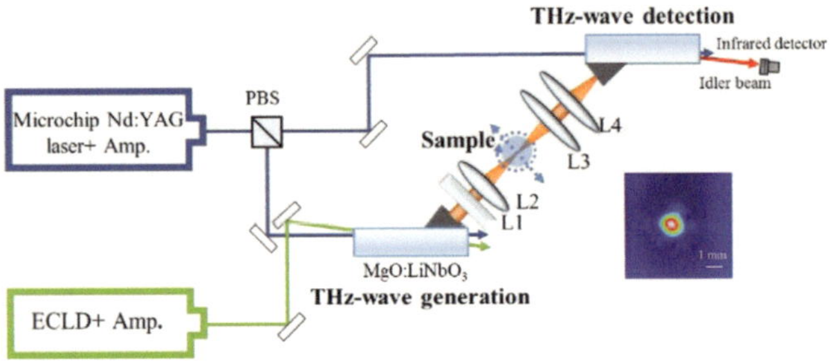

Figure 3. A 3D THz wave CT configuration. A THz image shown in inset is taken at the focal point using a THz imager (IR/VT0831, NEC Corp.). ECLD, external cavity laser diode. Reprinted with permission from Ref. [51].

Besides the above highlight, there are more examples. We discuss some here. (i) A THz InGaAs Schottky barrier diode array detector with 20 elements is built for real-time, compact and portable scanners in a TPI system [42]. (ii) Another real-time TPI system is built using a palm-size THz camera that contains a microbolometer array and a compact quantum cascade laser (QCL) [52]. (iii) A real-time transmission-type THz microscope is proposed by employing a THz penetration-enhancing agent, glycerol, to improve the THz penetration depth in tissues because the glycerol has low absorption of THz waves compared with water [53].

The use of novel THz sources, waveguides and detectors is valuable for THz imaging systems. The QCL in THz frequency is a compact source of THz radiation with high power and high spectral purity. As such, the source is useful for many TPI applications such as long-range imaging and materials analysis. The QCL-based THz imaging approaches and their key advancements are reviewed [54]. Another example is that by utilizing a split tapered waveguide with a subwavelength aperture, near-field THz imaging is accomplished [55].

4.2. Biomedical applications

Interest in biomedical THz research is growing rapidly [56, 57]. THz radiation has very low photon energy and thus does not cause any ionization hazard for biological tissues. Unique absorption spectra over the THz band have been found in different biological tissues. The feature makes THz attractive for biomedical applications because THz can provide complementary information to existing techniques. TPI or THz-TDS is one essential approach in biomedical THz research. In the following, we will describe TPI in various biomedical imaging applications.

Studies of cancer by THz imaging or spectroscopy have gained increasing interests in recent years [58]. First, the presence of cancer often induces increased blood supply to the affected tissues and a local increase in tissue water content, which can be utilized as contrast for THz imaging of cancer. For instance, a sample of dehydrated human colon tissues embedded in paraffin is studied [59]. The results demonstrate the potential of THz imaging to distinguish adenocarcinoma-affected colon areas and the ability to image dehydrated tissues. Second, the structural changes in affected tissues can also be an important sign for diagnosis and can be observed by THz imaging. As another example, a reflection THz imaging system is used to identify tumors in freshly excised whole brain tissue from normal brain tissue based on structural observation [60]. Because the THz reflection intensity is higher in brain tumors than in normal tissue, the difference in the THz reflection intensity between the normal and tumor brain tissues can be adopted for the diagnosis of cancer. **Figure 4** shows the visual, THz and MR images of fresh whole brain tissues with and without tumors. **Figure 4(a)–(c)** shows the brain tissue with tumors and **Figure 4(d)** shows the normal brain tissue. The tumor boundaries in the THz images agree well with those visible images, indicating that the THz imaging technique could be useful for diagnosing brain tumors. Potentially, THz imaging could be employed as a complementary label-free technique allowing surgeons to determine tumor margins in real time. In addition, THz image contrast differences are also observed in the normal brain tissue image in **Figure 4(d)** and these correspond to the gray and white matter areas, showing the ability of THz imaging to study brain structure. Furthermore, this study shows that the THz signals correspond to the cell density when water was removed. In other words, the THz contrast between normal and cancerous brain tissues can be distinguished not only by differences in the water content but also by differences in the cell density.

One other promising application of TPI is the study of breast cancer, which has been explored extensively. (i) Breast tumor phantoms that match the refractive indices and absorption coefficients in the THz band are developed to facilitate the study of breast cancer THz imaging [61]. Phantom properties are verified through THz-TDS. (ii) THz-TDS and TPI are used for characterization of paraffin-embedded breast cancer tissue [62]. (iii) A pulsed THz system is used to enact a quick and reasonable estimation of the breast cancer margin thickness of embedded breast cancer tissue [63]. (iv) A similar work investigates TPI for the application of surgical margin assessment of breast cancer in 3D, where the depth information is retrieved using time-of-flight analysis [64]. (v) A linear sampling algorithm can be applied to TPI data for identifying breast cancer tumor margins [65]. (vi) THz measurement of normal and breast cancer tissue in the range of 0.1–4 THz is presented [66], showing the ability of THz technology for characterization of cancerous and normal breast tissue. (vii) A clinical study that fifty-one samples from patients in Cambridge and Guy's Hospital in London is conducted [67], showing the ability of THz technology to classify tumor and normal breast tissue with good accuracy.

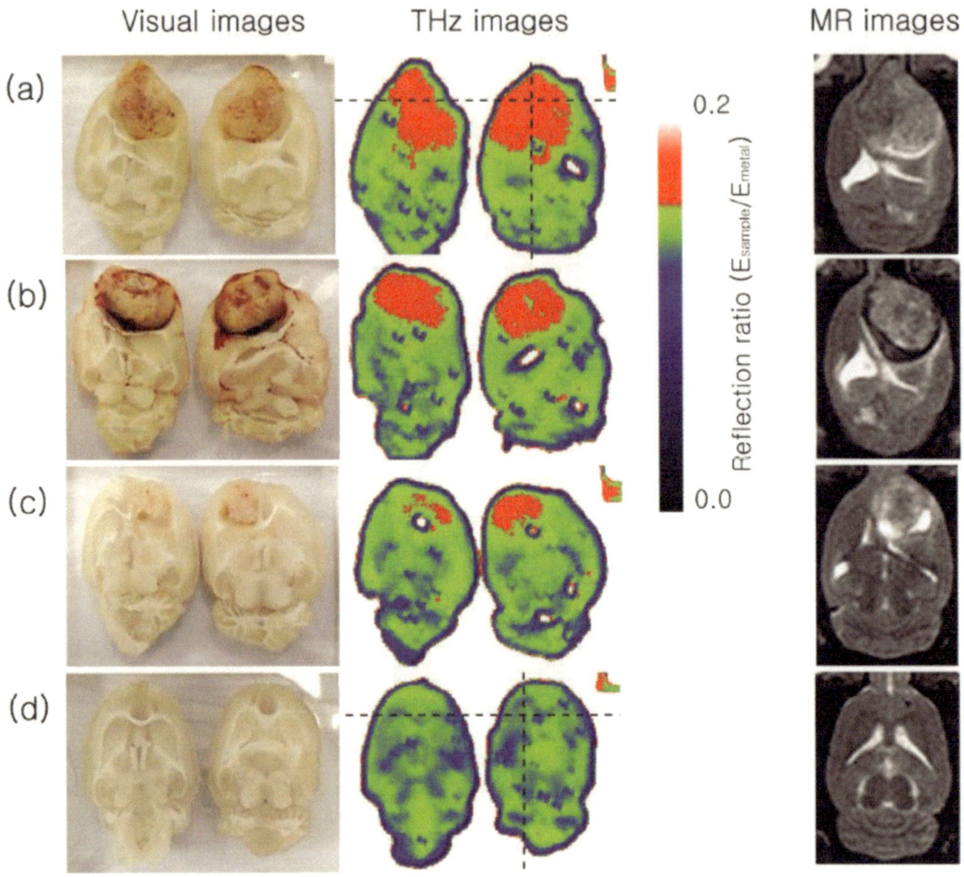

Figure 4. Whole brain images with (a–c) and without (d) tumors by visual, THz, and MR imaging. For THz images, the size is 4 × 3 cm^2 and scanning resolution is 250 μm. Reprinted with permission from Ref. [60].

4.3. Pharmaceutical applications

Pharmaceutical applications are one of the emerging opportunities offered by TPI and THz pulsed spectroscopy [68]. Solid dosage forms are the pharmaceutical drug delivery systems of choice for oral drug delivery [69]. These solid dosage forms are often coated to modify the properties of the active pharmaceutical ingredients, in order to help release kinetics [69, 70]. The critical coating attributes such as coating thickness, uniformity and density have chief influence on the tablet performance; advanced quality control techniques are required. TPI is an emerging nondestructive method to quantitatively characterize coating quality. Compared with established imaging techniques, e.g., near-infrared and Raman spectroscopy, TPI has the advantage to enable structural features of coated solid dosage forms at depth by the ability of THz radiation to penetrate many pharmaceutical materials. A typical THz time-domain waveform used for characterization of a single-layer coated tablet is illustrated in **Figure 5** [69].

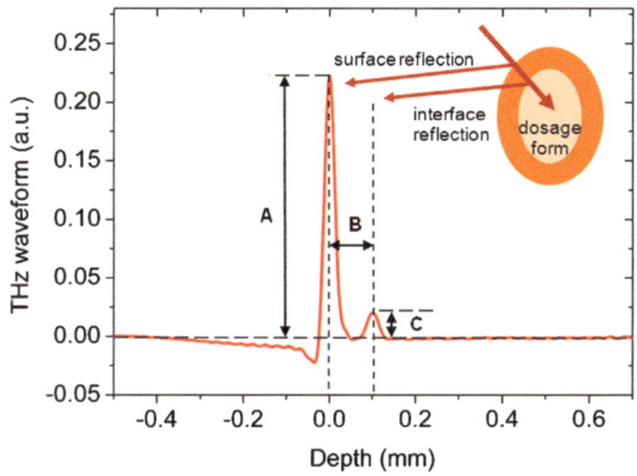

Figure 5. Characterization of coating attributes of a tablet by a THz time-domain waveform. At a depth of 0 mm in the THz waveform, the positive peak with a magnitude of A represents the difference in the refractive indices of the air and the outer surface of the coating layer (Interface 1). At a depth of ~0.1 mm, the next positive peak with a magnitude of C is caused by the boundary between the inner surface of the coating layer and the tablet core (Interface 2). One can retrieve parameters from the THz waveform: (1) the changes of the refractive index due to Interface 1, and the signal scattering can be obtained from the magnitude "A"; (2) the coating thickness can be calculated from the time delay, "B," between the reflection of Interface 1 and that of Interface 2; (3) the changes of the refractive index due to Interface 2 can obtained from the magnitude "C." Reprinted with permission from Ref. [69].

A number of studies have been done recently. (i) TPI is used for nondestructive evaluation of film-coated tablets by deriving parameters such as film thickness, film surface reflectance and interface density differences between the film layer and core tablets [71]. (ii) Spectral domain optical coherence tomography (OCT) and TPI for quantifying film coating thickness of tablets are studied [72]. The finding shows that OCT is suitable for characterizing pharmaceutical dosage forms with thin film coatings, whereas TPI is suited for thick coatings. (iii) To ensure robust measurements, the evaluation of film coating thickness using TPI should take some factors into account, such as signal processing of the raw data and signal distortions that can occur at tablet edges or areas with defects [73]. (iv) Enteric coatings in tablets are widely used to reduce gastrointestinal side effects and to control the release properties of oral medications. TPI is used to identify structural defects within enteric coating tablets with poor acid resistance [74]. (v) TPI is used for evaluation of the intra tablet and inter tablet coating uniformity and identification of critical process parameters in a coating process [75, 76]. (vi) TPI can also be used in evaluating the effect of coating equipment on tablet film quality [77]. (vii) TPI is used to quantify the hardness and surface density distribution of tablets [78]. (viii) TPI is a feasible and rapid tool to characterize ribbon density distributions [79], which are important parameters in dry granulation process in pharmaceutical industries. (ix) Tablet dissolution is crucial in medication and is strongly affected by swelling and solvent penetration into its matrix. A reflection mode TPI is used to measure swelling and solvent ingress in pharmaceutical compacts [80]. (x) Layer separation is a crucial defect in many bilayer tablets. TPI is used to provide a precise estimate of the layer separation risk [81].

4.4. Art and archeology

Examination of art and archeology is important for cultural heritage scientists to understand artistic materials and to devise better conservation procedures [82]. THz presents a number of valuable features specifically for the investigation of art and archeology such as no radiation risk with deep penetration (**Figure 6**), low power and noncontact mode. Recent progress shows that THz technology for art investigation is an efficient, convenient and affordable approach. We introduce several examples. (i) TPI is used to image apsidal wall painting in 3D to provide subsurface features at depths up to 1 cm from the surface [83]. Characterization of subsurface features is useful in conservation of art history as well as in building archeology. (ii) TPI is used as a technique to image obscured mural paintings [84]. Image processing can be used to solve the issue due to an uneven surface, enabling the visualization of the obscured painting. (iii) THz reflective tomography is used to identify the preset defects in a plaster [85]. (iv) A portable THz-TDS system is used to image panel paintings from a lab and a museum, offering useful information on the internal structure of the paintings and on their conditions [86]. (v) TPI is also used to image an oil canvas painting by Pablo Picasso and the multilayer structure is clearly revealed [87]. (vi) An artwork attributed to the Spanish artist Goya painted in 1771 is imaged and analyzed by THz time-domain system [88]. The study indicates that THz images present features that cannot be seen with optical inspection.

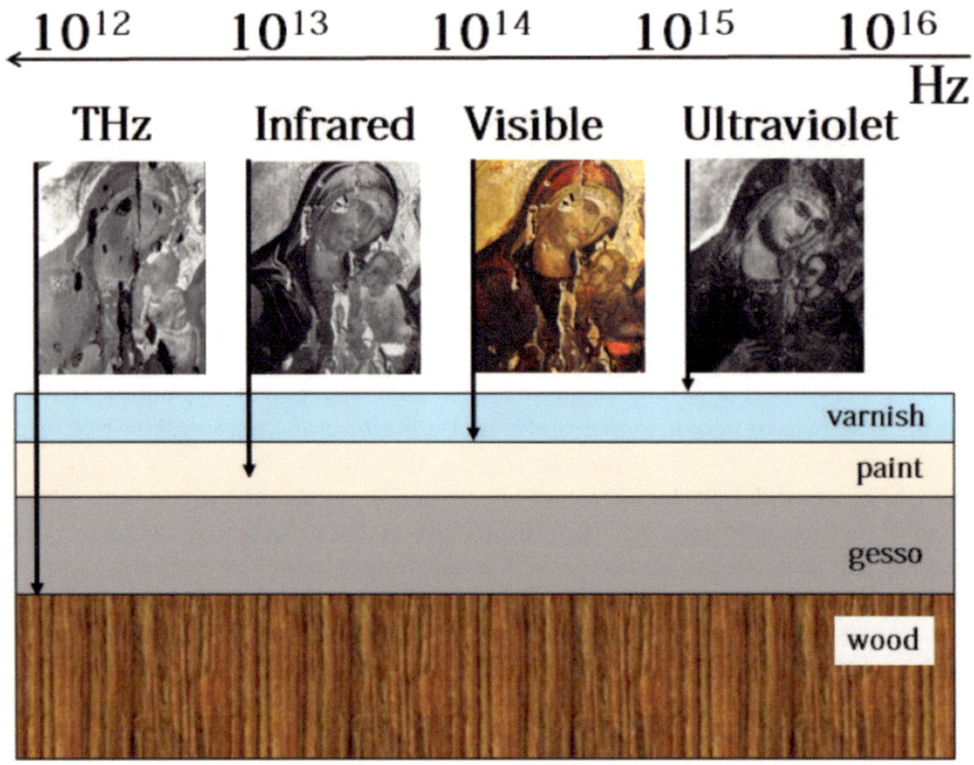

Figure 6. Paintings studied by THz radiation. For painting inspection, a panel painting can be penetrated deeper using THz and X-ray radiography than using other traditional approaches such as infrared radiation. Although THz and X-ray radiations can provide deep penetration into paintings, only THz promises safe and nonionizing 3D imaging of paintings. Reprinted with permission from Ref. [82].

4.5. Other applications

There are still a variety of TPI applications in the fields of architecture, chemistry, material science, environmental protection and homeland security. (i) TPI and THz-TDS are employed in characterization of construction and building materials [89]. Different types of thermal building insulation materials are analyzed [90]. (ii) TPI and THz-TDS are used to identify wheat grains at different stages of germination [91]. Specifically, the inner chemical structure during germination can be revealed from the THz spectra. (iii) TPI and THz-TDS are also explored to identify features such as coating, pores and cracks in polymer materials [92]. Another study employs TPI to study glass fiber-reinforced composite laminates in polyetherimide resin [93]. (iv) Reflective pulsed THz tomography can be a tool to monitor oil pollution. A cup of water covered by a layer of sesame oil with different densities is devised to simulate oil spills [94]. The results show that different densities can be determined by THz images. (v) The spectral fingerprint of high explosive material is investigated using THz-TDS [95]. Besides, the explosive material concealed in an opaque envelope can be identified using TPI.

5. Conclusions and future prospects

In summary, a number of detection techniques of THz pulse radiation and applications based on TPI and THz-TDS are presented in this chapter. We first elaborate the significance of THz pulses. Pulsed THz technologies bring about useful information from the interaction between THz radiation and matters. The applications based on pulsed THz technologies in part heavily rely on the technological advancement in THz pulse detection. We review the recent rapid development of THz pulse detection techniques in various aspects such as using novel materials, adopting innovative designs of detectors and even employing new detection mechanisms. These will open up new fields of applications or carry out particular tasks that cannot be attained previously. Of particular interest are the applications using TPI as well as THz-TDS. Development and test of TPI systems have shown steady progress in recent years. In addition, a wide range of TPI or THz-TDS applications including the fields of medical imaging and diagnosis, pharmaceutics, art and archeology, material science, architecture and so on has been intensively investigated by researchers and scientists. The exploration of THz pulsed detection technology and imaging applications will eventually lead to commercial products and systems for specific purposes, which is expected to have a profound impact on our lives. The substantial improvements over the past few years lay foundation for extending THz technology to new and potentially groundbreaking realms.

Acknowledgements

This work was supported by the National Natural Science Foundation of China (no. 61405112), the National High Technology Research and Development Program of China (863 Program) (no. 2015AA020944) and the Shanghai Pujiang Program (no. 14PJ1404400).

Author details

Sung-Liang Chen[1]* and L. Jay Guo[2]

*Address all correspondence to: sungliang.chen@sjtu.edu.cn

1 University of Michigan-Shanghai Jiao Tong University Joint Institute, Shanghai Jiao Tong University, Shanghai, China

2 Department of Electrical Engineering and Computer Science, University of Michigan, Ann Arbor, Michigan, United States

References

[1] Jepsen PU, Cooke DG, Koch M. Terahertz spectroscopy and imaging—modern techniques and applications. Laser & Photonics Reviews 2011;**5**:124–166.

[2] Zhang W, Azad Abul K, Grischkowsky D. Terahertz studies of carrier dynamics and dielectric response of n-type, free standing epitaxial GaN. Applied Physics Letters 2003;**82**:2841.

[3] Grischkowsky D, Keiding S, Exter Mv, Fattinger C. Far-infrared time-domain spectroscopy with terahertz beams of dielectrics and semiconductors. Journal of the Optical Society of America B 1990;**7**:2006–2015.

[4] Wegener M. Extreme Nonlinear Optics: An Introduction. Springer-Verlag Berlin Heidelberg; 2005. ISBN 978-3642060908.

[5] Hebling J, Yeh KL, Hoffmann MC, Nelson KA. High-power THz generation, THz nonlinear optics and THz nonlinear spectroscopy. IEEE Journal of Selected Topics in Quantum Electronics 2008;**14**:345–353.

[6] Balogh E, Kovacs K, Dombi P, Fulop JA, Farkas G, Hebling J, Tosa V, Varju K. Single attosecond pulse from THz-assisted high-order harmonic generation. Physical Review A 2011;**84**:023806.

[7] Fleischer S, Zhou Y, Field RW, Nelson KA. Molecular orientation and alignment by intense single-cycle THz pulses. Physical Review Letters 2011;**107**:163603.

[8] Wong LJ, Fallahi A, Kärtner FX. Compact electron acceleration and bunch compression in THz waveguides. Optics Express 2013;**21**:9792.

[9] Palfalvi L, Fülop JA, Toth G, Hebling J. Evanescent-wave proton postaccelerator driven by intense THz pulse. Physical Review ST-Accelerators and Beams 2014;**17**:031301.

[10] Ezdi K, Heinen B, Jördens C, Vieweg N, Krumbholz N, Wilk R, Mikulics M, Koch M. A hybrid time-domain model for pulsed terahertz dipole antennas. Journal of European Optical. Society-Rapid Publications 2009;**4**:09001.

[11] Dietz RJB, Gerhard M, Stanze D, Koch M, Sartorius B, Schell M. THz generation at 1.55 μm excitation: six-fold increase in THz conversion efficiency by separated photoconductive and trapping regions. Optics Express 2011;**19**:25911–7.

[12] Takayanagi J, Jinno H, Ichino S, Suizu K, Yamashita M, Ouchi T, Kasai S, Ohtake H, Uchida H, Nishizawa N, Kawase K. High-resolution time-of-flight terahertz tomography using a femtosecond fiber laser. Optics Express 2009;**17**:7549–7555.

[13] Fitzgerald AJ, Wallace VP, Jimenez-Linan M, Bobrow L, Pye RJ, Purushotham AD, Arnone DD. Terahertz pulsed imaging of human breast tumors. Radiology 2006;**239**:533–540.

[14] Shen YC, Taday PF. Development and application of terahertz pulsed imaging for nondestructive inspection of pharmaceutical tablet. IEEE Journal of Selected Topics in Quantum Electronics 2008;**14**:407–415.

[15] Fitzgerald AJ, Cole BE, Taday PF, Nondestructive analysis of tablet coating thicknesses using terahertz pulsed imaging. Journal of Pharmaceutical Sciences 2005;**94**:177–183.

[16] Maurer L, Leuenberger H. Terahertz pulsed imaging and near infrared imaging to monitor the coating process of pharmaceutical tablets. International Journal of Pharmaceutics 2009;**370**:8–16.

[17] Ho L, Müller R, Römer M, Gordon KC, Heinämäki J, Kleinebudde P, Pepper M, Rades T, Shen YC, Strachan CJ, Taday PF, Zeitler JA. Analysis of sustained-release tablet film coats using terahertz pulsed imaging. Journal of Controlled Release 2007;**119**:253–261.

[18] Tani M, Herrmann M, Sakai K. Generation and detection of terahertz pulsed radiation with photoconductive antennas and its application to imaging. Measurement Science and Technology. 2002;**13**:1739–1745.

[19] Auston DH, Cheung KP, Valdmanis JA, Kleinman DA. Cherenkov radiation from femtosecond optical pulses in electro-optic media. Physical Review Letters 1984;**53**:1555–8.

[20] Kim KY, Taylor AJ, Glownia JH, Rodriguez G. Coherent control of terahertz supercontinuum generation in ultrafast laser–gas interactions. Nature Photon 2008;**2**:605–609.

[21] Thomson MD, Blank V, Roskos HG, Terahertz white-light pulses from an air plasma photo-induced by incommensurate two-color optical fields. Optics Express 2010;**18**:23173–23182.

[22] Petersen PB, Tokmakoff A. Source for ultrafast continuum infrared and terahertz radiation. Optics Letters 2010;**35**:1962–1964.

[23] Fuji T, Suzuki T. Generation of sub-two-cycle mid-infrared pulses by four-wave mixing through filamentation in air. Optics Letters 2007;**32**:3330–3332.

[24] Dai J, Liu J, Zhang XC. Terahertz wave air photonics: terahertz wave generation and detection with laser-induced gas plasma. IEEE Journal of Selected Topics in Quantum Electronics 2011;**17**:183–190.

[25] Kostakis I, Saeedkia D, Missous M. Terahertz generation and detection using low temperature grown InGaAs-InAlAs photoconductive antennas at 1.55 μm pulse excitation. IEEE Transactions on Terahertz Science and Technology 2012;**2**:617–622.

[26] Dietz RJB, Globisch B, Roehle H, Stanze D, Göbel T, Schell M. Influence and adjustment of carrier lifetimes in InGaAs/InAlAs photoconductive pulsed terahertz detectors: 6 THz bandwidth and 90 dB dynamic range. Optics Express 2014;**22**:19411–19422.

[27] Peng K, Parkinson P, Fu L, Gao Q, Jiang N, Guo YN, Wang F, Joyce HJ, Boland JL, Tan HH, Jagadish C, Johnston MB. Single nanowire photoconductive terahertz detectors. Nano Letters 2015;**15**:206–210.

[28] Mitrofanov O, Brener I, Luk TS, Reno JL. Photoconductive terahertz near-field detector with a hybrid nanoantenna array cavity. ACS Photonics 2015;**2**:1763–1768.

[29] Berry CW, Wang N, Hashemi MR, Unlu M, Jarrahi M. Significant performance enhancement in photoconductive terahertz optoelectronics by incorporating plasmonic contact electrodes. Nature Communications 2013;**4**:1622.

[30] Muraviev A, Gutin A, Rupper G, Rudin S, Shen X, Yamaguchi M, Aizin G, Shur M. New optical gating technique for detection of electric field waveforms with subpicosecond resolution. Optics Express 2016;**24**:12730–12739.

[31] Vieweg N, Rettich F, Deninger A, Roehle H, Dietz R, Göbel T, Schell M. Terahertz-time domain spectrometer with 90 dB peak dynamic range. Journal of Infrared Millimeter and Terahertz Waves 2014;**35**:823–832.

[32] Urbanowicz A, Pačebutas V, Geižutis A, Stanionyte S, Krotkus A. Terahertz time-domain-spectroscopy system based on 1.55 μm fiber laser and photoconductive antennas from dilute bismides. AIP Advances 2016;**6**:025218.

[33] Tani M, Kinoshita T, Nagase T, Horita K, Que CT, Estacio E, Yamamoto K, Bakunov MI. Non-ellipsometric detection of terahertz radiation using heterodyne EO sampling in the Cherenkov velocity matching scheme. Optics Express 2013;21:9277–9288.

[34] Tani M, Horita K, Kinoshita T, Que CT, Estacio E, Yamamoto K, Bakunov MI. Efficient electro-optic sampling detection of terahertz radiation via Cherenkov phase matching. Optics Express 2011;**19**:19901–19906.

[35] Ibrahim A, Férachou D, Sharma G, Singh K, Kirouac-Turmel M, Ozaki T. Ultra-high dynamic range electrooptic sampling for detecting millimeter and sub-millimeter radiation. Scientific Reports 2016;**6**:23107.

[36] Oguchi K, Iwasaki H, Okano M, Watanabe S. Polarization-sensitive electro-optic detection of terahertz wave using three different types of crystal symmetry: toward broadband polarization spectroscopy. Applied Physics Letters 2016;**108**:011105.

[37] Kovalev SP, Kitaeva GK. Two alternative approaches to electro-optical detection of terahertz pulses. JETP Letters 2011;**94**:91–96.

[38] Chen SL, Chang YC, Zhang C, Ok JG, Ling T, Mihnev MT, Norris TB, Guo LJ. Efficient real-time detection of terahertz pulse radiation based on photoacoustic conversion by carbon nanotube nanocomposite. Nature Photonics 2014;**8**:537–542.

[39] Cai X, Sushkov AB, Suess RJ, Jadidi MM, Jenkins GS, Nyakiti LO, Myers-Ward RL, Li S, Yan J, Gaskill DK, Murphy TE, Drew HD, Fuhrer MS. Sensitive room-temperature terahertz detection via the photothermoelectric effect in graphene. Nature Nanotechnology 2014;**9**:814–819.

[40] Minamide H, Hayashi S, Nawata K, Taira T, Shikata J, Kawase K. Kilowatt-peak terahertz-wave generation and sub-femtojoule terahertz-wave pulse detection based on nonlinear optical wavelength-conversion at room temperature. Journal of Infrared Millimeter and Terahertz Waves 2014;**35**:25–37.

[41] Otsuji T, Watanabe T, Tombet SAB, Satou A, Knap WM, Popov VV, Ryzhii M, Ryzhii V. Emission and detection of terahertz radiation using two-dimensional electrons in III–V semiconductors and graphene. IEEE Transactions on Terahertz Science and Technology 2013;**3**:63–71.

[42] Han SP, Ko H, Park JW, Kim N, Yoon YJ, Shin JH, Kim DY, Lee DH, Park KH. InGaAs Schottky barrier diode array detector for a real-time compact terahertz line scanner. Optics Express 2013;**21**:25874–25882.

[43] Probst P, Scheuring A, Hofherr M, Rall D, Wünsch S, Il'in K, Siegel M, Semenov A, Pohl A, Hübers HW, Judin V, Müller AS, Hoehl A, Müller R, Ulm G. YBa 2 Cu 3 O 7−δ quasioptical detectors for fast time-domain analysis of terahertz synchrotron radiation. Citation: Applied Physics Letters 2011;**98**:043504.

[44] Shin J, Jin Z, Nosaka Y, Nakazawa T, Kodama R. Probe beam-free detection of terahertz wave by electroluminescence induced by intense THz pulse. Journal of Physics: Conference Series 2016;**688**:012108.

[45] http://arxiv.org/abs/1602.05739.

[46] Abraham E, Cahyadi H, Brossard M, Degert J, Freysz E, Yasui T. Development of a wavefront sensor for terahertz pulses. Optics Express 2016;**24**:5203–5211.

[47] Glyavin MY, Goykhman MB, Gromov AV, Palitsin AV, Panin AN, Rodin YV, Fil'chenkov SE. A waveguide high-pass filter system for measuring the spectrum of pulsed terahertz sources. Infrared Physics & Technology 2016;**76**:11–20.

[48] Pawar AY, Sonawane DD, Erande KB, Derle DV. Terahertz technology and its applications. Drug Invention Today 2013;**5**:157–163.

[49] Yin X, Ng BWH, Abbott D. Terahertz Imaging for Biomedical Applications. Springer-Verlag New York; 2012. ISBN: 978-1-4614-1820-7.

[50] Guillet JP, Recur B, Frederique L, Bousquet B, Canioni L, Manek-Hönninger I, Desbarats P, Mounaix P. Review of terahertz tomography techniques. Journal of Infrared Millimeter and Terahertz Waves 2014;**35**:382–411.

[51] Tripathi SR, Sugiyama Y, Murate K, Imayama K, Kawase K. Terahertz wave three-dimensional computed tomography based on injection-seeded terahertz wave parametric emitter and detector. Optics Express 2016;**24**:6433–6440.

[52] Oda N, Lee AWM, Ishi T, Hosako I, Hu Q. Proposal for real-time terahertz imaging system with palm-size terahertz camera and compact quantum cascade laser. Proceedings of SPIE 8363, Terahertz Physics, Devices, and Systems VI: Advanced Applications in Industry and Defense; 2012. 83630A.

[53] Oh SJ, Kim SH, Jeong K, Park Y, Huh YM, Son JH, Suh JS, Measurement depth enhancement in terahertz imaging of biological tissues. Optics Express 2013;**21**:21299–21305.

[54] Dean P, Valavanis A, Keeley J, Bertling K, Lim YL, Alhathlool R, Burnett AD, Li LH, Khanna SP, Indjin D, Taimre T, Raki'c AD, Linfield EH, Davies AG. Terahertz imaging using quantum cascade lasers—a review of systems and applications. Journal of Physics D: Applied Physics 2014;**47**:374008.

[55] Liu S, Mitrofanov O, Nahata A. Near-field terahertz imaging using sub-wavelength apertures without cutoff. Optics Express 2016;**24**:2728–2736.

[56] Ouchi T, Kajiki K, Koizumi T, Itsuji T, Koyama Y, Sekiguchi R, Kubota O, Kawase K. Terahertz imaging system for medical applications and elated high efficiency terahertz devices. Journal of Infrared Millimeter and Terahertz Waves 2014;**35**:118–130.

[57] Fan S, He Y, Ung BS, Pickwell-MacPherson E. The growth of biomedical terahertz research. Journal of Physics D: Applied Physics 2014;**47**: 374009.

[58] Yu C, Fan S, Sun Y, Pickwell-MacPherson E. The potential of terahertz imaging for cancer diagnosis: a review of investigations to date. Quantitative Imaging in Medicine and Surgery 2012;**2**:33–45.

[59] Faustino F, Kasalynas I, Venckevicius R, Seliuta D, Valusis G, Urbanowicz A, Molis G, Carneiro F, Silva CDC, Granja PL. Terahertz absorption and reflection imaging of carcinoma-affected colon tissues embedded in paraffin. Journal of Molecular Structure 2016;**1107**: 214–219.

[60] Oh SJ, Kim SH, Ji YB, Jeong K, Park Y, Yang J, Park DW, Noh SK, Kang SG, Huh YM, Son JH, Suh JS. Study of freshly excised brain tissues using terahertz imaging. Biomedical Optical Express 2014;**5**:2837–2842.

[61] Walter A, Bowman T, El-Shenawee M. Development of breast cancer tissue phantoms for terahertz imaging. Proceedings of SPIE 9700, Design and Quality for Biomedical Technologies IX; 2016. 970003.

[62] Bowman T, El-Shenawee M, Campbell L. Regional spectroscopy of paraffin-embedded breast cancer tissue using pulsed terahertz transmission imaging. Proceedings of SPIE 9706, Optical Interactions with Tissue and Cells XXVII; 2016. 97061W.

[63] Bowman T, El-Shenawee M, Campbell L. Time of flight estimation for breast cancer margin thickness using embedded tumors. Proceedings of SPIE 9706, Optical Interactions with Tissue and Cells XXVII; 2016. 97061V.

[64] Bowman T, Walter A, El-Shenawee M. Margin assessment of three-dimensional breast cancer phantoms using terahertz imaging. Proceedings of SPIE 9700, Design and Quality for Biomedical Technologies IX; 2016. 97000J.

[65] Bowman T, Hassan A, El-Shenawee M. Terahertz imaging of breast cancer margin using the linear sampling method. 2013 IEEE Antennas and Propagation Society International Symposium (APSURSI); 2013. 538–539.

[66] Bowmanl T, EI-Shenaweel M, Sharma SG. Terahertz spectroscopy for the characterization of excised human. 2014 IEEE MTT-S International Microwave Symposium (IMS2014); 2014. 1–4.

[67] Fitzgerald AJ, Pinder S, Purushotham AD, O'Kelly P, Ashworth PC, Wallace VP. Classification of terahertz-pulsed imaging data from excised breast tissue. Journal of Biomedical Optics 2012;**17**:016005.

[68] Shen YC. Terahertz pulsed spectroscopy and imaging for pharmaceutical applications: a review. International Journal of Pharmaceutics 2011;**417**:48–60.

[69] Haaser M, Gordond KC, Strachane CJ, Radesf T. Terahertz pulsed imaging as an advanced characterisation tool for film coatings—a review. International Journal of Pharmaceutics 2013;**457**:510–520.

[70] Haaser M, Karrout Y, Velghe C, Cuppok Y, Gordon KC, Pepper M, Siepmann J, Rades T, Taday PF, Strachan CJ. Application of terahertz pulsed imaging to analyse film coating characteristics of sustained-release coated pellets. International Journal of Pharmaceutics 2013;**457**: 521–526.

[71] Dohi M, Momose W, Yoshino H, Hara Y, Yamashita K, Hakomori T, Sato S, Terada K. Application of terahertz pulse imaging as PAT tool for non-destructive evaluation of film-coated tablets under different manufacturing conditions. Journal of Pharmaceutical and Biomedical Analysis 2016;**119**:104–113.

[72] Lin H, Dong Y, Shen Y, Zeitler JA. Quantifying pharmaceutical film coating with optical coherence tomography and terahertz pulsed imaging: an evaluation. Journal of Pharmaceutical Sciences 2015;**104**:3377–3385.

[73] Brock D, Zeitler JA, Funke A, Knop K, Kleinebudde P. Critical factors in the measurement of tablet film coatings using terahertz pulsed imaging. Journal of Pharmaceutical Sciences 2013;**102**:1813–1824.

[74] Niwa M, Hiraishi Y, Terada K. Evaluation of coating properties of enteric-coated tablets using terahertz pulsed imaging. Pharmaceutical Research 2014;**31**:2140–2151.

[75] Brock D, Zeitler JA, Funke A, Knop K, Kleinebudde P. Evaluation of critical process parameters for intra-tablet coating uniformity using terahertz pulsed imaging. European Journal of Pharmaceutics and Biopharmaceutics 2013;**85**:1122–1129.

[76] Brock D, Zeitler JA, Funke A, Knop K, Kleinebudde P. Evaluation of critical process parameters for inter-tablet coating uniformity of active-coated GITS using terahertz pulsed imaging. European Journal of Pharmaceutics and Biopharmaceutics 2014;**88**: 434–442.

[77] Haaser M, Naelapaa K, Gordon KC, Pepper M, Rantanen J, Strachan CJ, Taday PF, Zeitler JA, Rades T. Evaluating the effect of coating equipment on tablet film quality using terahertz pulsed imaging. European Journal of Pharmaceutics and Biopharmaceutics 2013;**85**:1095–1102.

[78] May RK, Su K, Han L, Zhong S, Elliott JA, Gladden LF, Evans M, Shen Y, Zeitler JA. Hardness and density distributions of pharmaceutical tablets measured by terahertz pulsed imaging. Journal of Pharmaceutical Sciences;**102**:2179–2186.

[79] Zhang J, Pei C, Schiano S, Heaps D, Wu CY. The application of terahertz pulsed imaging in characterising density distribution of roll-compacted ribbons. European Journal of Pharmaceutics and Biopharmaceutics 2016;**106**:20–25.

[80] Yassin S, Su K, Lin H, Gladden LF, Zeitler JA. Diffusion and swelling measurements in pharmaceutical powder compacts using terahertz pulsed imaging. Journal of Pharmaceutical Sciences 2015;**104**:1658–1667.

[81] Niwa M, Hiraishi Y, Iwasaki N, Terada K. Quantitative analysis of the layer separation risk in bilayer tablets using terahertz pulsed imaging. International Journal of Pharmaceutics 2013;**452**:249–256.

[82] Cosentino A. Terahertz and cultural heritage science: examination of art and archaeology. Technologies 2016;**4**:6.

[83] Dandolo CLK, Jepsen PU. Wall painting investigation by means of non-invasive terahertz time-domain imaging (THz-TDI): inspection of subsurface structures buried in historical plasters. Journal of Infrared and Millimeter and Terahertz Waves 2016;**37**:198–208.

[84] Walker GC, Bowen JW, Matthews W, Roychowdhury S, Labaune J, Mourou G, Menu M, Hodder I, Jackson JB. Sub-surface terahertz imaging through uneven surfaces: visualizing Neolithic wall paintings in Çatalhöyük. Optics Express 2013;**21**:8126–8134.

[85] Yuan M, Sun W, Wang X, Wang S, Zhang Q, Ye J, Zhang Y. Investigating murals with terahertz reflective tomography. Proceedings of SPIE 9625, 2015 International Conference on Optical Instruments and Technology: Terahertz Technologies and Applications; 2015. 962507.

[86] Picollo M, Fukunaga K, Labaune J. Obtaining noninvasive stratigraphic details of panel paintings using terahertz time domain spectroscopy imaging system. Journal of Cultural Heritage 2015;**16**:73–80.

[87] Fukunaga K, Ikari T, Iwai K. THz pulsed time-domain imaging of an oil canvas painting: a case study of a painting by Pablo Picasso. Applied Physics A 2016;**122**:106.

[88] Seco-Martorell C, López-Domínguez V, Arauz-Garofalo G, Redo-Sanchez A, Palacios J, Tejada J. Goya's artwork imaging with terahertz waves. Optics Express 2013;**21**:17800–17805.

[89] Abina A, Puc U, Jeglič A, Zidanšek A. Applications of terahertz spectroscopy in the field of construction and building materials. Applied Spectroscopy Reviews 2015;**50**:279–303.

[90] Abina A, Puc U, Jeglič A, Zidanšek A. Structural characterization of thermal building insulation materials using terahertz spectroscopy and terahertz pulsed imaging. NDT&E International 2016;**77**:11–18.

[91] Jiang Y, Ge H, Lian F, Zhang Y, Xia S. Early detection of germinated wheat grains using terahertz image and chemometrics. Scientific Reports 2016;**6**:21299.

[92] Pastorelli G, Trafela T, Taday PF, Portieri A, Lowe D, Fukunaga K, Strlič M. Characterisation of historic plastics using terahertz time-domain spectroscopy and pulsed imaging. Analytical and Bioanalytical Chemistry 2012;**403**:1405–1414.

[93] Dong J, Kim B, Locquet A, McKeon P, Declercq N, Citrin DS. Nondestructive evaluation of forced delamination in glass fiber-reinforced composites by terahertz and ultrasonic waves. Composites Part B 2015;**79**:667–675.

[94] Sun W, Wang X, Zhang Y. A method to monitor the oil pollution in water with reflective pulsed terahertz tomography. Optik 2012;**123**:1980–1984.

[95] Liu J, Fan WH, Chen X, Xie J. Identification of high explosive RDX using terahertz imaging and spectral fingerprints. Journal of Physics: Conference Series 2016;**680**:012030.

Chapter 12

Semiconductor THz Lasers and Their Applications in Spectroscopy of Explosives

Mykhailo Klymenko, Oleksiy V. Shulika and
Igor A. Sukhoivanov

Additional information is available at the end of the chapter

http://dx.doi.org/10.5772/67625

Abstract

Recently, applications of THz spectroscopy for detecting explosive agents have attracted much attention due to following reasons: many CBRNE agents have fingerprint-like features in the THz wavelength range; the THz spectroscopy provides an ability for remote and non-destructive identification of explosives; the THz radiation penetrates through many covering dielectric materials including paper, leather, fabric and so on. One of the most important components of THz spectroscopy setups is the source of THz radiation, which has to be high-power, tunable, low-cost and to have compact sizes. In this chapter, we are going to overview recent progress of wide variety of THz emitters considered as candidates for that role. We will pay a special attention to recent trends in engineering of spectral characteristics of THz quantum-cascade lasers and their tunability. Also we will describe the advantages and difficulties that accompany a THz spectroscopy of explosives.

Keywords: terahertz time-domain spectroscopy, quantum-cascade laser, photoconductive antenna, CBRNE agents, gain spectrum

1. Introduction

The term CBRNE is an acronym for chemical, biological, radiological, nuclear, and explosive agents that represent warfare hazards [1]. The terahertz spectroscopy in the range 0.1–10 THz has become an important tool for studying the low frequency vibrational properties of molecules in the condensed phase. Many CBRNE agents exhibit characteristic spectral features in the THz frequency range, mainly in 0.5–10 THz [2, 3]. Therefore, THz spectroscopy has shown strong potential for security applications related to the detection of CBRNE materials [4, 5].

Additionally, in recent years, there has been significant progress in imaging of metallic objects under dielectric covering materials by using the millimeter wave/lower THz frequency range 50 GHz – 1 THz [6]; many imaging techniques have already found their applications in UK airports [7]. Both active and passive imaging modalities have been proposed; the former involves the analysis of radiation directed at the target by an operator, while the latter relies on naturally occurring ambient radiation emitted by, or reflected off, the target.

One of main advantages of THz spectroscopy and imaging is based on the fact that many dielectric materials such as paper, leather and woven, cotton and synthetic fabric are transparent in THz frequency range. Differences in the electromagnetic properties of skin, explosives, metal and so on can allow the detection of weapons or concealed packages. Another advantage of the THz frequency radiation is that it has a relatively low photon energy and, thus, it is nonionizing in comparison with X-rays that makes it potentially more suitable for human-involved scanning and detections. Furthermore, the radiation power obtainable from typical THz-TDS systems is rather small (less than 1 mW) and is not considered to pose a health risk [8], although QCL pulse and CW sources, which can be characterized by the radiation power greater than tens of milliwatts, may pose some risks for operators and personnels.

The most widely used THz-based technique for the spectroscopic identification of CBRNE agents is based on the THz time domain spectroscopy (THz-TDS). One of the most important components of THz spectroscopy setups is a source of radiation which has to be high-power, tunable, and low-cost having compact sizes. In Section 2, we are going to review a recent progress of wide verity of semiconductor THz lasers considered as candidates for that role. We will pay special attention on the recent trends in the tunability of THz quantum-cascade lasers. In Section 3, we will describe a basic THz-TDS setup and its application in spectroscopy of CBRNE.

2. THz emitters

2.1. Variety of THz sources

According to Tonouchi's review paper [9], THz sources can be classified into three categories: THz sources based on the emission induced by an optical radiation, quantum-cascade lasers, and sources based on solid-state electronic devices. The first category includes devices that involves generating ultrafast photocurrents in a photoconductive antenna [10, 11] with the carrier acceleration by the external electric-field, by the surface-depletion field, or by means of the photo-Dember effect in semiconductors [12]. This category also includes devices whose operation is based on non-linear optical effects induced by an incident CW optical radiation (optical rectification, difference-frequency generation, and optical parametric oscillations). The second category covers devices based on semiconductor quantum-cascade structures where THz radiation results from the drift electrical current passing through a multi-layered electrically-biased semiconductor structure. The third category includes devices such as uni-travelling-carrier photodiodes and Gunn diodes. In this chapter, we focus on the first two categories of devices which are the most suitable for the THz time-domain spectroscopy. Also, we shrink variety of considered devices to those based on semiconductor materials only. For instance, we exclude from the overview pretty efficient THz emitters based on an optically excited plasma in atmosphere [13] which allow to provide remote spectroscopic measurements for distant objects. Moreover, we do not aim to provide a comprehensive overview of all aspects of those THz generators but rather focus on the properties related to applications in the THz spectroscopy and chemical sensing.

2.2. Photoconductive antennas

One of the most common ways to generate THz radiation exploits semiconductor photoconductive antennas made of semiconductor materials and metallic stripes decomposed on its surface. Exposed to light with the photon energy overcoming the band gap of the semiconductor, the active region of the photoconductive antenna changes its electrical conductivity due to optically generated electron-hole pairs under the surface (see **Figure 1**). The free charge carriers are accelerated by the electrostatic field created by metallic stripes. The most commonly used materials for THz photoconductive antennas are radiation-damaged silicon-on-sapphire and low-temperature grown GaAs. The materials are such that their carrier lifetimes are in the subpicosecond range. The short lifetimes are determined by a high concentration of defects, at which carriers are trapped and recombined. The defects are caused by O^+ ions implanted in silicon by the ion bombardment [14]. The density of defects is controlled by the intensity of and timing of the ion beam. The low-temperature grown gallium arsenide is grown by molecular beam epitaxy at low substrate temperatures with followed rapid thermal annealing. The material contains a high density (>1018 cm^3) of point defects such as As antisites, As interstitials, and Ga-related vacancies [15]. The effective carrier mobilities of radiation-damaged silicon-on-sapphire and low-temperature grown GaAs are reported as 10–100 cm^2/V s [16] and 200–400 cm^2/V s, respectively [17]. Since the mobility of holes in low-temperature grown GaAs is lower by an order of magnitude than the mobility of electrons, carrier transport in the THz frequency range is dominated by electrons.

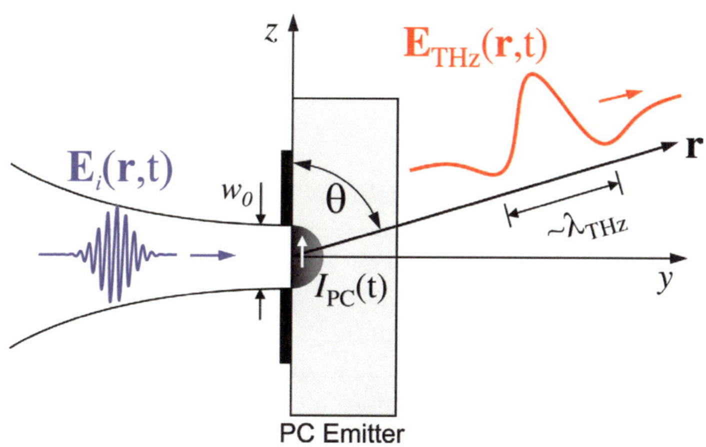

Figure 1. Generating THz radiation from the photoconductive antenna (adopted from [10]).

The output power of a photoconductive emitter depends on the bias voltage and the optical pump power. The amplitude of the radiation field increases with both parameters [11]. The maximum radiation power, however, is limited by the breakdown voltage of its substrate material. The breakdown field of the low-temperature grown GaAs is reported as ~300 kV/cm. The THz output power saturates at high optical pump powers due to the screening of the bias field by photocarriers. This saturation effect is more pronounced in dipole antennas comparing to other types of antennas. This is caused by the small size of the gap between electrodes in the dipole antennas. Nowadays, the output power of the commercial available photoconductive emitters is limited by tens of microwatts in the pulse operating mode [18, 19].

The photoconductive antennas are broadband (with the bandwidth of several THz [20]) that is beneficial for the THz time-domain spectroscopy.

The photoconductive emitters may exploit nonlinear optical effects as well. In this case, the optical radiation interacts with the non-linear medium of a semiconductor material generating THz radiation. Several non-linear effects may be employed: the optical rectification, which is a second-order non-linear processes [12], and photomixing (see Ref. [10] for instance). Such methods of THz emission allow for the continuous-wave generation; however, the emission power of such type of emitters is not large enough for most of the spectroscopic applications.

In principle, even a pure surface of a bulk semiconductor material without metallic stripes and applied external electrostatic fields can radiate terahertz radiation after being excited by an ultra-fast optical pulse. In this case, photo-generated carriers are accelerated by means of the surface depletion layer electric field or photo-Dember effect [21]. The photo-Dember effect is related to spikes of a photocurrent that arises as a result of the difference between diffusion velocities of electrons and holes. Because electrons respond to applied electric field with a larger acceleration due to their higher mobility comparing to holes, electrons diffuse more rapidly from the surface inside the semiconductor material. That creates an effective surge current normal to the surface. Although the emission of THz radiation from pure semiconductor surfaces represents an interest as a diagnostic tool for semiconductor surface physics, it is not widely exploited in applications due to its extremely low output power.

2.3. Quantum-cascade lasers

The photoconductive antennas are pretty bulk devices containing the generator of ultra-shot optical pulses as their integral part. Also, they are characterized by low converting efficiency. A good candidate to overcome these problems is the quantum-cascade structure, a compact electrically driven THz source operating in both pulse and cw mode. QCLs are unipolar devices where a laser emission is achieved through a series of the intersubband transitions in a repeated stack of semiconductor multiple quantum-well heterostructures. The idea of such heterostructures was first proposed by Kazarinov and Suris in 1971 [22], and later implemented for the mid-infrared radiation by Jerome Faist, Federico Capasso, Deborah Sivco, Carlo Sirtori, Albert Hutchinson, and Alfred Cho at Bell Laboratories in 1994 [23].

Band diagrams, band structures, and joint density of states for both interband and intersubband quantum-well devices are schematically shown in **Figure 2**. In the interband structures, the optical radiation is a result of electron transitions from the conduction subband to the valence subband. As a result, the minimal quantum of the energy is limited by the band gap of the quantum-well material. Curvatures of the bands involved in the transition have very different magnitudes and, what is more important, different senses of curvature. It results in the joint density of states, which is step one in this case.

Optical transitions in the quantum-cascade heterostructures occur between subbands within an allowed band (see **Figure 2(b)**). In contrast to the energy gap between two bands, the subband structure is governed by the conduction band offset and width of the quantum-well layer. Minimal transition energy is not limited by the fundamental band gap and can be tailored by a material composition of the quantum well and the thickness of the quantum-well layer. Since there is no a fundamental limit exposed on the energy gap between two subbands, the intersubband transitions may be exploited to achieve lasing in THz range. The curvature of dispersion curves characterized by the effective masses is usually almost the same for all subbands. It results in the joint density of states being localized around a constant transition energy E_{12}, **Figure 2(b)**. Such a joint density of states results in narrow-band optical characteristics of QCLs which are farther shrunk due to the cavity effect.

A single cascade of QCL usually consists of two parts: the active region and the injector (see **Figure 3**). The radiative intersubband transitions occur in the active region, while the injector serves to extract electrons from the lowest subband in a cascade of QCL and to transfer them to the upper subband in the next cascade providing thus the electron pumping on the upper laser levels. Also, another role of the injector is to block leakage of electrons from the upper subband to the next cascade.

There are generally two different approaches adopted for the injector of terahertz QCLs: chirped superlattice [24] and resonant phonon depopulation [25]. The chirped superlattice is based on fast intraminiband scattering that is engineered in a way to achieve a population inversion between two minibands. This approach shows the lowest threshold current density. The resonant-phonon design relies on fast resonant electron-phonon scatterings which serve to provide an electron extraction from one cascade to another (see **Figure 3**). Because of the relatively large phonon energy of 36 meV compared to the superlattice miniband width of

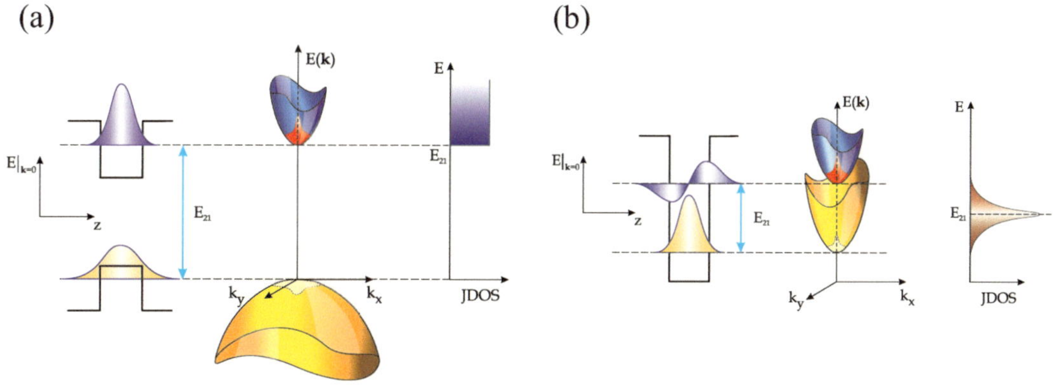

Figure 2. (a) Interband and (b) intersubband radiative transitions in semiconductor quantum-well structures.

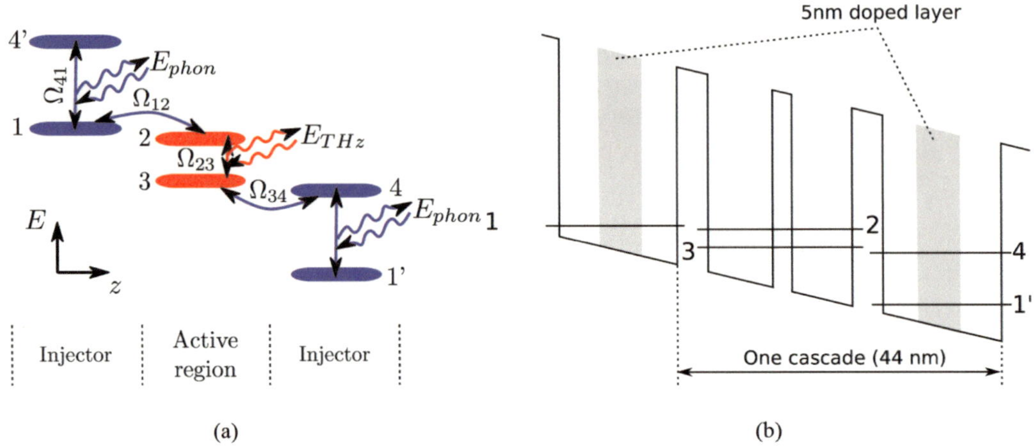

Figure 3. (a) Energy versus coordinate structure of the subband edges in the three-well QCL and (b) the energy diagram of the three-well QCL reported in [31]: the layer thicknesses are of 43/89/24.6/81.5/41/160 angstroms starting from injector barrier (the barriers are indicated in bold fonts). The electrical bias equal 12.2 kV/cm.

15 meV , the resonant-phonon design has less thermal backfilling, and an enhancement in operating temperature has been demonstrated [25]. In the first resonant-phonon design, one cascade of the structure consisted of four quantum wells: two wells to form the pair of lasing states (active region) and two wells for the phonon depopulation and carrier injection (these wells are often called the phonon and injector wells, respectively). That QCL was able to operate in both pulse regime [26] and cw mode [27]. It has been lately optimized to enhance the maximum operating temperature from 137 to 164 K for the pulse mode and from 93 to 117 K in cw mode [28].

In a simpler case, the phonon and the injector wells may be combined [29]. When the design with a single quantum well per injector was first proposed [23], it has slightly less maximum operating temperature of 142 K in pulse mode with the lasing frequency of 3.4 THz. However lately [30], the three-well design has been optimized and it has been proven that this design combines simplicity with a good performance (at the lasing frequency of 3.9 THz, 63 mW of peak optical power was measured at 5 K, and approximately 5 mW could still be detected at 180 K). The most recent advances on the three-well design has been reported in Ref. [31] and reviewed inRef. [32]. According to Ref. [31], the new temperature record for the three-well QCL is 199.5 K with lasing frequency of 3.22 THz and detectable output power. The structure may operate also at 8 K with the emission power of approximately 38 mW and a threshold current density of 1 kA/cm^2.

Below we describe the design of the three-well QCL showing a record operating temperature reported in [31]. Although terahertz QCLs based on InGaAs/InAlAs have been demonstrated [33], the majority of THz QCLs reported are made from GaAs/AlGaAs system because of its flexibility in barrier height that offers an additional degree of freedom for designing. The design is based on the GaAs/$Al_{0.15}Ga_{0.85}As$ material system. The laser structure was grown by molecular beam epitaxy on a semi-insulating GaAs substrate with a 10 μm-thick active region and a sheet electron density of 3×10^{10} cm^{-2} per period using a 3D Si-doping within the middle 5 nm of the injector well. The active region is placed between 100 nm bottom n+ GaAs layer, doped with impurity concentration of 5×10^{18} cm^{-3}, and a top stack consisting of 50 nm of 5×10^{18} cm^{-3} and 10 nm of low temperature-grown 5×10^{19} cm^{-3} n+ GaAs layers. The wafers were first processed into THz QCL structures with Au double metal with wide top Ti/Au metallization forming a Schottky contact, and 1 mm-long Fabry-Perot resonator. In this configuration, the device was lasing with the maximal temperature of 180 K. Additional enhancing of the maximal operating temperature has been achieved by lowering the waveguide loss and improving heat dissipation. That has been achieved by using Cu-Cu-based process with lower waveguide loss and better heat dissipation. Also, the 100 nm thick top n+ contact layer was removed that has further lowered the waveguide loss. The waveguide of the THz QCLs is based on a Cu-Cu double metal geometry. It was fabricated by using Cu-Cu wafer bonding and standard photolithography. Such a metal waveguide provides a large confinement factor close to unity; however, it has relatively high losses caused by the free-carrier absorption. The contacts are represented by bottom and top metal stacks: the first one is made of Ta/Cu (10/600 nm) and the second one is made of Ta/Cu/Ti/Au (10/300/20/150 nm). Wet etching was performed ($H_3PO_4/H_2O_2/H_2$:3/1/25) to etch through the entire thickness of the 10 μm-thick active region. The ridge waveguide of fabricated THz QCLs is ~170 μm wide. The substrate of the samples was thinned down to ~150 μm [31] and then cleaved into laser bars with a 1.8 mm-long Fabry-Perot resonator. The laser bars were then gold plated on the backside, indium soldered on a copper package, and then mounted in a closed-cycle cryostat for measurements [31].

At the electrical bias of 12.2 kV/cm, each cascade of QCL has four subbands with following energy gaps between their edges: E_{23} = 11.5 meV, E_{13} = 14.2 meV, and $E_{41'}$ = 34.6 meV (see **Figure 3(b)** to understand subband numbering). The lasing occurs between subbands 2 and 3. The injection of electrons on the upper subband is provided from the subband 1 by tunnel coupling, while the extraction passes from the lower subband 3 to the subband 4. The transition from the subband 4 to the subband 1′ in the injector involves the emission of LO phonons with energy E_{phon}. The design provides such combination of oscillator strengths Ω_{ij} for diagonal transition in the active region and rates of phonon-assisted scatterings in the injector that a large inversion of populations between lasing states is maintained ensuring large emission power.

2.4. QCL as THz amplifiers

In Ref. [34], a new approach to the generation of high-power broadband THz pulses has been proposed exploiting the effect of the material gain. According to this work, a semiconductor quantum-cascade structure can be used to amplify THz pulses generated via conventional photoconductive antennas. To provide sufficient amplification of THz-pulsed radiation, a quantum-cascade structure should have a large gain coefficient, whose value is generally limited by the gain clamping effect. To overcome this limitation, ultrafast switching of the material gain has been used. The gain coefficient is increased for a short time interval while the input THz pulse is propagating through a QCS. The modulation is implemented using an Auston switch.

In Ref. [35], *by* a series of numerical simulations based on the density matrix formalism, authors have demonstrated that, instead of using the Auston switch, a large instant gain coefficient can be achieved in QCLs by a direct exposition of the structure by ultra-short optical pulses (see **Figure 4(a)**).

The optical pulse is applied to the three-well QCL operating in the pre-threshold regime. As illustrated in **Figure 4(b)**, the design exploits the valance band to provide an interband pumping of electrons into the conduction band. As a result the considered three-well GaAs/AlGaAs QCL demonstrates THz gain increasing up to an order of magnitude under pumping by 100 fs optical pulses with the intensity of 100 MW/cm^2. The gain switching process is characterized by 1 ps rise time and 8 ps recovery time.

The emission spectra from QCL are narrow-band. Therefore, in order to apply them in the THz spectroscopy, one should either make the lasers frequency-tunable or extend their gain/emission spectral band. In the first case, a wideband frequency tuning in QCL can be achieved using a MEMS-based movable silicon plunger [36, 37]. A solution for the second case relies on a multi-stack QCL [38]. The spectral gain of the multi-stack terahertz QCL, composed of three active regions with emission frequencies centered at 2.3, 2.7, and 3.0 THz, is characterized by spectral gain full width at half-maximum of 1.1 THz. Bandwidth and spectral position of the measured gain match with the broadband laser emission. As the laser action ceases with increasing operating temperature, the gain at the dominant lasing frequency of 2.65 THz degrades sharply.

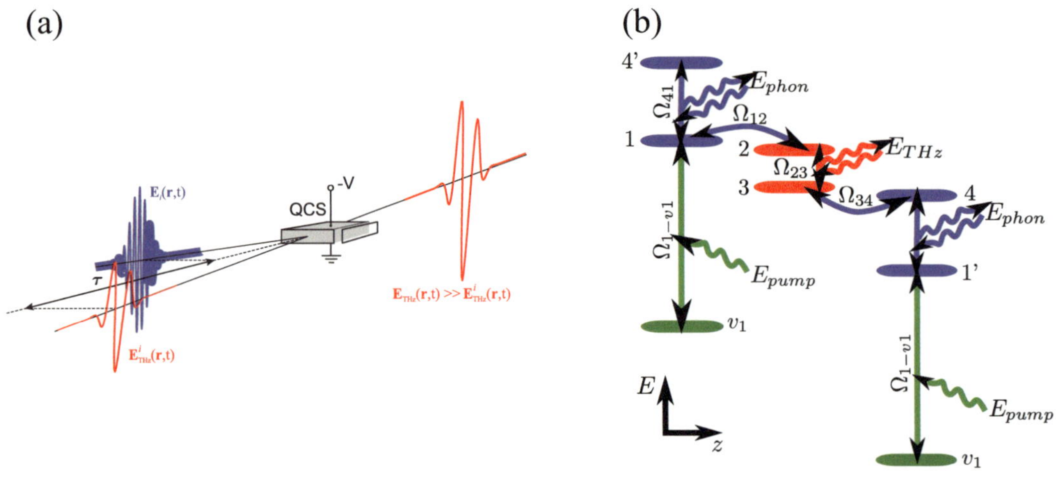

Figure 4. (a) Schematic diagram of the gain switching via ultrafast interband optical pumping and (b) energy versus coordinate structure of the subband edges in the three-well QCL with interband optical pump for gain switching.

3. THz spectroscopy of CBRNE agents

3.1. THz-TDS setups

In **Figure 5**, we show a typical experimental THz-TDS setup with the GaAs photoconductive antenna operating as a THz emitter. The THz-TDS setup starts with a femtosecond laser producing an optical-pulse train. Each pulse separates into two paths. One reaches the THz emitter where the optical pulses are transformed into ultrashort electromagnetic pulses. The THz pulse then propagates in free space and is focused onto an ultrafast detector. The other part of the pulse is also delivered onto the detector after passing through a time-delay stage. The detector measures the electric-field amplitude of the electromagnetic waves. Although the experimental details vary between different systems, in this example taken from Ref. [6], the emitter contains two vacuum-evaporated NiCr/Au electrodes separated by a submillimeter-wide gap. The electrodes are placed on a doped 1 μm-thick GaAs layer grown on an undoped GaAs substrate. A bias voltage applied across the electrodes changes around ±100 V, modulated at a few kilohertz. The pumping fs pulse generated by Ti:sapphire laser is typically of a few hundred milliwatts average power (~10–100 fs pulse width, 800 nm center wavelength). In this arrangement, a bandwidth of THz pulses is extended up to 3 THz. In an alternative arrangement, the THz radiation is collected "backwards" (in the direction of the reflected pump laser beam). As a result, the absorption and dispersion of the THz pulses in the substrate are minimized that enhances the bandwidth up to 20–25 THz [39]. The emitted THz pulses are collimated and focused onto the sample by a pair of parabolic mirrors. Samples can be scanned across the focus to build a two-dimensional image, each pixel contains spectral information.

There are several ways to provide a coherent detection of the transmitted or reflected THz radiation. Most common technique is based on the ultrafast Pockels effect, where the THz radiation is collected by an electro-optical crystal, for example, ZnTe [40, 41]. The THz field induces an instantaneous birefringence in the electro-optic medium, which is probed with a second visible or near-IR laser beam. Such a beam is split from the pump source. The birefringence modulates the ellipticity of the probe and is detected by a setup consisting of $\lambda/4$ waveplate, a Wollaston prism, and two balanced photodiodes. Also, a lock-in technique can be used to measure a signal from the photodiode using the modulated bias field of the

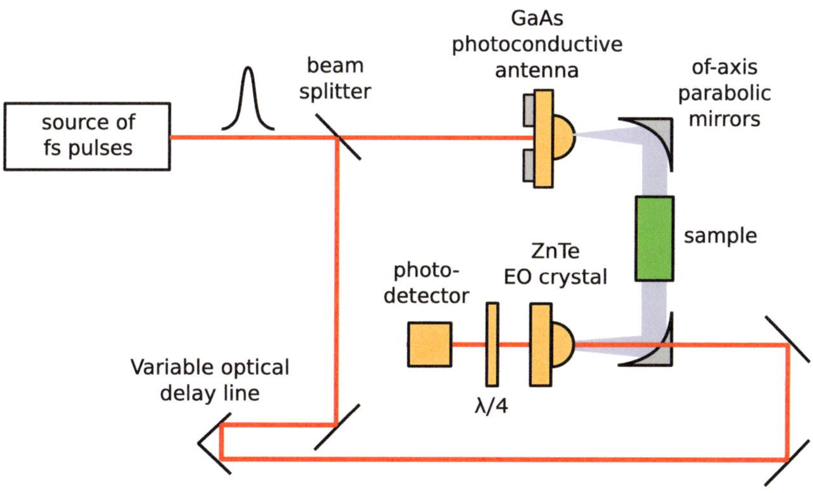

Figure 5. Schematic of a THz-TDS setup.

photoconductive emitter as a reference. In order to obtain the electric field in the time domain, one measures the signal as a function of the time delay between the THz pulse and probe pulses at the electro-optic crystal using a variable delay line. The frequency spectrum of the THz radiation is then obtained by the Fourier transform.

To demonstrate significantly higher spectral resolution and ability to resolve closely spaced lines, the sample may be placed in a parallel plate waveguide with a subwavelength gap between the two plates [42]. Here, a polycrystalline film of a target analyte is formed on one of the inner surfaces of the waveguide plate. The confinement of THz waves ($\lambda/6$ for a typical 50 micron gap) together with a long interaction pathlength (~2–3 cm) allows the THz response of the film to be measured in a sensitive manner over a broad bandwidth from 0.2 to 4 THz. A key idea here is that using relatively simple film formation methods, it is possible to form a highly crystalline quality thin film, which minimizes inhomogeneous line broadening that plagues traditional THz sample preparation methods such as pellet formation. That leads to the resolution of fingerprint lines in threat solids, which are not observed in the conventional free space THz spectroscopy of corresponding pellet samples.

3.2. Explosive materials and their THz spectra

THz-TDS has been shown to be a particularly sensitive technique for studying the structural dynamics of crystal forms, providing additional information for sample analysis. This spectroscopy technique can provide a spectral information being complementary to Raman spectroscopy, owing to the different selection rules governing IR and Raman transitions. Despite the relatively similar chemical structure, the THz frequency spectra are clearly distinct. THz-TDS dominates over other spectroscopic far-IR techniques, such as Fourier transform infrared spectroscopy, since it is insensitive to the thermal background (and hence benefits from a high signal-to-noise ratio). Also, THz-TDS does not require a cryogenically cooled bolometer detector. Another benefit of THz-TDS follows from the fact that both the absorption coefficient and refractive index can directly be extracted without requiring a Kramers-Kronig analysis.

One of the most attractive features of THz spectroscopy is that the THz radiation can penetrate though many nonmetallic and nonpolar materials. This allows to provide spectral analysis of materials concealed within dry packaging, such as paper, natural and synthetic fabrics, plastics, that enables imaging of concealed metallic objects.

First systematic experimental data on THz spectra of a wide range of explosives have been reported by in Ref. [43] and overviewed in Ref. [2]. Their spectra covered the frequency range 0.1–2.8 THz. Later in Ref. [3], the spectroscopic measurements have been extended up to 6 THz (see **Figure 6**). Although many explosives have distinct vibrational absorption features at room temperature, these absorption features tend to be rather broad and overlapping with Q factors ($f/\Delta f$) is less than 10. The broad absorption features are caused by the homogeneous and inhomogeneous line broadening. The resolution of the underlying THz resonances of an explosives solid is a challenging problem and is necessary to achieve a full understanding of the THz properties of molecules of explosives.

http://dx.doi.org/10.5772/67625

Up to date, a wide range of explosive materials have been investigated by THz-TDS, including both pure explosive compounds and plastic explosives, including explosive compounds and mixtures combined with a variety of plasticizers, desensitizers, dyes, waterproof coatings, and fabrics to aid storage and use. All these adulterants may potentially have their own spectral signatures in the THz region, complicating or obscuring the analysis. The pure explosive materials, whose THz spectra are well studied, include ammonium nitrate 2,4,6-trinitrotoluene (TNT) and its degradation product 2,4-dinitrotoluene, 1,3,5,7-tetranitro-1,3,5,7-tetrazocane (HMX), 1,3,5-trinitroperhydro-1,3,5-triazine (RDX), and 1,3-dinitrato-2,2-bis(nitratomethyl)propane (PETN). A comprehensive overview of the THz spectra of a wide range of explosive materials can be found in Ref. [2]. Studies of the plastic explosives Metabel (PETN-based), SX2 (RDX-based), C-4 (RDX-based), PBX (predominantly containing HMX), and Semtex-H (containing both RDX and PETN) have also been reported [44–46]. Most of these studies have focused on THz transmission spectroscopy, as this allows extraction of the absorption coefficient and refractive index of the sample, and the sample thickness/dilution can be optimized to maximize the measurement signal-to-noise ratio. However, there is growing interest in developing THz reflection spectroscopy geometries (both specular and diffuse), as this may be more natural arrangement for standoff detection.

RDX possesses an unusual six-membered heterocyclic ring comprising three nitrogen and three methylene groups, with a nitrite group attached to each of the nitrogen atoms in the ring. This high nitrogen and oxygen content causes explosive properties of this material. A typical THz-TDS spectrum of RDX is shown in **Figure 6(b)**. RDX has a number of characteristic absorption features. The sensing of RDX through covering materials such as plastic, leather, and cotton has been investigated in Ref. [2]; the same has been done using a reflection geometry [47]. In both cases, the spectral features of RDX (especially the absorption peak at 0.8 THz) can be identified. The HMX spectrum (see **Figure 6(a)**) exhibits spectral features below 3 THz, which concur with previously published work and additionally shows six features between 3 and 5 THz. The features between 3 and 5 THz are distinctly different from those present in RDX, allowing improved spectral identification. The data for HMX samples in the 5–6 THz range also have several spectral features; however, they are placed very close to each other [3]. PETN is more sensitive to shock or friction than RDX (or TNT) and is also an extremely powerful explosive. Its THz spectra has two broad bands between 1.5 and 2.5 THz and between 2.5 and 3.5 THz [3]. The TNT spectrum (see **Figure 6(b)**) has a feature near 2 THz as well as an increase in absorption with increasing frequency in the 0.5–3 THz range [3, 43]. Above 3 THz, there are also several significant features.

Most explosive materials are not used in their pure molecular crystalline form, but are mixed with other agents to make plastic explosives such as SX2, Semtex, and Metabel. Metabel shares three absorption features with PETN as well as contributions from each other compound. Also, one observes a clear correspondence between the features of the SX2 spectra and its constituent component RDX. Semtex consists of both RDX and PETN. Comparison of the spectrum of Semtex with those of its components shows a correspondence in a number of peaks): peaks in the Semtex spectrum are caused by a contribution from vibrational modes of

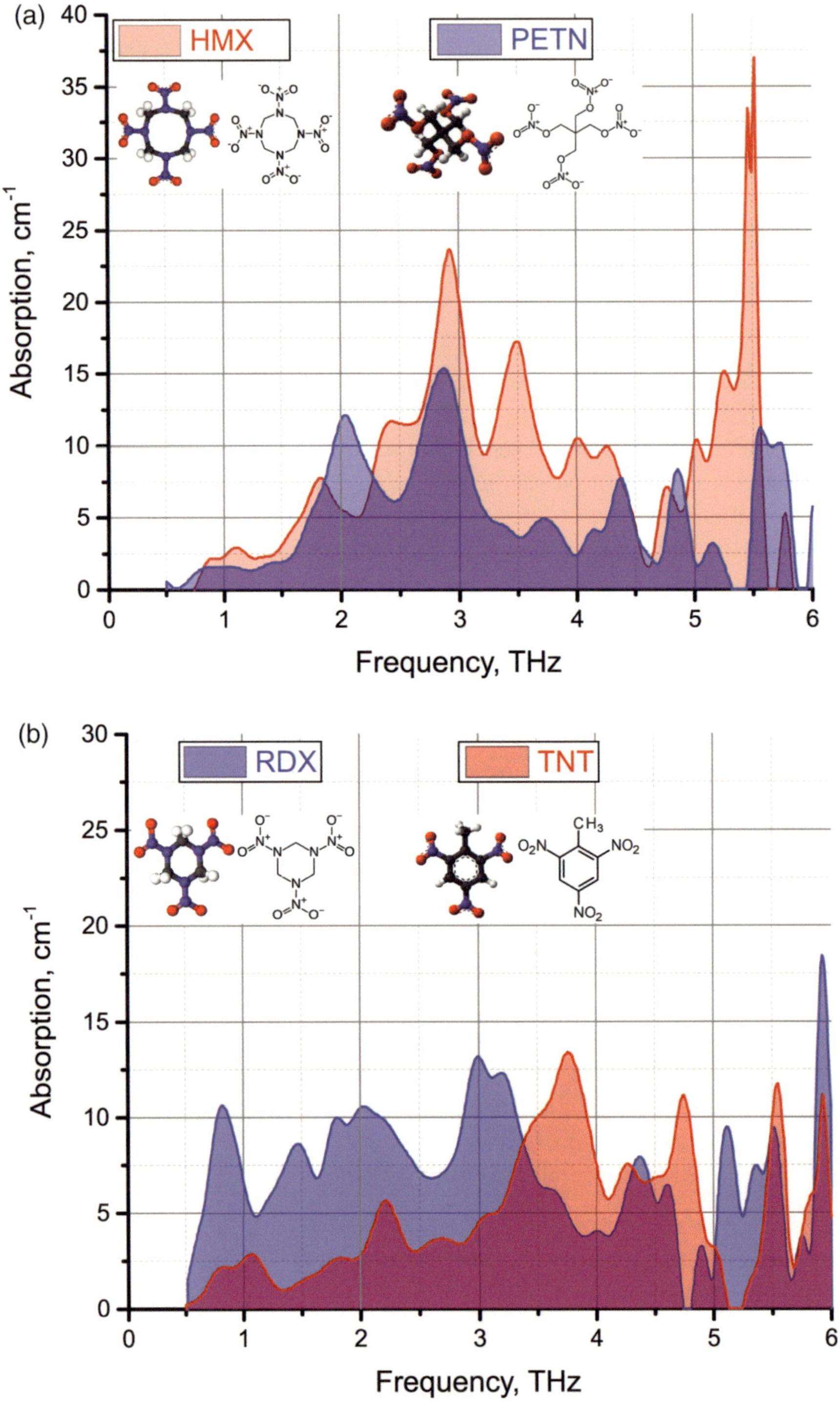

Figure 6. THz absorption spectra of the explosives HMX, PETN, RDX and TNT (the results are adopted from [3]). The chemical structures of the explosives are also shown.

both RDX and PETN, as well as a number of other components. In general, the spectra of the plastic explosives are dominated by the signatures arising from their constituent explosives.

4. Summary

In summary, THz radiation possesses a set of attractive unique features, such as absorption with strongly non-uniform spectral characteristics in most of CBRNE agents, ability to penetrate through many dielectric materials, and nonionizing effect on organic matter. All these features have induced fast recent development of the hardware for THz spectroscopy, including both THz sources and detectors. The photoconductive antennas are most promising sources for the THz time domain spectroscopy. The THz quantum-cascade lasers and emitter based on photomixing phenomena seem to be the best candidates for the coherent broadband continuous-wave terahertz spectroscopy. The quantum-cascade lasers are more compact and produce higher power (tens of miliwatt in low-temperature regime). However, the output power drops with increasing of temperature. Recent developments in the field [31] evidence that the quantum cascade laser can produce a detectable power of several miliwatt at the temperature of 199.5 K. Also, they can be spectral-tunable using MEMS-based movable silicon plunger [36, 37]. The emitters based on the photomixing phenomena do not have issues with operating temperature; however, their output power is limited since their operation is based on high-order optical non-linear effects.

Most explosives have very non-uniform and non-monotonic spectral characteristics, which may serve as fingerprint-like markers for their detection. The absorption spectra of a mixture of explosives contain recognizable features of the constituents that allow for the detection of the components of the mixture. Also, their spectroscopic features are not strongly affected by most of the possible dielectric covering materials. Thus, CBRNE agents can be identified through packaging.

Acknowledgements

Partial support from the University of Guanajuato under projects DAIP-1021/2016 and DAIP-932/2016 is greatly acknowledged. OVS conveys his appreciation to Mauro F. Pereira (Sheffield Hallam University, UK) for continuous support and encouragement.

Author details

Mykhailo Klymenko, Oleksiy Shulika* and Igor Sukhoivanov

*Address all correspondence to: oshulika@ugto.mx

Department of Electronic Engineering, Division of Engineering, Campus Irapuato-Salamanca, University of Guanajuato, Mexico

References

[1] http://doh.dc.gov/service/cbrne-chemical-biological-radiological-nuclear-and-explosives

[2] Chen J., Chen Y., Zhao H., Bastiaans J.G., Zhang X.-C. Absorption coefficients of selected explosives and related compounds in the range of 0.1–2.8 THz. Opt. Exp., 15 (19), 12060 (2007)

[3] Leahy-Hoppa M.R., Fitch M.J., Zheng X., Hayden L.M., Osiander R. Wideband terahertz spectroscopy of explosives. Chem. Phys. Lett., 434, 227–230 (2007)

[4] Kemp M.C., Taday P.F., Cole B.E., Cluff J.A., Fitzgerald A.J., Tribe W.R. Security applications of terahertz technology. Proc. SPIE-Int. Soc. Opt. Eng., 5070, 44–52 (2003)

[5] Federici J.F., Schulkin B., Huang F., Gary D., Barat R., Oliveira F., Zimdars D. THz imaging and sensing for security applications: explosives, weapons, and drugs. Semicond. Sci. Technol., 20 (7), S266–S280 (2005)

[6] Davies A.G., Burnett A.D., Fan W., Linfield E.H., Cunningham J.E. Terahertz spectroscopy of explosives and drugs. Mater. Today, 11 (3), 18–26 (2008)

[7] Appleby R., Wallace H.B. Standoff detection of weapons and contraband in the 100 GHz to 1 THz region. IEEE Trans. Antennas Prop., 55, 2944 (2007)

[8] Berry E., et al. Do in vivo terahertz imaging systems comply with safety guidelines? J. Laser Appl., 15, 192 (2003)

[9] Tonouchi M. Cutting-edge terahertz technology. Nat. Photon., 1, 97–105 (2007)

[10] Lee Y.-S. Principles of terahertz science and technology. New York: Springer Science + Business Media; 2009. 340 p. DOI: 10.1007/978-0-387-09540-0

[11] Tani M., Matsuura S., Sakai K., Nakashima S. Emission characteristics of photoconductive antennas based on low-temperature-grown GaAs and semi-insulating GaAs. Appl. Opt., 36 (30), 7853–7859 (1997)

[12] Sakai K. editor. Terahertz optoelectronics (Topics in applied physics, vol. 97). Heidelberg: Springer-Verlag Berlin Heidelberg; 2005. 387 p. DOI: 10.1007/b80319

[13] Cook D.J., Hochstrasser R.M. Intense terahertz pulses by four-wave rectification in air. Opt. Lett., 25, 1210–1212 (2000)

[14] Doany F.E., Grischkowsky D., Chi C.-C. Carrier lifetime versus ion-implantation dose in silicon on sapphire. Appl. Phys. Lett., 50, 460 (1987)

[15] Gupta S., Frankel M.Y., Valdmanis J.A., Whitaker J.F., Mourou G.A. Subpicosecond carrier lifetime in GaAs grown by molecular beam epitaxy at low temperatures. Appl. Phys. Lett., 59, 3276 (1991)

[16] Auston D. H. Ultrafast optoelectronics. In: Kaiser W. editor. Ultrashort laser pulses: generation and applications. 2nd ed. New York: Springer-Verlag Berlin Heidelberg; 1993. p. 183-234. DOI: 10.1007/BFb0070977

[17] Look D.C., Walters D.C., Robinson G.D., Sizelove J.R., Mier M.G., Stutz C.E. Annealing dynamics of molecular-beam epitaxial GaAs grown at 200 °C. J. Appl. Phys., 74, 306 (1993)

[18] Ferguson B., Zhang X.-C. Materials for terahertz science and technology. Nat. Mater., 1, 26–33 (2002)

[19] Zhao G., Schouten R.N., van der Valk N., Wenckebach W.T., Planken P.C.M. Design and performance of a THz emission and detection setup based on a semi-insulating GaAs emitter. Rev. Sci. Instrum., 73, 1715–1719 (2002)

[20] Katzenellenbogen N., Grischkowsky D. Efficient generation of 380 fs pulses of THz radiation by ultrafast laser pulse excitation of a biased metal–semiconductor interface. Appl. Phys. Lett., 58, 222–224 (1991)

[21] Dekorsy T., Auer H., Waschke C., Bakker H.J., Roskos H.G., Kurz H., Wagner V., Grosse P. Emission of submillimeter electromagnetic waves by coherent phonons. Phys. Rev. Lett., 74, 738 (1995)

[22] Kazarinov R.F., Suris R.A. Possibility of amplification of electromagnetic waves in a semiconductor with a superlattice. Fizika i Tekhnika Poluprovodnikov, 5 (4), 797–800 (1971)

[23] Faist J., Capasso F., Sivco D.L., Sirtori C., Hutchinson A.L., Cho A.Y. Quantum-cascade laser. Science, 264 (5158), 553–556 (1994)

[24] Tredicucci A., Köhler R., Mahler L., Beere H.E., Linfield E.H., Ritchie D.A. Terahertz quantum cascade lasers - first demonstration and novel concepts. Semicond. Sci. Technol., 20 (7), S222 (2005)

[25] Williams B.S., Callebaut H., Kumar S., Hu Q. 3.4-THz quantum cascade laser based on longitudinal-optical-phonon scattering for depopulation. Appl. Phys. Lett., 82, 1015 (2003).

[26] Williams B.S., Kumar S., Callebaut H., Hu Q., Reno J.L. Terahertz quantum-cascade laser operating up to 137 K. Appl. Phys. Lett., 83, 5142 (2003)

[27] Kumar S., Williams B.S., Kohen S., Hu Q., Reno J.L. Continuous-wave operation of terahertz quantum-cascade lasers above liquid-nitrogen temperature. Appl. Phys. Lett., 84, 2494 (2004)

[28] Williams B.S., Kumar S., Hu Q., Reno J.L. Operation of terahertz quantum-cascade lasers at 164 K in pulsed mode and at 117 K in continuous-wave mode. Opt. Exp., 13, 3331 (2005)

[29] Luo H., Laframboise S.R., Wasilewski Z.R., Aers G.C., Liu H.C., Cao J.C. Terahertz quantum-cascade lasers based on a three-well active module. Appl. Phys. Lett., 90, 041112 (2007)

[30] Kumar S., Hu Q., Reno J.L. 186 K operation of terahertz quantum-cascade lasers based on a diagonal design. Appl. Phys. Lett., 94, 131105 (2009)

[31] Fathololoumi S., Dupont E., Chan C.W.I., Wasilewski Z.R., Laframboise S.R., Ban D., Mátyás A., Jirauschek C., Hu Q., Liu H.C. Terahertz quantum cascade lasers operating up to ~ 200 K with optimized oscillator strength and improved injection tunneling. Opt. Exp., 20 (4), 3866–3876 (2012)

[32] Vitiello M.S., Scalari G., Williams B., De Natale P. Quantum cascade lasers: 20 years of challenges. Opt. Exp., 23 (4), 5167–5182 (2015)

[33] Ajili L., Scalari G., Hoyler N., Giovannini M., Faist J. InGaAs–AlInAs/InP terahertz quantum cascade laser. Appl. Phys. Lett., 87 (14), 141107–141107 (2005)

[34] Jukam J., Dhillon S.S., Oustinov D., Madeo J., Manquest C., Barbieri S., Sirtori C., Khanna S.P., Linfield E.H., Davies A.G., Tignon J. Terahertz amplifier based on gain switching in a quantum cascade laser. Nat. Photon., 3 (12), 715–719 (2009)

[35] Shulika O.V., Klymenko M.V., Sukhoivanov I.A. Ultrafast interband pumping of quantum-cascade structures: a feasibility study of a THz pulse amplifier. Laser Photon. Rev., 8 (1), 188–195 (2014)

[36] Qin Q., et al. Tuning a terahertz wire laser. Nat. Photon., 3, 732–737 (2009)

[37] Sirtori C., Barbieri S., Colombelli R. Wave engineering with THz quantum cascade lasers. Nat. Photon., 7, 691–701 (2013)

[38] Bachmann D., et al. Spectral gain profile of a multi-stack terahertz quantum cascade laser. Appl. Phys. Lett., 105, 181118 (2014)

[39] Shen, Y.-C., et al. Ultrabroadband terahertz radiation from low-temperature-grown GaAs photoconductive emitters. Appl. Phys. Lett., 83, 3117 (2003)

[40] Wu Q., Zhang X.-C. 7 terahertz broadband GaP electro-optic sensor. Appl. Phys. Lett., 70, 1784 (1997)

[41] Leitenstorfer A., et al. Detectors and sources for ultrabroadband electro-optic sampling: experiment and theory. Appl. Phys. Lett., 74, 1516 (1999)

[42] Theuer M., Sree Harsha S., Grischkowsky D. Flare coupled metal parallel-plate waveguides for high resolution THz time-domain spectroscopy. J. Appl. Phys., 108, 113105, (2010)

[43] Tribe W.R., et al. Hidden object detection: security applications of terahertz technology. Proc. SPIE, 5354, 168 (2004)

[44] Burnett A., et al. Broadband terahertz time-domain and Raman spectroscopy of explosives. Proc. SPIE, 6549, 654905 (2007)

[45] Yamamoto K., et al. Noninvasive inspection of C-4 explosive in mails by terahertz time-domain spectroscopy. Jpn. J. Appl. Phys., 43, L414 (2004)

[46] Funk D.J., et al. THz transmission spectroscopy and imaging: application to the energetic materials PBX 9501 and PBX 9502. Appl. Spectrosc., 58, 428 (2004)

[47] Liu H.-B., et al. Detection and identification of explosive RDX by THz diffuse reflection spectroscopy. Opt. Exp., 14, 415 (2006)

Chapter 13

Terahertz Spectroscopy for Gastrointestinal Cancer Diagnosis

Faustino Wahaia, Irmantas Kašalynas,
Gintaras Valušis, Catia D. Carvalho Silva and
Pedro L. Granja

Additional information is available at the end of the chapter

http://dx.doi.org/10.5772/66999

Abstract

In this chapter, we present a number of sensitive measurement modalities for the study and analysis of human cancer-affected colon and gastric tissue using terahertz (THz) spectroscopy. Considerable advancements have been reached in characterization of bio-tissue with some accuracy, although too dawn, and still long and exhaustive work have to be done towards well-established and reliable applications. The advent of the THz-time-domain spectroscopy (THz-TDS) test modality at a sub-picosecond time resolution has arguably fostered an intensive work in this field's research line. The chapter addresses some basic theoretical aspects of this measurement modality with the presentation of general experimental laboratory setup diagrams for THz generation and detection, sample preparation aspects, samples optical parameters calculation procedures and data analysis.

Keywords: spectroscopy, absorption, ATR mode, reflection and transmission, carcinomas

1. Introduction

Terahertz (THz) (10^{12} Hz) frequency band is a small section of the electromagnetic spectrum lying between the microwave and infrared (IR) regions sometimes referred to as THz gap or T-rays. There is no standard definition for the THz band, but it has most often come to refer to frequencies in the range of 0.1–10 THz [1–3]. The results of several studies [4–7] using terahertz techniques started the breakneck race of even more studies towards biomedical applications of terahertz technology.

1.1. Terahertz technology advantages

The energies of THz radiation are significantly lower such that at 1 THz, the energy is about 4.1 meV [8]; compared to X-rays (with photon energy typically in order of few keV [9], i.e., few million times higher than that for THz photon energy), it is considered non-ionizing. In the absorption processes where THz waves interact with biological media, the Gibbs free energy conveyed in the THz photons beam is not sufficient to induce any chemical reactions [10]; therefore, any measurement technique operating at this frequency range and at reasonably low power levels could be considered as non-invasive and offers higher contrast spectral features; almost all dielectric materials are transparent to THz radiation; it gives an unambiguous information concerning big molecules; it offers higher spatial resolution compared to infrared one, therefore much better for imaging purposes; meanwhile, IR frequency band brings only local information on molecules, gives only information on chemical binding between shortest neighbours, and THz radiation brings global information on molecules, gives information on the whole molecule and its specific rotation-vibrational modes [11]. So that the frequency band could be of paramount importance for the study and characterization of biological media.

The vibrational spectral fingerprint of biomolecules lies in this frequency range [12, 13] (and many materials of interest have unique spectral fingerprints in the terahertz range [14, 15]), therefore, it can be used for theirs identification, making THz spectroscopy a promising sensing tool for biomedical applications and disease diagnosis.

Due to space constraints, we avoided, in the present chapter, broad practical considerations on methods, and by the nowadays, vast applications of THz spectroscopy. Being confident in having counted with the contribution of leading experts in the field of THz technology, we realize that the chapter will be concise and comprehensive. We aim at providing a representative overview of the current state of art of THz spectroscopy for cancer detection, as well as for colon and stomach cancer particularly.

The development of ultrafast lasers and the discovery of the Auston switch in the 1970s [16] led to a new generation of THz spectrometers in the early 1990s [17] that were able to generate and detect pulses of coherent terahertz radiation with unprecedented ease and sensitivity. The generation and detection diagram are presented in **Figure 1**.

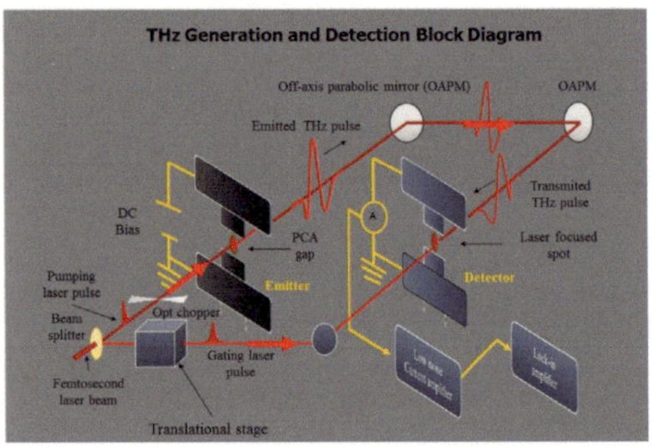

Figure 1. Diagram of THz generation and detection. The emitter is on the left; the receiver is on the right. The THz pulse is recorded as a function of time delay and the obtained time-domain signal is Fourier-transformed numerically.

http://dx.doi.org/10.5772/66999

2. Terahertz time-domain spectroscopy

Since 1990s, the research towards the application of THz spectroscopy to probe and characterize various biomolecules has advanced considerably. The THz time-domain spectroscopy (THz-TDS) is actually a spectroscopic technique in which the properties of a material are investigated using THz short pulses. Their generation and detection scheme is sensitive to the material's effect on both the amplitude and the phase. The technique can provide more information than conventional Fourier-transform spectroscopy (FTS), which is only sensitive to the amplitude. Therefore, it could be a useful analytical tool for materials study and characterization.

An ultrashort optical pulse (normally femtosecond [fs] in duration) is used to pump (illuminate) a photoconductive (PC) semiconductor creating pairs of photocarriers (electron-hole pairs). The photocunductive (PC) material changes suddenly from being an insulator into being a conductor. Then the conduction state leads to an abrupt electrical current across a biased kind of dipole antenna stuck on a semiconductor substrate referred to as photoconductive antenna (PCA). This changing current emits short pulse (~2 picoseconds [ps]) [18] THz electromagnetic field.

Typically, there are two electrodes stuck on a low temperature grown gallium arsenide (LTG-GaAs), semi-insulating gallium arsenide (SIGaAs), indium phosphate (InP) substrate or other semiconductor material. The electrodes are made in the form of a simple dipole antenna with a gap (G) of a few microns (~5 μm) and have a bias voltage up to 40 V between them (**Figure 2**).

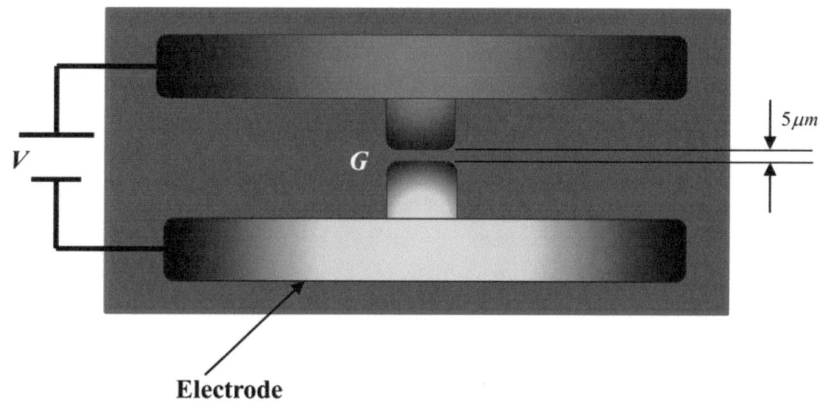

Figure 2. Diagram of a PCA for THz emission and detection.

And, the other one is the probe pulse, and goes through a translational stage to provide a relative periodical time delay [19] (typically controlled by changing the relative path between the pump and probe beams with a linear stage). Both the pump and probe pulses are derived from the same optical beam and, therefore, have the same duration, which typical has a range between 10 and 150 fs. Usually, another PCA is used as detector, which, interacting with the probe, generates an electrical signal proportional to the amplitude of the THz pulse for that particular time delay, that is, it samples the THz signal in the detector. There are other methods besides that based on photoconductive antennas, such as based on optical rectification, wherein, by passing high-intensity femtosecond laser pulses through a transparent crystal such as zinc telluride (ZnTe), gallium phosphide (GaP), and gallium selenide (GaSe), a terahertz short pulse is generated in this case biasing is not needed. The process is nonlinear where anyone of above mentioned crystals is suddenly electrically polarized at high laser intensities (amplified). The so changing electrical polarization emits terahertz radiation.

The detection process is similar with that for generation (PC detection). To do that, the bias electrical field across the detecting PCA is generated by the THz electric field pulses this time. The THz electric field generates current across the receiving PCA connections wires. A low-noise amplifier is used for the signal amplification. And, finally, the amplified current is the measured parameter, which is proportional to the THz field transient. A lock-in amplifier (LIA) is also used to demodulate the signal, and this avoids $1/f$ (flicker) noise problems that are present in the detector-limited measurement scheme.

The THz signal is directly measured as a function of time, and the frequency spectra of both sample material signal and reference (without sample) one are obtained by a numerical Fourier transformation. Further calculations of the obtained spectra yield the spectroscopic information of the sample material under study. Since the measurements are made on electric field instead of intensity, both amplitude and phase can be determined at once, thus leading to the calculation of the sample's frequency-dependent optical constants such as absorption coefficient and the refractive index. This is an advantage compared to the well-established Fourier transform spectroscopy (FTS), which is based on the intensity detection with recourse to Kramers-Kronig [20] data treatment, with all the uncertainties associated.

Terahertz spectroscopy for cancer detection is arguably among the most active research topics within the research groups in the field of T-rays' technology. The utmost importance of the information obtained by THz spectroscopic technique has incentivized the researchers to keep seeking for efficient source and sensitive detectors. As was stated previously, this chapter will not provide a full review of the works performed in this field. It otherwise intends to show (on the base of selected few previous works) the potential of this technique for gastrointestinal (GI) cancer detection particularly.

A THz-TDS is a pump-probe-like technique [21] since the signal and reference are measured by sampling using a delayed probe optical pulse, taking the form of a time trace with sub-picosecond resolution. The signal-to-noise ratio (SNR) is much higher than that of Fourier transform infrared spectroscopy (FTIR) [22]. The Coherent-gated detection gives a noise equivalent power of ~10–16 W/√Hz, which is six orders of magnitude better than the pyrodetectors, normally used in Fourier transform spectrometers. A typical setup diagram for THz-TDS measurements is shown in **Figure 3**.

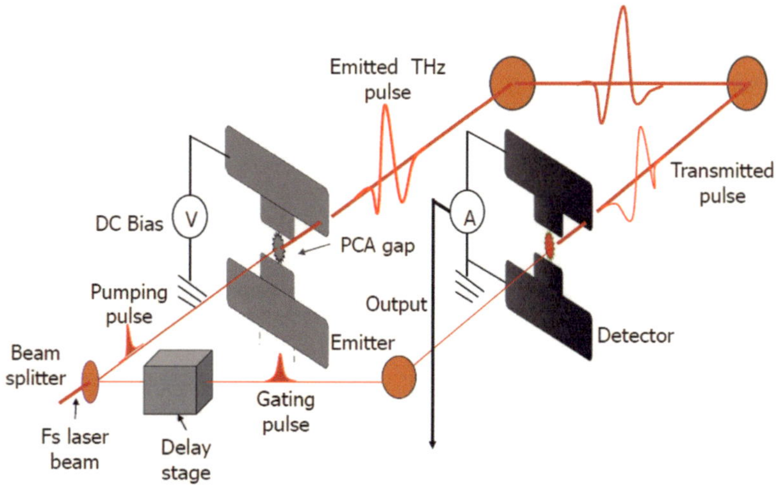

Figure 3. Typical experimental setup for THz spectrometer using photoconductive antennas (PCAs) in the transceiver.

3. Time-domain terahertz spectroscopy in reflection geometry

In this modality, the measurements can be performed by reflection geometry (**Figure 4**) as well as by attenuated total reflection (ATR) [23–25]; the use of an evanescent wave (i.e., a THz wave illuminates the interface of two media, with different refractive indices n_1 and n_2, at a certain critical angle θ_c, and the wave is totally reflected at the interface). Then, part of the wave enters and propagates through the studied medium at a short distance (known as evanescent wave). This method is required in measurements involving strongly absorbing materials and in transmission geometry involving no or slightly absorbing ones.

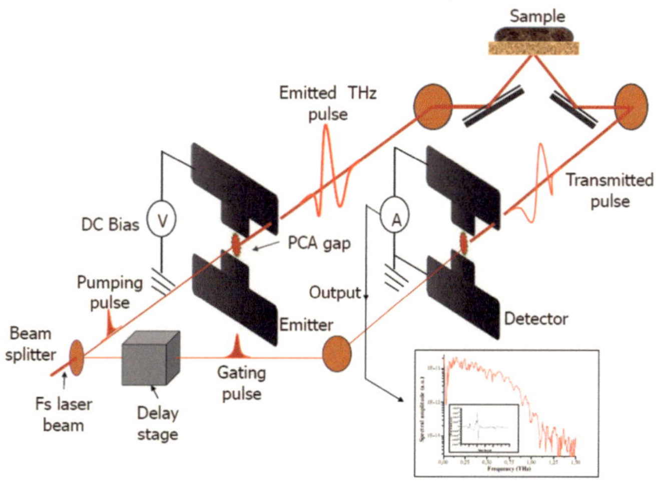

Figure 4. General experimental setup for THz-TDS in reflection geometry. The sample (colon or gastric tissue) is sandwiched in between two high density polyethylene (HDPE) or other window material sheets, normally.

As the THz signal is extremely weak in the order of >10 μW [26], an lock-in amplifier (LIA) is used to extract the signal from the thermal background. Due to the coherent and gated detection nature of the THz-TDS technique, the signals are almost unmasked by the thermal background noise. A computer controls the delay line, reads the LIA values at each position and records the THz pulse waveforms. The amplitude and phase information of the THz field, as foretold, are accounted at once [27]. The THz properties of samples are measured with the difference between the generated THz pulse, E_{ref}, and the transmitted sample THz pulse, E_s: using Fast Fourier Transform, these signals provide complex-valued optical properties such as refractive indices, absorption coefficients, dielectric constants and conductivities.

The technique provides more information if compared with a conventional Fourier-transform spectroscopy (FTS), which is only sensitive to the amplitude. THz radiation has several distinct advantages over other forms of spectroscopy: Examples which have been demonstrated include several different types of explosives [28], polymorphic forms of many compounds used as active pharmaceutical ingredients (API) [29, 30] in commercial medications as well as several illegal narcotic substances [31]. Since many materials are transparent to THz radiation, these items of interest can be observed through visually opaque intervening layers, such as packaging and clothing. Though not strictly a spectroscopic technique, the ultrashort width of the THz radiation pulses allows for measurements (e.g., thickness, density, location of defects) on difficult to probe materials (e.g., foam) [32, 33].

3.1. Data analysis: reflection geometry

The optical parameters of a sample may be evaluated using reflection or transmission as mentioned above. For high THz absorption media, such as fresh biological tissues, there is a limit of sample thickness for THz-TDS in transmission geometry. Over that limit, reflection geometry must be used (**Figure 3**).

Once temporal measurements are made and profiles of the sample signal and reference one are obtained, the optical parameters, such as refractive index and absorption coefficient, are then calculated. They are extracted using the following expression [34, 35].

$$\frac{E_s(\omega)}{E_{ref}(\omega)} = \frac{t_{w,s}\, r_{w,s}}{t_{w,s}}(\omega)\exp\left[i\frac{4\pi\, n_w\, \mathrm{d}\nu}{c}\right] \quad (1)$$

Where $E_s(\omega)$ is the measured electric field of the pulse reflected at the interface window-sample, $E_{ref}(\omega)$ is the measured electric field of the pulse at the front surface (air-window) of the sample holder, $r_{w,s}$ is the complex reflection coefficient for the window-sample interface to be calculated and $t_{a,w}$ and $t_{w,a}$ are, respectively, the known transmission coefficients for air-window and window-air interfaces.

Since the absorptions of the air and window are negligible (normally), the above transmission coefficients are assumed to be real; the n_w is the known refractive index of the medium used as reference (window material) and d is the effective thickness of the window, which can be calculated from

$$d = \frac{d_w}{\cos\phi} \tag{2}$$

where d_w is the thickness of the sample holder's window and φ is the reflection angle on the back side of the window (in contact with the sample). The known and measured quantities in Eq. (2) allow the determination of $r_{w,s}(\omega)$, which can be written in its complex form

$$r_{w,s}(\omega) = A\, e^{i\phi} \tag{3}$$

where A is the amplitude and ϕ is the phase. Considering the case of normal incidence, where the reflection and transmission coefficients have simpler forms, the complex reflection coefficient becomes

$$r_{w,s}(\omega) = \frac{n_w - n_s(\omega)}{n_w + n_s(\omega)} \tag{4}$$

where $n_s(\omega) = n_s + i\,\alpha_s(\omega)\,c/4\pi\, v_{THz}$ is the complex refractive index of the sample; n_s is the real part of $n_s(\omega)$, $\alpha_s(\omega)$ is the absorption coefficient; c is the speed of light in vacuum; and ν_{THz} is the frequency of THz radiation. Substituting Eq. (3) in Eq. (4), we obtain for the sample index of refraction the following expression [36]:

$$n_w(\omega) = \frac{n_w(1 - A^2)}{1 + A^2 + 2A\cos\phi} \tag{5}$$

And for absorption coefficient

$$\alpha_s(\omega) = \frac{4\pi\, n_s\, v}{c}\frac{-2A\sin\phi}{1 + A^2 + 2A\cos\phi} \tag{6}$$

In reflection geometry, the extraction of these frequency-dependent parameters is a bit more complicated than in transmission one, as can be seen in the following section. Similarly, for the transmission, both reference and sample signals are recorded. However, apart from a temporal shift of the sample and the reference pulse due to the reference window material, an additional spatial shift occurs due to refraction and the absorption data are contained within the phase information of the measurement, which is sensitive to system artefacts. Additionally, there is one more difficulty arising due to polarization change due to reflection under a specific angle that causes the reference and sample pulses to be differently polarized. This is relevant, since the sensitivity of the detector is polarization dependent.

4. Time-domain terahertz spectroscopy in transmission geometry

Measurement in transmission geometry (**Figure 5**), the laser beam is also split into a pump and a probe beam and so on. The signal is first enhanced by a low-noise current amplifier and then sent to the LIA, which sends reference frequency to the chopper.

A delay stage with dc-motor actuator, driven by a motion controller, moves a set of mirrors either continuously or stepwise with adjustable waiting time between steps. The mechanical precision of the movement is typically about 0.1 μm corresponding to time shift of about 0.66 fs. In THz-TD spectrometer, three parameters are of interest: the delay stage waiting-time (t_{delay}); the LIA's time constant, the integration time (t_{LIA}); and the frequency of the chopper ($\nu_{chopper}$). For a higher SNR and narrower amplifier bandwidth noise, longer t_{LIA} is needed [37].

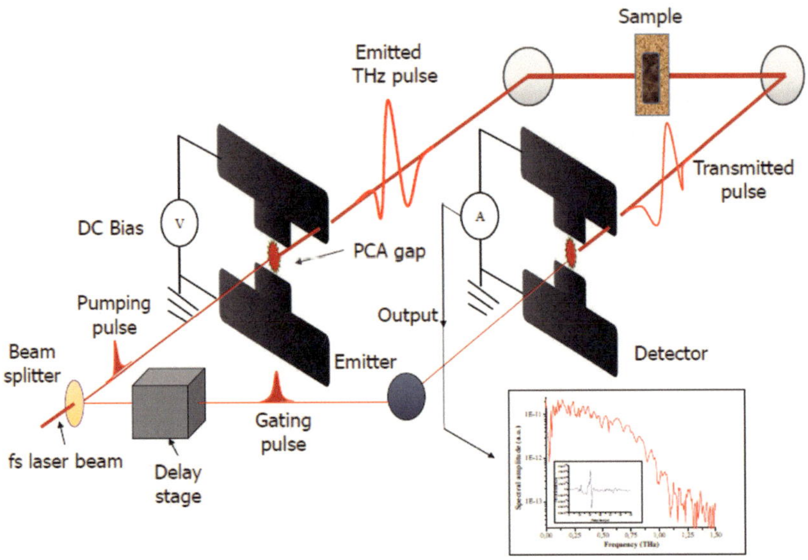

Figure 5. General experimental setup for THz-TDS in transmission geometry. The sample (colon or gastric tissue) is sandwiched in between two high density polyethylene (HDPE) or other window material sheets, normally.

4.1. Data analysis: transmission geometry

In THz-TDS measurements, the following assumptions are considered: (1) the sample under measurement is a homogeneous dielectric slab with parallel and flat surfaces, where the scattering of THz rays is negligible; (2) the incident angle of the THz beam is normal to the sample surfaces; (3) the transverse dimension of the sample is larger than the incident beam waist, so there is no diffraction; (4) the reference signal is measured under the same conditions as the sample signal; (5) the resolution of the measuring instrument is sufficiently high so that the quantization error is negligible, unless stated otherwise; (6) the measuring instruments are well calibrated; and (7) there is no human error in the measurements.

Figure 6 shows a diagram of typical sample mounting in a container (sample cell) for in transmission THz-TDS. The THz pulse propagates through a sandwich formed by two windows and an inner space containing the sample.

The determination of frequency-dependent optical constants of a sample comprehends several steps, as illustrated in **Figure 7**. Since the quantity provided by a THz-TDS measurement is a time-domain signal, then a physical model is required to relate the measured signal to the optical properties of the sample.

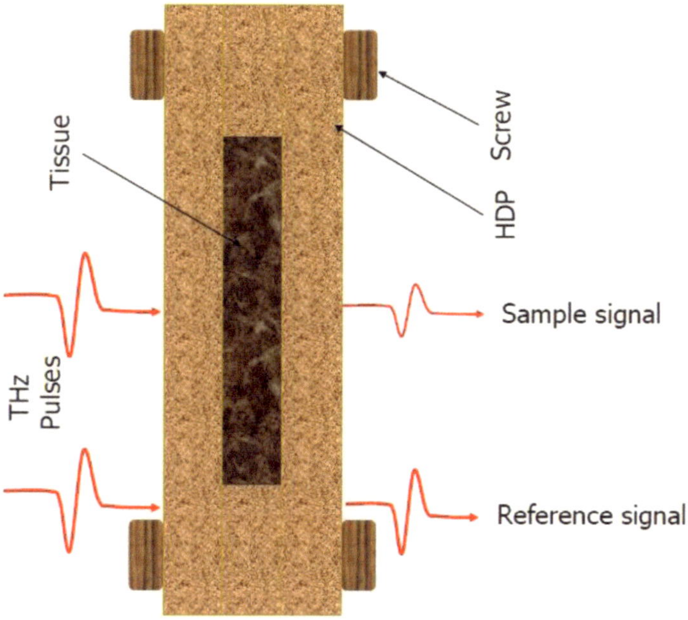

Figure 6. Sample mounting for THz-TDS in transmission geometry.

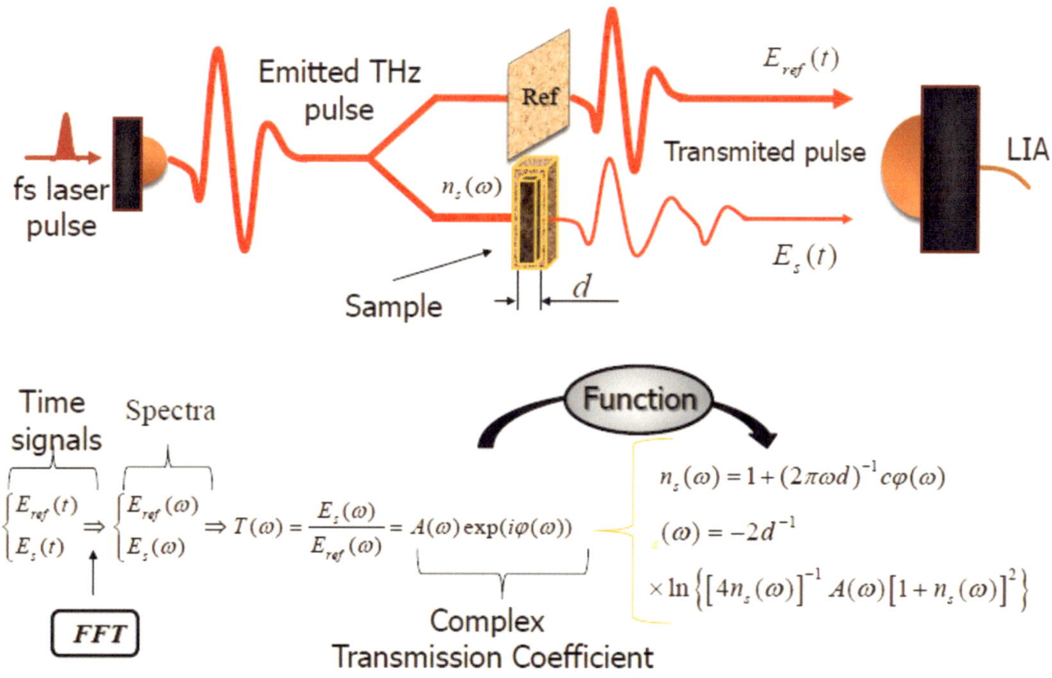

Figure 7. THz-TDS in transmission. Extraction of the real and imaginary refractive index, absorption coefficient of a sample through a fast Fourier transformation.

Normally, the waveforms of the THz pulses have very narrow time widths (~0.25 ps). As already stated, two pulses are recorded in THz-TDS: one with an empty sample holder and the other with the sample placed. The measured signal is proportional to the electric field of the THz pulse. The pulse that propagates through the sample is called sample pulse, E_s, and that propagating through the reference medium (with known dielectric constants) is the reference pulse, E_{ref}

Knowing the time evolution of those pulses, it is possible to get the information about the frequency-dependent dielectric constants (absorption coefficient (α_s) and the refractive index (n_s) of the sample). It can be noticed that the THz signal propagates through both the reference medium and the sample, over a distance d (the thickness of the sample). **Figure 7** presents a diagram representation of a THz-TDS in transmission geometry and data processing.

The Fresnel amplitude transmission coefficients through the interfaces, reference medium-sample and sample-reference medium, are expressed, respectively, as

$$t_{M,s} = \frac{2}{M + n(\omega)} \text{ and } t_{s,M} = \frac{2n(\omega)}{M + n(\omega)} \tag{7}$$

Where $n_s(\omega) = n_s - ik(\omega)$ is the complex refractive index of the sample. The real part of this expression denotes the refractive index of the sample and the imaginary one is related with the absorption coefficient.

Taking the Fourier transform of $E(t)$,

$$E(\omega) = \frac{1}{2\pi} \int_{-\infty}^{\infty} e^{-i\omega t} E(t)\, dt \tag{8}$$

The Fourier transform of the THz signal transmitted through the sample can be written as follow:

$$E_s(\omega) = E_0(\omega)\, t_{w,s}\, P h_s(\omega)\, t_{s,w} \tag{9}$$

Where $E_0(\omega)$ is the THz electric field shining on the first interface window-sample, $t_{w,s}$ represents the transmission coefficient of that interface, $t_{s,w}$ is the transmission coefficient of the second interface sample window and $P h_s(\omega) = e^{-i\phi_s(\omega)}$ represents the phase factor, where $\phi_s = k_s(\omega)\, d$ and $k_s(\omega)$ are the wave vector of the sample pulse. The phase factor represents the phase gained when the pulse propagates through the sample with thickness d.

The Fourier transform of the THz signal transmitted through the reference can be written as

$$E_{ref}(\omega) = E_0(\omega)\, \phi_{ref}(\omega) = E_0\, e^{-i\frac{M\omega d}{c}} \tag{10}$$

Comparing with the equivalent expression for the sample signal in Eq. (9), we observe that the transmission coefficients are absent since they are equal to unit. The ratio of the two signals, $E_s(\omega)$ and E_{ref}, gives the complex transmission of the sample, in some literatures, called transfer function. For normal incidence, this function can be expressed as [34]

$$T(\omega) = \frac{E_s(\omega)}{E_{ref}(\omega)} = \frac{4\,n_s(\omega)}{(n_s(\omega)+M)^2} \exp\left[i\,\tfrac{\omega d}{c}(n_s(\omega)-M)\right] \tag{11}$$

The effect of multiple reflections in the interfaces, reference medium-sample and sample-reference medium (i.e., Fabry-Pérot effect [38, 39]), may be accounted for, by incorporating the factor $FP(\omega)$ as follows [40],

$$T(\omega) = \frac{E_s(\omega)}{E_{ref}(\omega)} = \frac{4\,n_s(\omega)}{(n_s(\omega)+M)^2} \exp\left[i\,\tfrac{\omega d}{c}(n_s(\omega)-M)\right] FP(\omega) \tag{12}$$

where

$$FP(\omega) = \frac{1}{M - \left[\frac{n_s(\omega)-M}{n_s(\omega)+M} \exp(-2i\,n_s(\omega)\,\tfrac{\omega d}{c})\right]} \tag{13}$$

This effect could be, however, neglected in case the samples to be analysed are thick enough so that the Fabry-Pérot echo is considerably retarded, therefore, easily discriminable from the main signal. The aim is at find the absorption coefficient and the real refractive index of the samples. As was stated before, the ratio between the sample and the reference signal may be expressed as [41],

$$\frac{E_s(\omega)}{E_{ref}(\omega)} = A(\omega)\,exp(i\phi(\omega)) = \frac{4\,n_s(\omega)}{(n_s(\omega)+M)^2} \exp\left[i\,\tfrac{\omega d}{c}(n_s(\omega)-M) - n_i\,i\,\tfrac{\omega d}{c}\right] \tag{14}$$

If in the experiments, THz transparent material be used as a reference medium then, $M = n_w$, where n_w represents the window's refractive index. Then for the refractive index and absorption coefficient of the sample, the following expressions can be written [42, 43]

$$n_s(\omega) = \frac{c\phi(\omega)}{2\pi\nu d} + n_w \tag{15}$$

and

$$\alpha_s(\omega) = \frac{2}{d} \ln\left[\frac{4\,n_s(\omega)}{A\,(n_s(\omega)+n_w)^2}\right] \tag{16}$$

respectively.

In the process of extraction of the optical frequency-dependent parameters of materials, the dynamic range (DR) of a THz-TDS [44] is also taken in account. Due to the typical single-cycle regime of the THz pulse, the spectral amplitude is strong at low frequencies, and normally, a characteristic gradual roll-off occurs at high frequencies until the detected THz signal approaches the noise level at the experiment.

The DR of a THz-TDS setup is defined as the signal above the noise level of the spectrum. It is determined by the Fourier transformation of the measured temporal waveform (i.e., the spectrum of the THz radiation extracted from its time waveform in the scanning period, 0– T) above the noise level. The noise N includes the noise due to fluctuations in the THz field, $N_{THz}(t) = R(t)\,E_{THz}(t)$, where $T(t)$ is a dimensionless random ratio and N_{OPB} is the background noise from the detection of the THz. The background noise includes the optical probe-beam (OPB) shot noise in the detector, which is proportional to the recorded current and all noises in the detector gathered together, such as Johnson noise, amplification noise, thermal noise, the zero-THz-field photocurrent, etc.

Analytically, the DR can be expressed by the expression [45].

$$\mathrm{DR}(\omega) = \frac{\kappa(\omega)}{\sqrt{\frac{\delta t}{2\pi}\left(\frac{\kappa}{S}\right)^2 + \frac{T\delta t}{2\pi}\frac{1}{D^2}}} \tag{17}$$

where $k(\omega) = E(\omega)/A_{\mathrm{THz}}$ with A_{THz} representing the maximum amplitude of the THz field in time-domain is a factor dependent upon the actual waveform, δt is the temporal intervals at which the temporal waveform of the THz pulse is sampled, $S = 1/\sigma_R$ is the temporal measurement signal-to-noise ratio (SNR), T is the sampling range, $D = A/S_B$ where A represents the maximum amplitude of the THz field in time-domain, is the temporal dynamic range, and

$$\kappa = A^{-1}\sqrt{\int_{-\infty}^{\infty} |E(t)|^2 \, \mathrm{d}t} \tag{18}$$

Actually, the DR limits the maximum magnitude of the absorption coefficient that could be observed along the higher frequencies. The effect of the DR may wrongly be perceived as an absorption peak in samples whose absorption rises with the rise of the frequencies, since the rise and the beginning of the rolling-off resemble a signal peak.

Thus, the maximum absorption coefficient, which can be measured reliably, corresponds to the situation where the sample signal is attenuated to a level approaching the noise level, that is, the maximum of absorption coefficient data reliable can be obtained only in the range within the DR of the experiment [44]:

$$\alpha_{\max}(\omega)\, d = 2\ln\left(\mathrm{DR}(\omega)\frac{4n_s(\omega)}{(n_s(\omega)+1)^2}\right) \tag{19}$$

Values of α larger than $\alpha_{\max}$ could cause the detector saturation and therefore cannot be measured.

5. Terahertz spectroscopy of colonic and gastric carcinoma

The intestine is part of the digestive system—a tube that begins at mouth, forms stomach, small intestine, the large bowel (colon and rectum) and ends with the anus. The colon and rectum are located in this gastrointestinal tract (GIT) [46] system [47]. It is made up of two main parts, the colon and the rectum, which are the lower part of the digestive tract measuring about 80–100 cm and 12–15 cm, respectively [48]. The colon absorbs large amount of water and salts from broken down food and the rectum stores the waste material until it is removed from our body through the anus.

http://dx.doi.org/10.5772/66999

Colon and rectum cancers are cancers starting in the colon and rectum, respectively. Since they have many features in common, they are very often called with a unique name of colorectal cancer. The colon and the rectum cancers develop slowly along several years. Before cancer development, a growth of tissue or benign tumour normally starts as a non-cancerous polyp (non-cancerous tumour) on the inner colon or rectum epithelium. With time, it becomes an adenomatous polyps or adenomas that are polyps potentially transiting into cancer stage.

The colon and rectum tracts are sections of the large intestine, a tube-like structure characterized by a wall consisting of layers such as mucosa (the inner lining layer), *muscularis* mucosa (a thin muscle layer outer limit of the mucosa), submucosa (the fibrous tissue beneath the *muscularis*), *muscularis propria* (a thick muscular layer) and subserosa and serosa (the outmost layer of connective tissue covering most of the colon but not the rectum) (**Figure 8**).

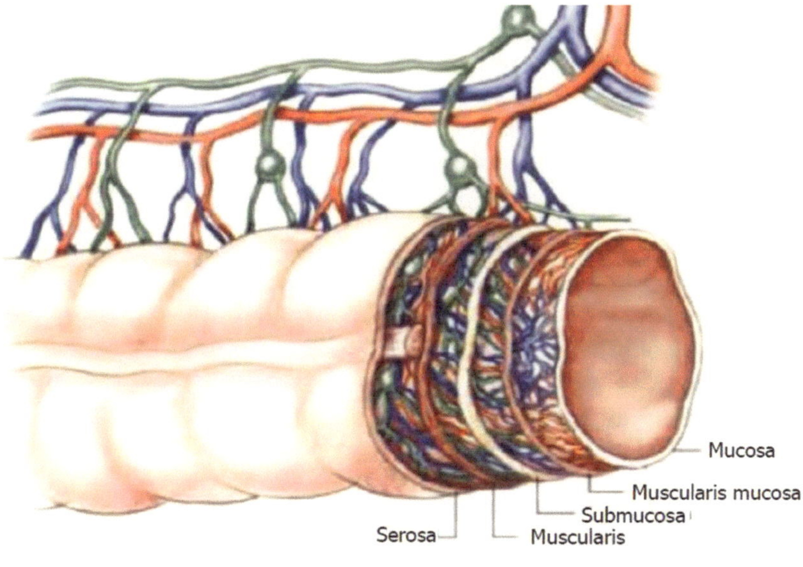

Figure 8. A section of colorectal layered wall [49].

Colon and gastric cancers are among the most commonly diagnosed cancers and death cause worldwide [50]. Early diagnosis is of utmost importance for in-time treatment. THz-TDS has demonstrated the capability to distinguish between normal and cancerous tissue as was stated and well referenced previously in this chapter. Presently, several THz-TDS systems are commercially available [51]; some of them being directly derived from laboratory setups, while others are compact and integrate with more sophisticated man-machine interfaces. There are, however, still many challenging issues to overcome, such as better understanding of THz wave—biological tissue interaction, which could enable the development of reliably and powerful THz diagnostic systems even for early cancer diagnosis.

6. Conclusions

Owing to the fact that THz waves have very low photon energy, that is, energy levels of few milli-electronvolt, which is well below the ionization energies of atoms and molecules [52], they do not pose any ionization hazard for biological tissues. The characteristic energies of biological media, because of molecular motions, such as rotational and vibrational ones, lie in the THz frequency band, allowing THz waves to directly detect their spectral signatures. Due to that, there is a worldwide interest in the exploitation of this frequency band and related techniques (spectroscopy and imaging) for biomedical application in the last two decades with much more terahertz spectra being reported in spectroscopic studies of cancer, particularly.

Water is essential in biological systems [53], and it plays a key role as the solvent in molecular reactions. It shows characteristic absorption features mainly the infrared (IR) and THz bands, due to which its resonance is related to symmetric stretch, bending, libration and rotation, which are modified by hydrogen bonding in the liquid state. THz waves are very sensitive to water content and strongly attenuated by water in which biological molecules reside [54–57]. The presence of cancer often causes increased blood supply to affected tissues and a local increase in tissue water content may be observed [58–60]: this fact acts as a natural contrast mechanism for terahertz spectroscopy of cancer. Furthermore, the structural changes that occur in affected tissues have also been shown to contribute to terahertz spectroscopy contrast.

The THz techniques as THz-TDS, providing broadband information on biological tissue and which make possible the discrimination of tissue regions with different optical characteristics (e.g., neoplastic and non-neoplastic tissue) over working THz frequency range, are thus expected to bring a more comprehensive screening and diagnosis of human disease, particularly in the case of cancer.

The THz transmission spectroscopy has previously been used to obtain the THz optical characteristics of skin tissue [61–63]. THz-pulsed spectroscopy has also been used to successfully characterize DNA and proteins [64, 65], allowing intermolecular interactions to be probed. The THz-TDS combined with THz imaging could be used for macroscopic visualization of tumour margins in fresh tissues according to first published results on cancer tissue imaging using THz-pulsed radiation [66], and later confirmed by studies on various cancer types and organs [67].

This chapter presents some arguments that indicate that THz-TDS has the potential [68, 69] to be a superior complement to the techniques for cancer screening in use nowadays. In the work of several groups with freshly excised tissues, the differences on the refractive indices and absorption coefficients in dysplastic tissues have been mainly attributed to the presence of a higher water content. However, other possible factors have been pointed out by authors of numerous studies [70–72]. Additional studies are needed towards the determination of all of the contrast-contributing factors other than water towards an efficient, reliable and functional THz techniques and methods. Some cancer contrast factors are already known such as (1) the increase in the vasculature due to the release of growth factors that also lead to rapid cell division and higher cell densities; (2) the conditions within the tumour microenvironment that differ considerably from those of normal tissue, for instance, low oxygen levels (hypoxia). Angiogenesis is one of the main responses of the tumour to overcome the hypoxia; (3) the rapid and uncontrolled cell division that leads to an increased cell density and/or to the presence of certain proteins; and (4) tryptophan degradation in women with breast cancer [73], and it shows a resonance absorption peak at 1.435 THz [74], and other amino acids play a crucial role in the proliferation of tumour cells and their influence in the contrast is still in evaluation. There is a dependence of the tumour development upon the nutrients received through the blood. Tumour cells, like all other cells, need amino acids for their proliferation. The majority of the tumour cells have the capability to gather amino acids more than normal cells can [75].

In THz-TDS, the dysplastic tissue normally shows higher refractive indices and absorption coefficients that distinguish them from the normal ones. This fact could reinforce the feasibility of THz-TDS technique for gastrointestinal cancer detection. Furthermore, the works in this area demonstrate that the higher percentage of water in cancerous tissues is not the only factor contributing to the contrast of the observed through refractive indices and absorption coefficients spectra [76].

Author details

Faustino Wahaia[1,2]*, Irmantas Kašalynas[3], Gintaras Valušis[3], Catia D. Carvalho Silva[4] and Pedro L. Granja[1,2,5,6]

*Address all correspondence to: fwahaia@fc.up.pt

1 Institute for Innovation and Health Research (i3S), University of Porto, Porto, Portugal

2 Institute of Biomedical Engineering (INEB), University of Porto, Porto, Portugal

3 Department of Optoelectronics, Center for Physical Sciences and Technology, Vilnius, Lithuania

4 Centro Hospitalar São João, Porto, Portugal

5 Institute of Biomedical Sciences Abel Salazar (ICBAS), Universidade do Porto, Porto, Portugal

6 Faculty of Engineering, University of Porto (FEUP), Porto, Portugal

References

[1] Gallerano GP. Overview of Terahertz Radiation Sources. In: *FEL Conference*; 2004. Publ. JACoW.Org

[2] Mukherjee P, Gupta B. Terahertz (THz) Frequency sources and antennas—a brief review. *Int J Infrared Millimeter Waves*. 2008;29(12):1091–1102. doi:10.1007/s10762-008-9423-0.

[3] Nagatsuma T, Kaino A, Hisatake S, et al. Continuous-wave terahertz spectroscopy system based on photodiodes. *PIERS Online*. 2010;6:390–394. doi:10.2529/PIERS091029132808.

[4] Nishizawa J, Suto K, Sasaki T, et al. Spectral measurement of terahertz vibrations of biomolecules using a GaP terahertz-wave generator with automatic scanning control. *J Phys D Appl Phys*. 2003;36(23):2958–2961. doi:10.1088/0022-3727/36/23/015.

[5] Woodward RM, Wallace VP, Arnone DD, Linfield EH, Pepper M. Terahertz pulse imaging of skin cancer with time and frequency domain. *J Biol Phys*. 2003;29:257–261.

[6] Ferguson B, Zhang X-C. Materials for terahertz science and technology. *Nat Mater*. 2002;1(1):26–33. doi:10.1038/nmat708.

[7] Pickwell E, Wallace VP. Biomedical applications of terahertz technology. *J Phys D Appl Phys*. 2006;39(17):R301–R310.

[8] Kaindl RA, Carnahan MA, Hägele D, Lövenich R, Chemla DS. Ultrafast terahertz probes of transient conducting and insulating phases in an electron–hole gas. *Nature*. 2003;423(6941):734–738. doi:10.1038/nature01676.

[9] Seibert JA, Boone JM. X-ray imaging physics for nuclear medicine technologists. Part 2: X-ray interactions and image formation. *J Nucl Med Technol*. 2005;33(1):3–18. doi:33/1/3 [pii].

[10] Yin X-X, Hadjiloucas S, Zhang Y. Complex extreme learning machine applications in terahertz pulsed signals feature sets. DOI: http://dx.doi.org/10.1016/j.cmpb.2014.06.002

[11] Fan S, He Y, Ung BS, et al. The growth of biomedical terahertz research. *J Phys D Appl Phys*. 2014;47(37):374009. doi:10.1088/0022-3727/47/37/374009.

[12] Globus T. Sub-terahertz vibrational spectroscopy with high resolution for biological molecules and cells identification. *J Biomol Res Ther*. 2016;2016. doi:10.4172/2167-7956.1000E150.

[13] Yang X, Zhao X, Yang K, et al. Biomedical applications of terahertz spectroscopy and imaging. *Trends Biotechnol*. 2016;34(10):810–824. doi:10.1016/j.tibtech.2016.04.008.

[14] Schirmer M, Fujio M, Minami M, Miura J, Araki T, Yasui T. Biomedical applications of a real-time terahertz color scanner. *Biomed Opt Express*. 2010;1(2):354–366. doi:10.1364/BOE.1.000354.

[15] Fischer BM, Helm H, Jepsen PU. Chemical recognition with broadband THz spectroscopy. *Proc IEEE*. 2007;95(8):1592–1604. doi:10.1109/JPROC.2007.898904.

[16] Kimmit MF. Restrahlen to T-rays—100 years of terahertz radiation. *J Biol Phys*. 2003;29:75–85.

[17] Tonouchi M. Cutting-edge terahertz technology. *Nat Photonics*. 2007;1(2):97–105. doi:10.1038/nphoton.2007.3.

[18] Uhd JP, Jacobsen RH, Keiding SR. Generation and detection of terahertz pulses from biased semiconductor antennas. *Journal of the Optical Society of America* B. 1996;13(11):2424–2436.

[19] Wahaia F, Valusis G, Bernardo LM, et al. Detection of colon cancer by terahertz techniques. *J Mol Struct*. 2011;1006(1–3):77–82. doi:10.1016/j.molstruc.2011.05.049.

[20] Rolf Hulthén. Kramers–Kronig relations generalized: on dispersion relations for finite frequency intervals. A spectrum-restoring filter. *Journal of the Optical Society of America.* 1982;72(6):794–803.

[21] Mickan SP, Lee K-S, Lu T-M, Munch J, Abbott D, Zhang X-C. Double modulated differential THz-TDS for thin film dielectric characterization. *Microelectronics Journal.* 33 (2002) 1033–1042.

[22] Chen Y, Liu H, Deng Y, et al. THz spectroscopic investigation of 2,4-dinitrotoluene. *Chem Phys Lett.* 2004;400(4-6):357–361. doi:10.1016/j.cplett.2004.10.117.

[23] Un-ichi Shikata, Hiroyuki Handa et al. Terahertz ATR spectroscopy of liquids using THz-wave parametric sources. 2007 Conf Lasers Electro-Optics, IEEE - Pacific Rim. 2007:1–2.

[24] Hirori H, Yamashita K, Nagai M, et al. Attenuated total reflection spectroscopy in time domain using terahertz coherent pulses. *Jpn J Appl Phys.* 2004;43(No. 10A):L1287–L1289. doi:10.1143/JJAP.43.L1287.

[25] Nagai M, Yada H, Arikawa T, Tanaka K. Terahertz time-domain attenuated total reflection spectroscopy in water and biological solution. *Int J Infrared Millimeter Waves.* 2007;27(4):505–515. doi:10.1007/s10762-006-9098-3.

[26] Hindle F, Yang C, Mouret G, et al. Recent developments of an opto-electronic THz spectrometer for high-resolution spectroscopy. *Sensors (Basel).* 2009;9(11):9039–9057. doi:10.3390/s91109039.

[27] Schubert O, Hohenleutner M, Langer F, et al. Sub-cycle control of terahertz high-harmonic generation by dynamical Bloch oscillations. *Nature Photonics.* 2014;8:119–123.

[28] Liu H-B, Zhang X-C. Terahertz Spectroscopy for Explosive, Pharmaceutical, and Biological Sensing Applications. In: *Terahertz Frequency Detection and Identification of Materials and Objects.* Dordrecht: Springer Netherlands; 2007:251–323. doi:10.1007/978-1-4020-6503-3_17.

[29] Kissi EO, Bawuah P, Silfsten P, Peiponen K-E. A tape method for fast characterization and identification of active pharmaceutical ingredients in the 2–18 THz spectral range. *J Infrared, Millimeter, Terahertz Waves.* 2015;36(3):278–290. doi:10.1007/s10762-014-0123-7.

[30] Larkin PJ, Dabros M, Sarsfield B, Chan E, Carriere JT, Smith BC. Polymorph characterization of active pharmaceutical ingredients (APIs) using low-frequency Raman spectroscopy. *Appl Spectrosc.* 2014;68(7):758–776. doi:10.1366/13-07329.

[31] Dobroiu A, Sasaki Y, Shibuya T, Otani C, Kawase K. THz-wave spectroscopy applied to the detection of illicit drugs in mail. *Proc IEEE.* 2007;95(8):1566–1575. doi:10.1109/JPROC.2007.898840.

[32] Abina A, Puc U, Jeglič A, Zidanšek A. Applications of terahertz spectroscopy in the field of construction and building materials. *Appl Spectrosc Rev.* 2015;50(4):279–303. doi:10.1080/05704928.2014.965825.

[33] Mangeney J. THz photoconductive antennas made from ion-bombarded semiconductors. *J Infrared, Millimeter, Terahertz Waves*. 2012;33(4):455–473. doi:10.1007/s10762-011-9848-8.

[34] Duvillaret L, Garet F, Coutaz J-L. A reliable method for extraction of material parameters in terahertz time-domain spectroscopy. *IEEE J Sel Top Quantum Electron*. 1996;2(3):739–746. doi:10.1109/2944.571775.

[35] Krüger M, Funkner S, Bründermann E, Havenith M. Uncertainty and ambiguity in terahertz parameter extraction and data analysis. http://dx.doi.org/10.1007/. Accessed October 23, 2016.

[36] Obradovic J, Collins JHP, Hirsch O, Mantle MD, Johns ML, Gladden LF. The use of THz time-domain reflection measurements to investigate solvent diffusion in polymers. *Polymer (Guildf)*. 2007;48(12):3494–3503. doi:10.1016/j.polymer.2007.04.010.

[37] Mickan SP, Lee K-S, Lu T-M, et al. Thin film characterization using terahertz differential time-domain spectroscopy and double modulation Proceedings of SPIE Vol. 4591 (2001) © 2001 SPIE · 0277-786X/01/$15.

[38] Dorney TD, Baraniuk RG, Mittleman DM. Material parameter estimation with terahertz time-domain spectroscopy. *J Opt Soc Am A*. 2001;18(7):1562. doi:10.1364/JOSAA.18.001562.

[39] Scheller M. Data extraction from terahertz time domain spectroscopy measurements. *J Infrared, Millimeter, Terahertz Waves*. 2014;35(8):638–648. doi:10.1007/s10762-014-0053-4.

[40] Lee JW, Seo MA, Kim DS, et al. Fabry–Perot effects in THz time-domain spectroscopy of plasmonic band-gap structures. *Appl Phys Lett*. 2006;88(7):71114. doi:10.1063/1.2174104.

[41] Duvillaret L, Garet F, Coutaz J-L. A reliable method for extraction of material parameters in terahertz time-domain spectroscopy. *IEEE J Sel Top Quantum Electron*. 1996;2(3):739–746

[42] He M, Li M, Zhang W. Terahertz time-domain spectroscopy signature of animal tissues. 2008:274–277.

[43] Ying H, Huang P, Guo L, Wang, Zhang C. Terahertz spectroscopic investigations of explosives. *Phys Lett*. 2006;359:728–732.

[44] Jepsen PU, Fischer BM. Dynamic range in terahertz time-domain transmission and reflection spectroscopy. *Opt Lett*. 2005;30(1):29–31.

[45] Xu J, Yuan T, Mickan S, Zhang X-C. Limit of spectral resolution in terahertz time-domain spectroscopy. *Chines Phys Lett*. 2003;20(8):1266–1268.

[46] Gremel G, Wanders A, Cedernaes J, et al. The human gastrointestinal tract-specific transcriptome and proteome as defined by RNA sequencing and antibody-based profiling. *J Gastroenterol*. 2015;50(1):46–57. doi:10.1007/s00535-014-0958-7.

[47] B. ODE, J.C. Alcohol and the gastrointestinal tract. *Advances in Internal Medicine and Pediatrics*. 1980;45:1–75.

[48] Camilleri M, Linden DR. Measurement of gastrointestinal and colonic motor functions in humans and animals. *C Cell Mol Gastroenterol Hepatol*. 2016;2(4):412–428. doi:10.1016/j.jcmgh.2016.04.003.

[49] NCI. The Cancer of the Colon and Rectum. (Institute NC, ed.). NIH; 2006. ISBN: 0-309-65952-3, 340 pages, 6 x 9, (2006).

[50] Fitzmaurice C, Dicker D, Pain A, et al. The global burden of cancer 2013. *JAMA Oncol*. 2015;1(4):505. doi:10.1001/jamaoncol.2015.0735.

[51] Hochrein T. Markets, availability, notice, and technical performance of terahertz systems: Historic development, present, and trends. *J Infrared, Millimeter, Terahertz Waves*. 2015;36(3):235–254. doi:10.1007/s10762-014-0124-6.

[52] Ionization and Fragmentation of Complex Molecules and Clusters. Henrik A. B. Johansson, Stockholm 2011. ISBN 978-91-7447-399-5.

[53] Png GM, Choi JW, Ng B, Mickan SP, Abbott D. The impact of hydration changes in fresh bio-tissue on THz spectroscopic measurements. *Phys Med Biol*. 2008;53:3501–3517.

[54] Wahaia F, Valusis G, Bernardo LM, et al. Detection of colon cancer by terahertz techniques. *J Mol Struct*. 2011;1006(1–3):77–82. doi:10.1016/j.molstruc.2011.05.049.

[55] Wallace VP, Fitzgerald AJ, MacPherson EP, Taday RJPPF, Flanagan N, Ha T. Terahertz pulsed spectroscopy of human basal cell carcinoma. *Appl Spectrosc*. 2006;60 (10):1127–1133.

[56] MacPherson EP, Fitzgerald AJ, Taday PF, et al. Terahertz imaging and spectroscopy of skin cancer. *Biol Med Appl*. 2004:821–822.

[57] Sun Y, Fischer MB, MacPherson EP. Effects of formalin fixing on the THz properties of biological tissues. *J Biol Opt*. 2009;14(6):6401–6417.

[58] Reynolds TY, Rockwell S, Glazer PM. Genetic instability induced by the tumor microenvironment. *Cancer Res*. 1996;56:5754–5757.

[59] Mclntyre GI. Cell hydration as the primary factor in carcinogenesis: Unifying concept. *Med hypothesis*. 2006;66:518–526.

[60] Mcltyre GI. Increased cell hydration promotes both tumor growth and metastasis: A biochemical mechanism consistent with genetic signatures. *Med hypothesis*. 2007;69:1127–1130.

[61] Ashworth PC, Pickwell-MacPherson E, Provenzano E, et al. Terahertz pulsed spectroscopy of freshly excised human breast cancer. *Opt Express*. 2009;15:12444.

[62] Pickwell E, Cole B, FitzgeraldJ AJ, Pepper M, vivo study of human skin using pulsed terahertz radiation Phys. in Med. VPW 2004 I, 1595-1607 B 49. In vivo study of human skin using pulsed terahertz radiation. *Phys Med Biol*. 2004;49:1595–1607.

[63] Fitzgerald AJ, Berry E, Zinov'ev NN, et al. Catalogue of human tissue optical properties at terahertz frequencies. *J. Biol. Phys*. 2003;29(2–3):123–128.

[64] Tatiana Globus, Tatyana Khromova et al.,Terahertz characterization of dilute solutions of DNA. Proc. SPIE 6093, Biomedical Vibrational Spectroscopy III: Advances in Research and Industry, 609308 (2006).

[65] Xu J, Plaxco KW, Allen SJ. Probing the collective vibrational dynamics of a protein in liquid water by terahertz absorption spectroscopy. *Protein Sci.* 2006;15(5):1175–1181. doi:10.1110/ps.062073506.

[66] Wallace VP, Fitzgerald AJ, Pickwell E, Pye RJ, Taday PF, Flanagan N and Ha T. Terahertz pulsed spectroscopy of human basal cell carcinoma. *Appl Spectrosc.* 2006;60(10):1127–1133.

[67] Brun M-A, Formanek F, Yasuda A, Sekine M, Ando N, Eishii Y. Terahertz imaging applied to cancer diagnosis. *Phys Med Biol.* 2010;55(16):4615–4623. doi:10.1088/0031-9155/55/16/001.

[68] Yu C, Fan S, Sun Y, Pickwell-Macpherson E. The potential of terahertz imaging for cancer diagnosis: A review of investigations to date. *Quant Imaging Med Surg.* 2012;2(1):33–45. doi:10.3978/j.issn.2223-4292.2012.01.04.

[69] Doradla P, Alavi K, Joseph C, Giles R. Detection of colon cancer by continuous-wave terahertz polarization imaging technique. *J Biomed Opt.* 2013;18(9):90504. doi:10.1117/1.JBO.18.9.090504.

[70] Chung SH, Cerussi AE, Klifa C, et al. In-vivo water state measurements in breast cancer using broadband diffuse optical spectroscopy. *Phys Med Biol.* 2008;53:6713–6727.

[71] Wahaia F. Spectroscopic and imaging techniques using terahertz frequency band for biomedical applications. PhD Thesis (2011).

[72] Ashworth PC, Pickwell-MacPherson E, Pinder SE, et al. Terahertz spectroscopy of breast tumors. In: *Infrared and Millimeter Waves, 2007 and the 2007 15th International Conference on Terahertz Electronics. IRMMW-THz. Joint 32nd International Conference on*; 2007:603–605.

[73] Lyon DE, Walter JM, Starkweather AR, Schubert CM, McCain NL. Tryptophan degradation in women with breast cancer: A pilot study. *BMC Res Notes.* 2011;4:156. doi:10.1186/1756-0500-4-156.

[74] Yu B, Zeng F, Xing YYQ, et al. Torsional vibrational modes of tryptophan studied by terahertz time-domain spectroscopy. *Biophys J.* 2004;86:1649–1654.

[75] Cherkasova OP, Nazarov MM, Shkurinov AP. Terahertz spectroscopy of biological molecules. *Radiophys Quantum Electron.* 2009;52 (7):518–523.

[76] Wahaia F, Kasalynas I, Seliuta D, et al. Study of paraffin-embedded colon cancer tissue using terahertz spectroscopy. *J Mol Struct.* 2015;1079:448–453. doi:10.1016/j.molstruc.2014.09.024.

Chapter 14

Nanostructured Indium Tin Oxides and Other Transparent Conducting Oxides: Characteristics and Applications in the THz Frequency Range

Ci-Ling Pan, Chan-Shan Yang, Ru-Pin Pan,
Peichen Yu and Gong-Ru Lin

Additional information is available at the end of the chapter

http://dx.doi.org/10.5772/66344

Abstract

Transparent conductors are essential for optoelectronic components operating in the far-infrared or terahertz (THz) frequency range. Indium tin oxide (ITO), extensively used in the visible, is semi-transparent in the far-infrared frequency range. Other types of bulk transparent conducting oxides (TCOs), such as aluminum-doped zinc oxide (AZO) and aluminum and ytterbium-doped zinc oxide (AYZO), have not yet been explored for THz applications. Recently, biomimic nanomaterials have been shown to exhibit exotic optical properties, e.g., broadband, omnidirectional antireflective properties. Indeed, nanostructured ITO was found to exhibit the above desirable characteristics. In this chapter, we describe the fabrication and characterization of several TCOs, including ITO nanomaterials and several types of bulk TCO thin films, e.g., AZO and AYZO. Performance of THz phase shifters with ITO nanomaterials as transparent electrodes and liquid crystals for functionalities is presented.

Keywords: terahertz, spectroscopy, transparent conducting oxides, indium tin oxide, nanostructures, nanorods, nanowhiskers, liquid crystals, far-infrared, millimeter wave, sub-millimeter wave, phase shifter, optoelectonics, ITO, AZO, AYZO

1. Introduction

Materials that exhibit good electrical conductivity and high optical transmittance are important in many applications. An interesting group of materials with these properties is known as transparent conducting oxides (TCOs) [1]. Among the TCOs, indium tin oxide (ITO) is one of the most frequently investigated. It is widely used as transparent electrodes in optoelectronic devices such as solar cells [2, 3]. Zinc oxide (ZnO), a well-known TCO, is a semiconductor with wide band gap of about 3.3 eV at room temperature. Even after careful deposition procedure, the intrinsically doped ZnO films are still not well suited for applications because of their relatively high resistivity of about 10^{-2}–10^{-3} Ω cm and their instability at high temperature. Therefore, the extrinsic doping is necessary for lowering the resistivity. Zinc oxide films can be doped with metal ions such as aluminum (Al), chromium (Cr), gallium (Ga), and indium (In). Among these dopants, Al is relatively cheap, abundant and nontoxic. Thus, Al-doped ZnO (AZO) films are considered as low-cost substitutes for ITO films as transparent conducting films. AZO film has transparency higher than 80% and resistivity lower than 7×10^{-4} Ω cm and is comparable with the ITO film. Further, AZO films show high stability in hydrogen plasma during the fabrication process of amorphous silicon solar cells. In order to solve the degradation problem of AZO thin films, Al and ytterbium-doped ZnO (AYZO) is developed.

In view of emerging applications of terahertz (THz) technology, it is interesting to compare the THz performance of these TCO films and explore their potential use in THz optoelectronic devices. In the terahertz (THz) or far-infrared (FIR) frequency range, however, ITO exhibits high reflectance and strong absorption [2]. Recently, ITO nanocolumns have attracted a lot of attentions, because of its excellent optoelectronic and antireflection characteristics in the visible and near-infrared range [2, 3]. The successful development of ITO nanostructures as functional transparent electrodes relies on the clear understanding of their optical and electrical properties, which are linked to the structural properties. Unfortunately, previous studies mainly focused on preparation techniques and optical properties [4], except for our studies on mid- and far-infrared studies of the electrical characteristics of ITO thin films [5]. It was confirmed that this material is a good dichroic mirror for the FIR.

On the other hand, nanostructures of TCOs have attracted much attention. Recently, our group has confirmed that ITO nanomaterials, e.g., nanocolumn, nanorods, nanowires, and nanowhiskers (NWhs), exhibit superhydrophilicity as well as omnidirectional, broadband antireflective (AR) characteristics in the near-infrared and visible frequency range [6, 7]. These novel nanostructured TCOs have been employed successfully in devices such as light-emitting diodes (LEDs) and solar cells [8, 9]. Further, we showed that ITO NWhs are good conductors and highly transparent in the THz frequency range. The dielectric and conductive properties of these nanomaterials were investigated using THz time-domain spectroscopy (THz-TDS), which has been widely used to study the characteristics of novel materials, including metal thin films [10], nanostructure of ZnO [11] and CdSe quantum dots of various sizes [12].

In this chapter, we survey the frequency-dependent complex refractive indices and conductivities of several TCOs for their potential applications in the THz devices. TCOs studied include ITO nanomaterials, ITO sputtering films, AZO, and AYZO thin films. Besides, we will describe THz phase shifters employing ITO nanomaterials as transparent electrodes.

2. Preparation and characteristics of the transparent conducting oxides

2.1. ITO nanomaterials

Indium tin-oxide (ITO) nanorods and nanowhiskers were deposited on the high resistivity silicon substrate (499 μm) using glancing-angle electron-beam evaporation, and the nanostructures

http://dx.doi.org/10.5772/66344

were deposited on the microgrooved surface. The double-side-polished silicon substrate was attached to a holder, which was tilted at a deposition angle of 70° with respect to the incident vapor flux. There were seven holders which circle around the center of the chamber at a speed of 10 rpm during the deposition. The target source contained 5 wt % SnO_2 and 95 wt % In_2O_3. The chamber was first pumped down to a residue pressure of about 10^{-6} torr. Nitrogen gas is then injected at a flow rate at 1 sccm. This results in an oxygen-deficient atmosphere in the growth chamber. The working condition of the growth chamber was ~10^{-4} torr and 260°C. The tilted top view and cross-sectional scanning electron microscopy (SEM) images of the ITO nanorods are given in **Figure 1**) and those of nanowhiskers are given in **Figure 2**.

2.2. ITO sputtering thin films

Indium tin-oxide (ITO) thin films were grown on the high-resistivity silicon substrate at a temperature of 250°C using DC reactive magnetron sputtering. The target source composed of 5 wt.% SnO_2 and 95 wt.% In_2O_3. The magnetron operated at a power setting of about 300 W. While the film was being grown, argon and oxygen gas at flow rates of 20 and 0.4 sccm were injected such that the total pressure in the chamber was ~6 mtorr. All the

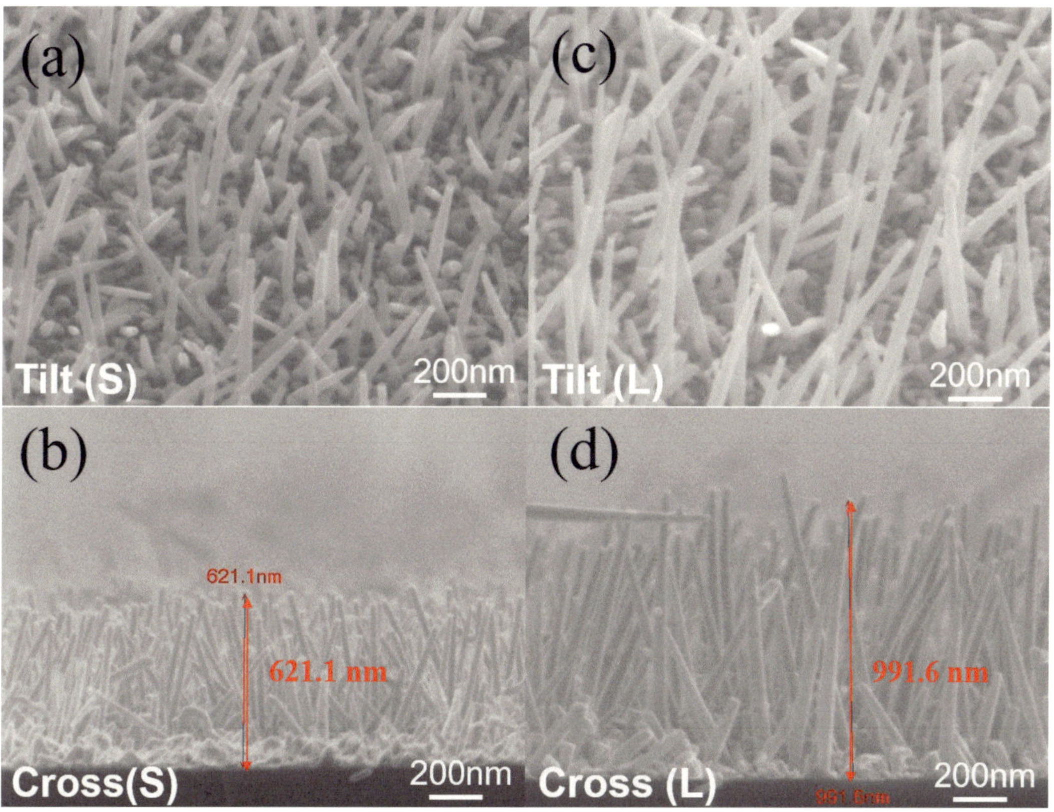

Figure 1. Tilted top view and cross-sectional scanning electron micrographs of ITO nanorods grown on silicon wafer with different lengths (a and b) 621.1 nm and (c and d) 991.6 nm. The scalar bars are 200 nm in length. Adapted from Ref. [13].

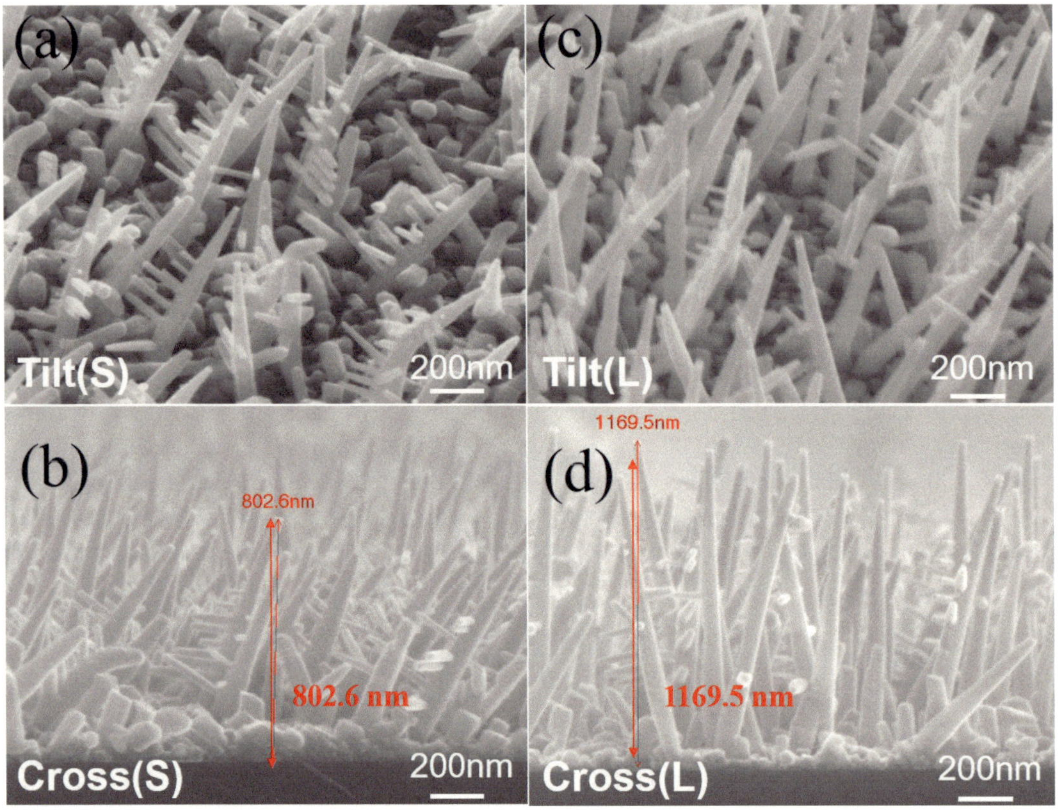

Figure 2. Tilted top view and cross-sectional scanning electron micrographs of ITO nanowhiskers grown on silicon wafer with different lengths (a and b) 802.6 nm and (c and d) 1169.5 nm. The scalar bars are 200 nm in length. Adapted from Ref. [13].

ITO thin films we studied were as-grown, not subject to further annealing. The thin films we studied were 333, 345, 615, and 1062 nm in thickness, as measured by an n&k Analyzer, model 1280.

2.3. AZO and AYZO thin films

All the AZO and AYZO films were deposited on undoped Si substrates by using the RF sputtering system. During the AZO process, the AZO target was composed of 98 wt.% ZnO and 2 wt.% Al. The chamber pressure and RF plasma power were controlled at 3 mtorr and 0.8 W/cm^2, respectively. The argon (Ar) fluence was fixed at 30 sccm during deposition. The AYZO target, on the other hand, was composed of 98 wt.% ZnO, 1 wt.% Al, and 1 wt.% Yb. The chamber pressure and magnetron plasma power were controlled at 3 mtorr and 0.896 W/cm^2, respectively. The chamber pressure and magnetron plasma power were set at 3 mtorr and 0.231 W/cm^2, respectively. The Ar and oxygen fluences were fixed at 12 and 1 sccm, respectively, during deposition process.

http://dx.doi.org/10.5772/66344

3. Experimental and analytical methods

3.1. Broadband THz time-domain spectroscopy (THz-TDS)

We employed several THz-TDS systems, including transmission and reflection types, among which the broadband system is shown in **Figure 3** as an example. In this system, we employed a commercial femtosecond Ti: sapphire laser system (Spectra Physics, Spitfire) as the light source, which provides 50 fs pulse duration at 815 nm central wavelength and 1.6 mJ pulse energy at 1 kHz. In the experimental setup, the incident beam is divided into the pump beam and probe beam by the beam splitter. Then, the collimated fundamental wave of the pump beam is focused on the focal point by the convex lens (with a focal length of $f = 15$ cm), and the intensity must be high enough to produce the plasma in the air. On the other hand, we use the 100-μm-thick β-barium-borate (BBO) crystal, which is cut so as to achieve type-I phase-matching for second-harmonic generation (SHG) of the 815 nm fundamental light, and the fundamental and second harmonic wave will form the asymmetric laser field. In other words, it means that electron is ionized by the fundamental and second-harmonic wave in the focal point. The ionized electrons in the plasma region are accelerated by the asymmetric laser field, which induces the electron oscillation. Subsequently, the THz wave is generated by the oscillation electron. The emitted THz radiation is collimated by an off-axis parabolic mirror with an effective focal length of $f = 50.8$ mm and then focused with a second one (also with $f = 100.16$ mm) onto the sample.The pump laser beam transmitted through the plasma region is blocked by the silicon wafer which has high transmissivity in the THz frequency region, while strongly absorbing the pump laser photons. Another pair of off-axis parabolic mirrors with the same focal length as that of the previous pair focus the THz signal into the detector. The resulting THz field was measured using an electro-optic sampling technique employing a 0.3-mm-thick GaP<110> crystal. The pellicle beam splitter which is transparent in THz frequency and the reflection is 40% in the 800 nm. In order to increase the signal-to-noise ratio (SNR), the chopper and lock-in amplifier are used. The frequency of chopper is about 500 Hz which is the most stable condition in our system. The translation stage scans the THz wave with 5 μm and total of 1024 steps. An antenna-based THz-TDS, as described in our previous works [2, 14, 15], was also employed. The spectroscopic characteristics of both systems are illustrated in Figure 2. Clearly, the two systems would allow us to study material properties in the band of 0.15–9.00 THz.

3.2. Extraction of optical parameters by THz-TDS

We extract the optical parameters of different kinds of materials by analyzing the THz waveform obtained from the THz-TDS. In this section, the method of extracting the frequency-dependent complex refractive index will be introduced. The process is achieved through the Fresnel equations and the multireflection approximation while numerical calculation is also applied.

The thin sample is grown on its substrate and its thickness is usually several micrometers or even several hundred nanometers. The time delays for thin samples whose thicknesses are of

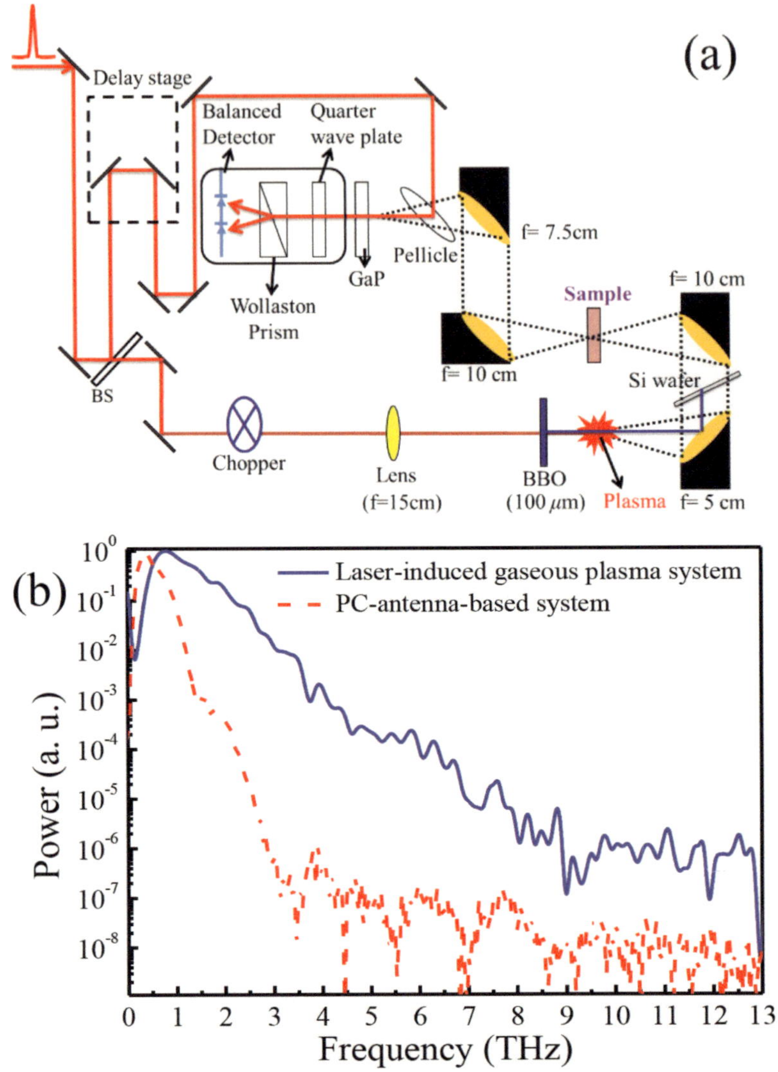

Figure 3. (a) Experimental setup of THz-TDS based on laser-induced gas plasma. (b) Typical performance of the THz-TDS systems based on laser-induced gaseous plasma (blue solid line) and PC antenna (red dash line) is shown. Adapted from Ref. [13].

the order of nanometers are smaller than 1 ps. Therefore, we are not able to distinguish the multireflected signal from the main signal. **Figure 4** shows that the main peak includes all multireflected signals which come from the thin sample layer and the second signal is caused by the reflection of the relatively thick substrate. In cases like this, we have to take multiple reflections into account in our analysis [5].

Schematic diagrams of electromagnetic model are shown in **Figure 5**. $E_0^*(\omega)$ is the incident THz field, and $E_{ref}^*(\omega)$ is the field which has passed through the substrate only. $E_{\text{film}}^*(\omega)$ is the field which has transmitted through the thin film sample including thin film layer and substrate

http://dx.doi.org/10.5772/66344

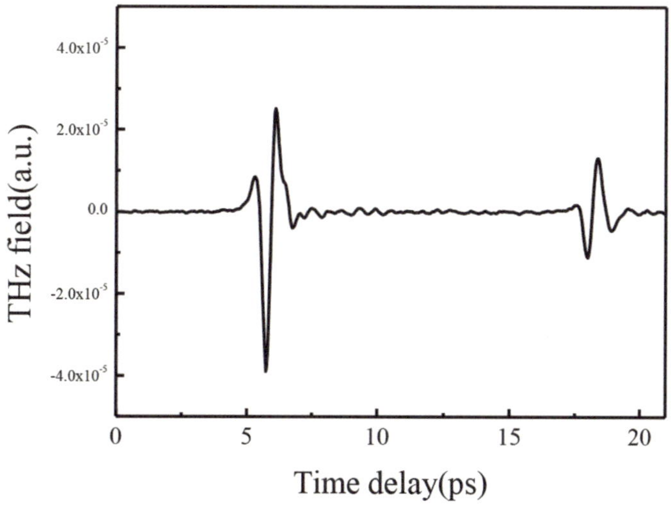

Figure 4. Transmitted THz waveform of 273 nm ITO thin film which is grown on fused silica substrate. The reflection signal appears behind the main signal for about 12 ps. The thickness of the substrate is 941 μm.

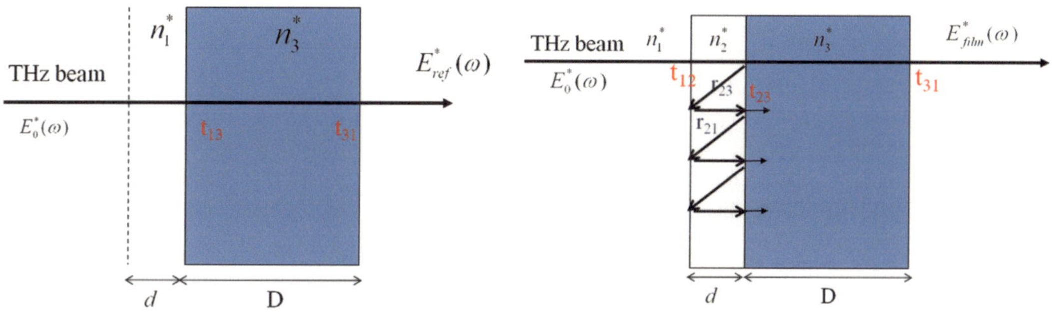

Figure 5. Schematic diagrams of electromagnetic model for reference and thin sample, respectively.

while the effect of multireflection is not considered here. $n_1^* = 1$, $n_2^* = n_2 + i\kappa_2$, and n_3^* are the refractive indices of air, thin film, and substrate, respectively. D is the thickness of the substrate and d is the sample layer. c_0 is the speed of light.

In our experimental setup, the THz pulse normally incidents the sample. Therefore, the well-known Fresnel equations of reflectance coefficient and transmittance coefficient can be expressed as the following equations:

$$t_{ij} = \frac{2n_i^*}{n_i^* + n_j^*} \tag{1}$$

$$r_{ij} = \frac{n_i^* - n_j^*}{n_i^* + n_j^*} \tag{2}$$

From Eq. (1) and (2), $E_{ref}^*(\omega)$ can be determined as the following equation.

$$E^*_{ref}(\omega) = E_0(\omega)\, t_{13}\, t_{31}\, e^{i\frac{n_1^*\omega d + n_3^*\omega D}{c}} \tag{3}$$

With the thin film layer, multireflection effect must be considered. The field which has passed through the thin film layer and is going to pass substrate layer can be noted as $E^*_{\text{sample}}(\omega)$. The field $E^*_{\text{sample}}(\omega)$ therefore has the relationship with $E^*_{\text{film}}(\omega)$ expressed as

$$E^*_{\text{film}}(\omega) = E^*_{\text{sample}}(\omega)\, t_{31}\, e^{i\frac{n_3^*\omega D}{c}} \tag{4}$$

$$\begin{aligned} E^*_{\text{sample}}(\omega) &= E_0^*(\omega)\, t_{12}\, t_{23}\, e^{i\frac{n_2^*\omega d}{c}} + E_0^*(\omega)\, t_{12}\, r_{23}\, r_{21}\, t_{23}\, e^{i\frac{3n_2^*\omega d}{c}} + E_0^*(\omega)\, t_{12}\, r_{23}{}^2\, r_{21}{}^2\, t_{23}\, e^{i\frac{5n_2^*\omega d}{c}} \\ &+\cdots + E_0^*(\omega)\, t_{12}\, r_{23}{}^q\, r_{21}{}^q\, t_{23}\, e^{i\frac{(2q+1)n_2^*\omega d}{c}} \end{aligned} \tag{5}$$

Here q is the number of multiple reflections. Assuming the number of multiple reflections is infinite ($q \to \infty$), the field $E^*_{\text{sample}}(\omega)$ can be simplified as

$$E^*_{\text{sample}}(\omega) = E_0^*(\omega) \frac{t_{12}\, t_{23}\, e^{i\frac{n_2^*\omega d}{c}}}{1 - r_{21}\, r_{23}\, e^{i\frac{2n_2^*\omega d}{c}}} \tag{6}$$

From Eqs. (3), (4), and (6), the theoretical complex transmittance can be determined.

$$T^*_{\text{theo}}(\omega) = \frac{E^*_{\text{film}}(\omega)}{E^*_{ref}(\omega)} = \frac{t_{12}\, t_{23}\, e^{i\frac{(n_2^*-1)\omega d}{c}}}{t_{13}\left(1 - r_{21}\, r_{23}\, e^{i\frac{2n_2^*\omega d}{c}}\right)} \tag{7}$$

Here t_{12}, t_{13}, t_{31}, t_{23}, r_{21}, and r_{23} are the Fresnel transmission and reflection coefficients which come from the Fresnel equations and are shown below.

$$t_{12} = \frac{2n_1^*}{n_1^* + n_2^*},\ t_{23} = \frac{2n_2^*}{n_2^* + n_3^*},\ t_{13} = \frac{2n_1^*}{n_1^* + n_3^*},\ r_{23} = \frac{n_2^* - n_3^*}{n_2^* + n_3^*},\ \text{and}\ r_{21} = \frac{n_2^* - n_1^*}{n_2^* + n_1^*} \tag{8}$$

From the above-mentioned formulas, the theoretical complex transmittance can be determined. This model is utilized for extractions of complex refractive index for thin film grown on thick substrate.

In order to avoid the effect of difference of substrate thickness D between reference and sample, we have to modify the equations we use in the theoretical model. The concept of thickness correction is described in **Figure 6**. By modifying Eqs. (4) and (6), the transmitted field of the sample $E^*_{\text{film}}(\omega)$ (Eq. 9) can be derived when thickness correction is taken into account. While the transmitted field of reference $E^*_{ref}(\omega)$ is the same, we can conclude the theoretical complex transmittance $T^*_{\text{theo}}(\omega)$ (Eq. (10) by combining Eq. (3) and Eq. (9).

$$E^*_{\text{film}}(\omega) = E_0^*(\omega) \frac{t_{12}\, t_{23}\, t_{31}\, e^{i\frac{n_2^*\omega d}{c}}}{1 - r_{21}\, r_{23}\, e^{i\frac{2n_2^*\omega d}{c}}} e^{i\frac{n_1^*\omega \Delta d + n_3^*\omega(D-\Delta d)}{c}} \tag{9}$$

$$T^*_{\text{theo}}(\omega) = \frac{E^*_{\text{film}}(\omega)}{E^*_{ref}(\omega)} = \frac{t_{12}\, t_{23}\, e^{i\frac{(n_2^*-1)\omega d}{c}}}{t_{13}\left(1 - r_{21}\, r_{23}\, e^{i\frac{2n_2^*\omega d}{c}}\right) e^{i\frac{(n_3^*-1)\omega \Delta d}{c}}} \tag{10}$$

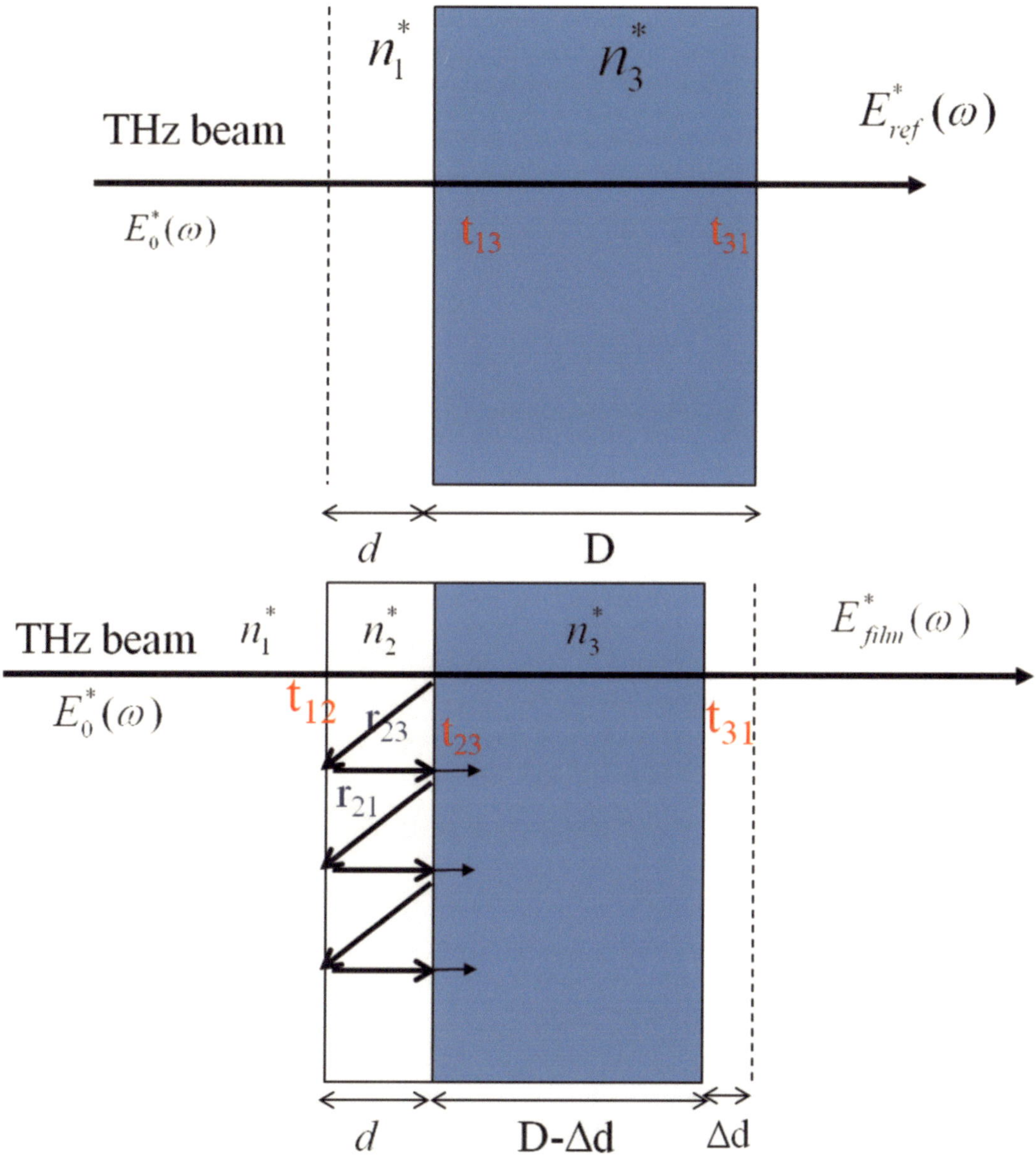

Figure 6. Schematic diagram of thickness correction.

We obtain the experimental data from THz time-domain spectroscopy. The experimental complex transmittance $T^*_{\exp}(\omega)$ can be therefore derived.

$$T^*_{\exp}(\omega) = \frac{E^*_{\text{film}}(\omega)}{E^*_{ref}(\omega)} \tag{11}$$

Here the complex fields $E^*_{\text{film}}(\omega)$ and $E^*_{ref}(\omega)$ are obtained from experimental data which are the results of transmitted THz pulse for sample and reference respectively. By comparing the complex transmittance of theoretical and experimental results, the unknown frequency-dependent complex refractive index n_2^* of the thin film can be extracted. First of all, the error function between the theoretical and experimental complex transmittance is defined as Eq. (12). Because of the unknown refractive index n_2^*, the transmittance is dependent both on frequency and n_2^*.

$$\text{Error}(\omega, n_2^*) = \left| T^*_{\text{exp}}(\omega, n_2^*) - T^*_{\text{theo}}(\omega, n_2^*) \right| \tag{12}$$

$$\text{Find Minimum} = \left[\text{Error}(\omega, n_2^*)\right] \tag{13}$$

With the help of mathematical program, we may make the theoretical transmittance match the experimental transmittance to a minimum difference value by finding a suitable complex refractive index n_2^*. The index $n_2^* = n_2 + i\,\kappa_2$ therefore can be extracted.

3.3. Terahertz conductivity

For a nonmagnetic medium, the frequency-dependent dielectric constant is the square of the complex refractive index. In the following equations, ε^* is the complex dielectric constant with real and imaginary parts, i.e., ε_r, n and ε_i, κ are the real and imaginary parts of the refractive index, respectively, all functions of frequency.

$$\varepsilon^* = \varepsilon_r + i\,\varepsilon_i = (n + i\kappa)^2 \tag{14}$$

$$\varepsilon_r = n^2 - \kappa^2\,,\ \varepsilon_i = 2n\kappa \tag{15}$$

From the relationship mentioned above, the real and imaginary parts of dielectric constant are derived. Furthermore, the dielectric constant is related to the optical conductivity which will be fully described by the following paragraph.

Secondly, with several approximations, the Maxwell's equation can be modified into the form related to conductivity. We assume that the flowing current density has the form of $\vec{J} = \sigma^* \vec{E}$ (Ohm's law) where σ^* is complex conductivity. Also, the harmonic assumption for electric field $\vec{E}(\omega, t) = \vec{E}_0 e^{-i\omega t}$, where $\vec{E}_0$ is the amplitude and ω and t represent frequency and time is applied. And the relation between displacement field and electric field $\vec{D} = \varepsilon_0 \varepsilon_\infty \vec{E}$ is also used.

$$\begin{aligned} \nabla \times \vec{H} &= \vec{J} + \frac{\partial \vec{D}}{\partial t} = \vec{J} - i\omega\,\varepsilon_0\,\varepsilon_\infty \vec{E} \\ &= -i\omega\,\varepsilon_0\left[\varepsilon_\infty - \frac{\sigma^*}{i\omega\,\varepsilon_0}\right]\vec{E} \\ &= -i\omega\,\varepsilon_0\,\varepsilon^* \vec{E} \end{aligned} \tag{16}$$

$$\varepsilon^* = \varepsilon_\infty + i\frac{\sigma^*}{\omega\,\varepsilon_0} \tag{17}$$

Here $\varepsilon_0 = 8.854 \times 10^{-12}$ (F/m) is the free-space permittivity. σ^* is complex conductivity, and ε^* and ε_∞ are the complex dielectric function and the dielectric constant, respectively. From Eq. (17), the complex conductivity can be derived:

$$\sigma^*(\omega) = (\sigma_r + i\,\sigma_i) = i\omega\,\varepsilon_0(\varepsilon_\infty - \varepsilon^*) \tag{18}$$

By substituting Eqs. (14) and (15) into Eq. (18), the complex conductivity's real and imaginary parts can be determined:

$$\sigma_r = \omega\,\varepsilon_0\,\varepsilon_i = 2n\kappa\omega\,\varepsilon_0 \tag{19}$$

$$\sigma_i = \omega\,\varepsilon_0(\varepsilon_\infty - \varepsilon_r) = \omega\,\varepsilon_0(\varepsilon_\infty - n^2 + \kappa^2) \tag{20}$$

The nanostructured ITO was considered as a composite material of ITO and air. The effective medium approximation (EMA) [16, 17] was assumed to be valid. We write the equivalent dielectric function of the nanostructured material of interest as

$$\varepsilon^*_{\text{Equivalent}} = f \times \varepsilon^*_{\text{m}} + (1 - f) \times \varepsilon^*_{\text{h}} \tag{21}$$

In Eq. (21), ε^*_{m} and ε^*_{h} are the dielectric constants of materials such as ITO and air, respectively, whereas f is the filling factor, i.e., the volume fraction of the nanomaterial in the composite. Similarly, the real (ε_{Re}) and imaginary (ε_{Im}) parts of the dielectric constants of pure nanostructured material can be written as

$$\varepsilon_{\text{Re}} = \left[n^2 - \kappa^2 - (1 - f)\right]/f \tag{22}$$

$$\varepsilon_{\text{Im}} = [2 \cdot n \cdot \kappa]/f \tag{23}$$

Finally, the real (σ_r) and imaginary (σ_i) parts of the complex conductivities of pure ITO nanostructures can be expressed in terms of the filling factors as

$$\sigma_r = \omega\,\varepsilon_0 \cdot (2n\kappa)/f \tag{24}$$

$$\sigma_i = \omega\,\varepsilon_0\left[\varepsilon_\infty - \left[n^2 - \kappa^2 - (1 - f)\right]/f\right] \tag{25}$$

Consequently, the complex conductivity can be acquired by THz-TDS and the analysis is just based on the most fundamental Fresnel equations and Maxwell's equation without any other complicated model.

3.4. Non-Drude behavior in terahertz conductivity

The Drude model was often used to analyze conductive properties of metals and bulk semiconductors. In the model, the conductivity is given by $\sigma^*(\omega) = \varepsilon_0 \omega_p^2 \tau / (1 - i\omega\tau)$ [18]. In this expression, τ and ω_p are scattering times of carriers in the material and plasma frequency, respectively. According to the model, the real part of conductivity, Re{σ}, reaches its highest value when $\tau = 0$. This is the so-called DC conductivity. At higher frequencies, Re{σ} decreases because of the $1/\tau$ dependence. On the other hand, the value of Im{σ} is always positive. Its maximum occurs when the frequency is in the vicinity of the inverse of the scattering rate [18]. The THz conductivities of ITO nanostructures, however, exhibit non-Drude-like behavior, e.g., depressed values of DC conductivity and negative values for Im{σ}.

Therefore, the Drude-Smith model, which takes into account the carrier localization effect, is applied to fit the experimentally deduced conductivity of the ITO material [19, 20]. In this model, we write

$$\sigma^*(\omega) = \varepsilon_0 \omega_p^2 \tau \cdot [1 + \gamma / (1 - i\omega\tau)] / (1 - i\omega\tau) \tag{26}$$

where ω_p and τ are defined as plasma frequency and scattering time, respectively. Also in Eq. (26), the parameter γ is equal to the expectation value of cosine of carrier scattering angle. It gauges the persistence of velocity of carriers. The values of γ vary from 0 to 1. If $\gamma \approx 0$, the material shows Drude-like behavior. That is, the scattering of carriers is isotropic. In the limit of $\gamma = -1$, on the other hand, the carriers fully backscattered. The carrier localization effect is strong. We have conducted a simulation study of the complex conductivity of ITO nanostructures based on Drude-Smith model. The plasma frequency (ω_p) and scattering time (τ) are fitting parameters. The best fitted values for these two are found to be equal to 363 rad THz and 49 fs, respectively. The cases of $\gamma = 0$, $\gamma = -0.25$, $\gamma = -0.75$, and $\gamma = -1$ are shown in **Figure 7(a–d)**, respectively. If $\gamma = 0$, see **Figure 7(a)**, the Drude-Smith model reduces to the Drude-like model, which has been described previously. In short, the curves for Re{σ} and Im{σ} versus frequency exhibit maxima. Further, the value of Re{σ} decreases with increasing frequency. For the case of $\gamma = -0.25$ (see **Figure 7(b)**), both maximum values will show a little shift toward higher frequency. For the cases of $\gamma = -0.75$ and -1 (see **Figure 7(c** and **d)**), the complex conductivities of the nanometer exhibit other features, e.g., a minimum for Im{σ}. Further, there is an angular frequency at which Im{σ} crosses zero. As a result, the Im{σ} is negative from DC to the zero-crossing frequency. The maxima of Re{σ} now occurs at a nonzero frequency. There are some studies that attribute the non-Drude-like behavior to backscattering of carriers by grain boundaries [21–25]. In addition, after releasing free electrons, the dopants in n-type TCOs will become positive ions. Due to the Coulomb interaction between free electrons and dopant ions, the latter will also act as scattering centers. Consequently, movement of the electrons will be further impeded [26, 27].

Armed with knowledge of ω_p and τ, the mobility (μ), carrier concentration (N_e), and DC conductivity (σ_0) of the material can then be calculated using the relations, $\mu = (1+\gamma)e\tau/m^*$, $N_e = \varepsilon_0\omega_p^2 m^*/e^2$, and $\sigma_0 = \varepsilon_0\omega_p^2\tau\,(1+\gamma)$, respectively. In these expressions, $e = 1.602 \times 10^{-19}$ C, is the electronic charge; the electron effective mass is given by $m^* = 0.3m_0$ [28], where $m_0 = 9.1094 \times 10^{-31}$ kg, is the electron's mass.

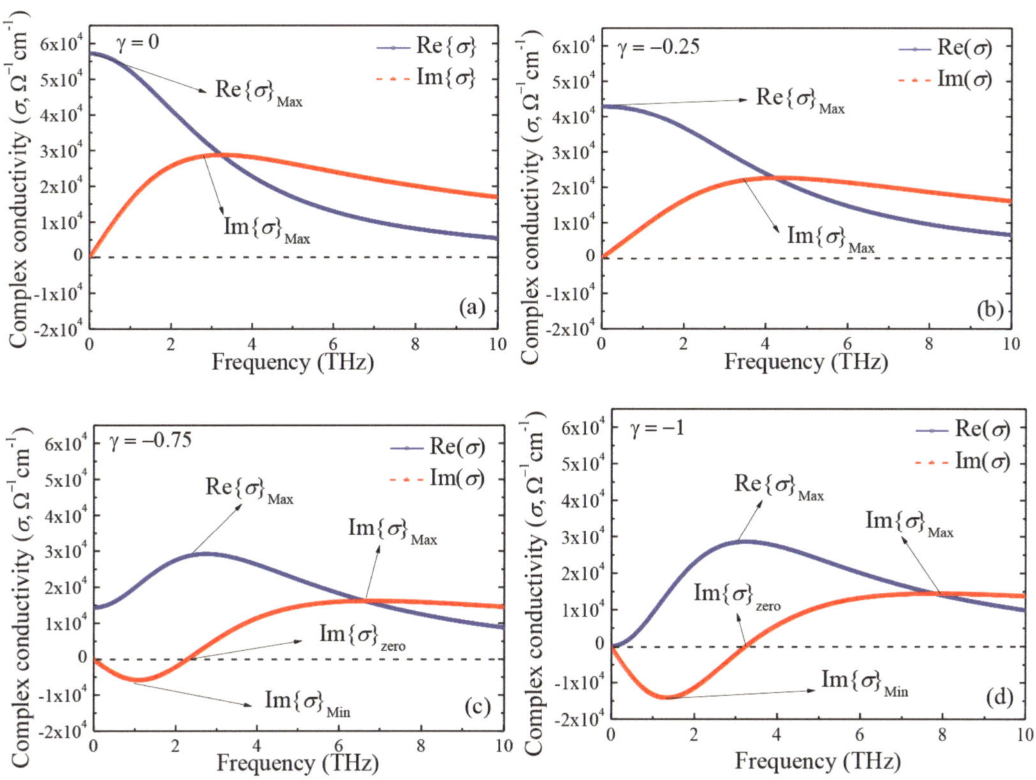

Figure 7. Simulation of the complex conductivities of ITO nanostructures using the Drude-Smith model for the cases of (a) $\gamma = 0$, (b) $\gamma = -0.25$, (c) $\gamma = -0.75$, and (d) $\gamma = -1$. From the fitting, the plasma frequency (ω_p) and scattering time (τ) are determined to be 363 rad THz and 49 fs, respectively.

4. THz optical constants and conductivities of selected TCOs

We have studied several TCOs using Fourier transform infrared spectroscopic (FTIR) and THz-TDS. First of all, the average transmittance of 100 nm thick ITO, AZO, and AYZO in the millimeter wave to submillimeter wave range was determined to be ~21%, ~48%, and ~28%, respectively. In **Figure 8(c)**, following the procedure detailed in Section 3 of the chapter, we found that the DC conductivity (σ_{DC}) and DC mobility (μ) for ITO thin films are 1.18–1.65 × 10^3 Ω^{-1} cm^{-1} and 96–120 cm^2 V^{-1} s^{-1}, respectively. In **Figure 8(a)** and **8(b)**, similarly, for AZO and AYZO thin films, σ_{DC} and μ are 0.41–1.06 × 10^3 Ω^{-1} cm^{-1} and 7–84 cm^2 V^{-1} s^{-1} versus 0.91–1.12 × 10^3 Ω^{-1} cm^{-1} and 26–85 cm^2 V^{-1} s^{-1}, respectively. Examining X-ray diffraction (XRD) data of these samples, we can immediately conclude that mobility of a particular sample is higher if the film exhibits larger grain size. On the other hand, if the carrier concentration (N_e) in the sample is high, the mobility of the sample would be lower as a result of the smaller distance between impurity ions. Our studies of these bulk TCO films to date thus indicate that AYZO has reasonable THz transmittance and electrical properties. Within this class, it may be the material of choice than ITO (lower THz transmittance) and AZO (lower mobilities) thin films for optoelectronic applications in the THz frequency range.

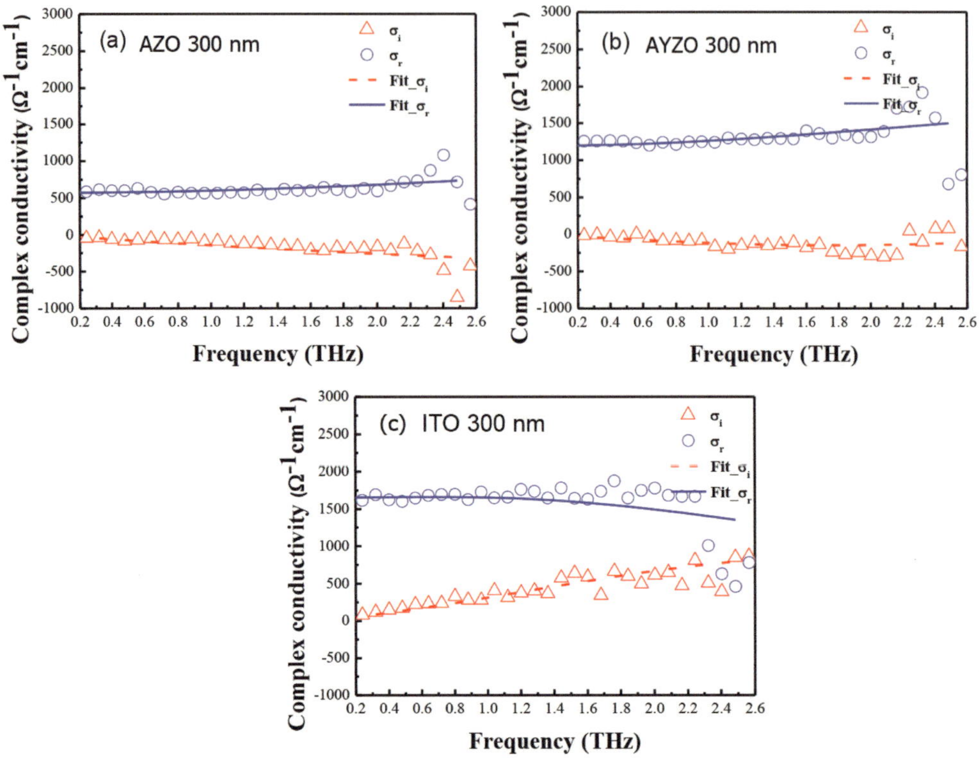

Figure 8. Real (blue circles) and imaginary (red triangles) parts of conductivity of (a) AZO, (b) AYZO, and (c) ITO thin films with thickness of 300 nm. Solid and dashed line corresponds to the fitting results based on the Drude-Smith model. Adapted from Ref. [29].

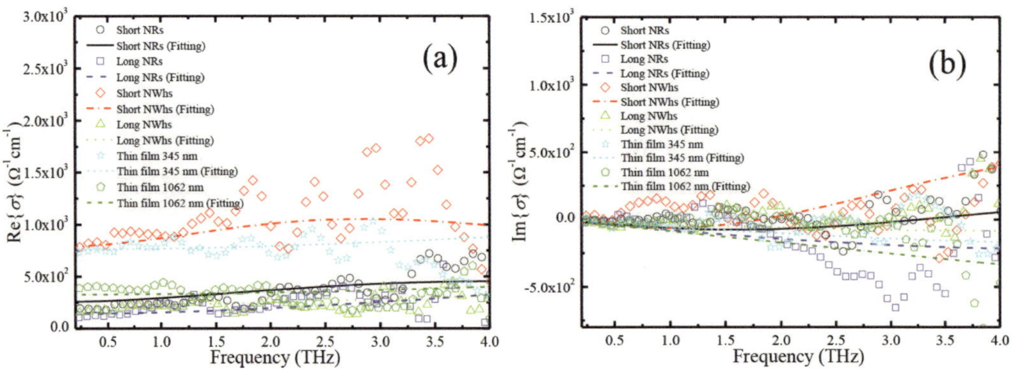

Figure 9. (a) Real and (b) imaginary conductivities of ITO nanomaterials (nanorods or NRs and nanowhiskers or NWhs) and thin films are plotted as a function of frequency (adapted from [13]). The symbols correspond to conductivities of samples extracted from THz-TDS measurements. The solid, dashed, dotted, and broken lines are fitting curves.

In **Figure 9**, we present complex conductivities of ITO nanomaterials and films studied by our group, reproduced from Ref. 13 by permission. The values of Re{σ} of both nanomaterials show a decreasing trend with frequencies in the THz band we studied. Regions of negative values of Im{σ} are also observed. These are typical features associated with the materials that exhibit characteristics of carrier localization [2]. That is, the Drude-Smith model would likely be the model of choice. Indeed, Re{σ} and Im{σ} of ITO nanomaterials can be fitted well by this model (see **Figure 9**). From the fit, we determine that, for NRs with heights of 621.1 and 991.6 nm, the carrier scattering times are 31.7 and 13.5 fs; the plasma frequencies are 561 and 1006 rad THz; while the parameters γ are −0.71 and −0.88, respectively. Similarly, for the NWhs with heights of 802.6 and 1169.5 nm, we find that the carrier scattering times are 39.6 and 13.2 fs; the plasma frequencies are 751 and 853 rad THz; the parameters γ are −0.60 versus −0.74, respectively. All γ values of nanostructured ITO studied by our group are nearly −1.00. Such values of γ indicate that carrier localization effects in ITO NRs and NWhs are very important [13].

5. Liquid crystal terahertz phase shifters with functional ITO Nanomaterials

5.1. Introduction

By now, the potential of THz science and technology to a host of applications ranging from fundamental sciences to national security is well known [30]. Consequently, key components, such as phase shifters, modulators, filters, and polarizers, need to be developed. Liquid crystals (LCs) can be used to add functionalities to various THz optical and optoelectronic components [31–33]. We have previously shown that phase shifts exceeding 2π at 1 THz can be achieved by using electrically controlled birefringence in a homeotropically aligned LC (E7, Merck) cell, 1.83 mm in thickness and biased at 100 V_{rms} [33]. As ITO films are opaque in the THz frequency range [2, 13], we employed two copper pieces separated by ~11 mm at two sides of LCs' cell as electrodes for biasing and spacers [33].

As we illustrated in previous sections, the ITO NWhs can be used as transparent electrodes for THz applications [2, 13]. Indeed, we demonstrated THz phase shifters with a phase shift of around $\pi/2$ at 1.0 THz with ITO NWhs used as transparent electrodes [34, 35]. Significantly, ITO NWhs can also be used to align the liquid crystal molecules. Therefore, traditional alignment layers and procedures are not required.

In this section of the chapter, we present several designs of THz phase shifters with maximum phase shifts exceeding 2π or 360°. This is significant for applications such as phased array radar in the submillimeter band. ITO nanomaterials described in this work were employed as transparent electrodes. These devices can be operated at relatively low voltage and exhibit outstanding transmittance in the THz frequency region.

5.2. Construction of the THz phase shifters

Three possible designs are presented in **Figure 10**. In all, the LC (MDA-00-3461 by Merck) layers were sandwiched between fused silica substrates that have the inner surface (facing the LC layer) either deposited with or without ITO NWhs. Polyimides were deposited and rubbed for LC alignment. Before turning on the voltage biasing the ITO NWhs, the LC molecules are aligned parallel to the substrates. As the bias is increased beyond the threshold field due to the Fréedericksz transition, LC molecules will be reoriented toward the direction

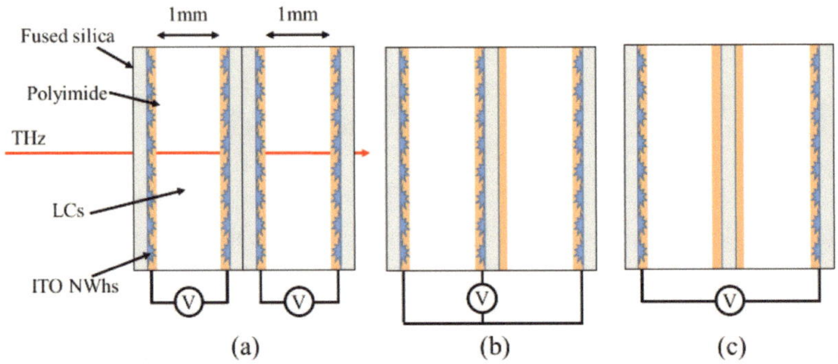

Figure 10. Schematic diagram of ITO NWhs 2π THz phase shifters in the (a) four-layer, (b) three-layer, and (c) two-layer design.

of the applied electric field. The threshold field is given by $E_{th} = \pi(k_1/(\varepsilon_0\Delta\varepsilon))^{1/2}/d$, where the free-space permittivity $\varepsilon_0 = 8.854 \times 10^{-12}$ F m^{-1} and d is thickness of a single cell or the distance between two electrodes as shown in **Figure 10**. The threshold field also depends on the splay elastic constant $k_1 = 12.6 \times 10^{-12}$ N and dielectric anisotropy, $\Delta\varepsilon = \varepsilon_{//} - \varepsilon_{\perp} = 11.2$ for this LC. After numerically finding the maximum tilt angle existing at every layer due to the weak boundary of substrates for LCs molecules, due to the effective birefringence, the phase shift can be given by $\delta = 2\pi f d\Delta n_{eff.Max}/c$, where f and c are the THz frequency and the speed of light in vacuum, respectively.

5.3. Results and discussions

All the three designs are found to be successful as THz phase shifters. In **Figure 11(a)**, we showed measured phase shifts of the devices as a function of the driving voltage. Phase shift exceeding 2π was demonstrated for all devices at 1.05 THz. The driving voltage needed is as low as 5 V_{rms}. Recalling performance of our previous work [33], the required voltage is lower by 20 times. Examining **Figure 1**, one immediately see that the threshold voltage is also quite low, less than 2 V_{rms}. Transmittance of all three phase shifters in the range of 0.2–1.2 THz is

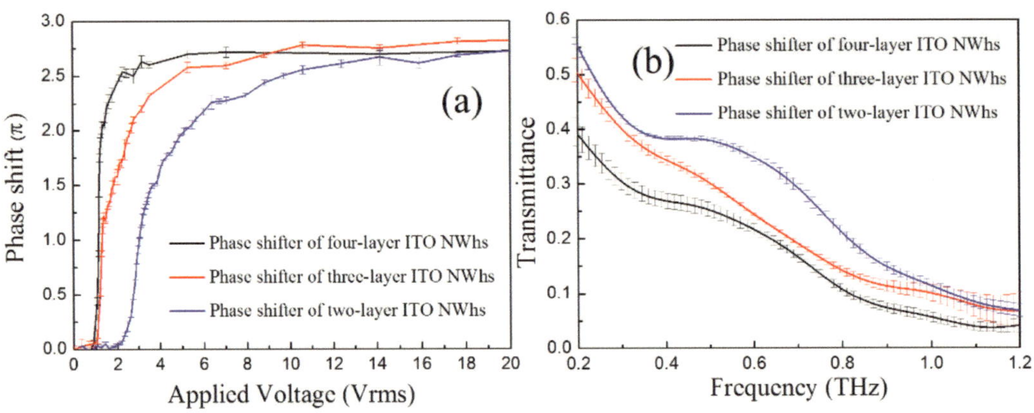

Figure 11. (a) Voltage-controlled phase shifts at 1 THz and (b) transmittance for three different 2π THz phase shifters based on multilayer ITO NWhs.

presented in **Figure 11(b)**. Note that the design with four layers of ITO NWhs exhibits the lowest threshold voltage and the lowest driving voltage needed to achieve 360° of phase shift. In contrast, the THz transmittance of the phase shifter with just two layers of ITO NWhs is significantly higher than others two. Although the loss of ITO NWhs is not so high in the THz frequency band, the accumulated loss of the device with the employment of more layers of electrodes is significant.

6. Summary

We have studied the frequency-dependent complex conductivities of several TCOs which have the potential in the THz applications. These include ITO nanomaterials, ITO sputtering films, AZO, and AYZO thin films. Methods for fabrication and characterization of these technologically important materials are described in detail. It is shown that ITO nanomaterials exhibit non-Drude-like electrical characteristics. Among the bulk films, AYZO has acceptable THz transmittance and good electrical properties such that it may be the better choice than the ITO and AZO thin films for optoelectronic applications in the THz band. Correlating with X-ray diffraction studies, we understood that the characteristics mentioned above can be explained by the extent of carrier localization in these materials. In comparison with the bulk films, ITO nanowhiskers are shown to be the material of choice as transparent electrodes in the THz band. To demonstrate, we show that the phase shifter using ITO NWhs exhibits transmittance as high as ~78% in the millimeter and submillimeter band. Several designs of the phase shifter using liquid crystal for tuning were demonstrated. Over 2π of phase shifts were realized at driving voltages that are over twenty times lower than previous designs.

Acknowledgements

This work was supported in part by the National Science Council, Taiwan, Republic of China, through grants 102-2622-E-007-021-CC2, 101-2221-E-007-103-MY3, and 101-2112-M-007-019-MY3, and a grant of the U.S. Air Force Office of Scientific Research, FA2386-13-1-4086.

Author details

Ci-Ling Pan[1]*, Chan-Shan Yang[1], Ru-Pin Pan[2], Peichen Yu[3] and Gong-Ru Lin[4]

*Address all correspondence to: clpan@phys.nthu.edu.tw

1 Department of Physics, National Tsing Hua University, Hsinchu, Taiwan

2 Department of Electrophysics, National Chiao Tung University, Hsinchu, Taiwan

3 Department of Photonics, National Chiao Tung University, Hsinchu, Taiwan

4 Graduate Institute of Photonics and Optoelectronics, National Taiwan University, Taipei, Taiwan

References

[1] G. Xu, Z. Liu, J. Ma, B. Liu, S.-T. Ho, L. Wang, P. Zhu, T. J. Marks, J. Luo, and Alex K. Y. Jen, "Organic electro-optics modulator using transparent conducting oxides as electrodes," Opt. Express 13, 7380–7385 (2005).

[2] C.-S. Yang, M.-H. Lin, C.-H. Chang, P. Yu, J.-M. Shieh, C.-H. Shen, O. Wada, and C.-L. Pan, "Non-Drude behavior in indium-tin-oxide nanowhiskers and thin films by transmission and reflection THz time-domain spectroscopy," IEEE J. Quantum Electron. 49(8), 677–690 (2013).

[3] P. Yu, C.-H. Chang, M.-S. Su, M.-H. Hsu, and K.-H. Wei, "Embedded indium-tin-oxide nanoelectrodes for efficiency and lifetime enhancement of polymer-based solar cells," Appl. Phys. Lett. 96, 153307 (2010).

[4] H. Yumoto, T. Sako, Y. Gotoh, K. Nishiyama, and T. Kaneko, "Growth mechanism of vapor–liquid-solid (VLS) grown indium tin oxide (ITO) whiskers along the substrate," Journal of Crystal Growth 203, 136–140 (1999).

[5] C.-W. Chen, Y.-C. Lin, C. H. Chang, P. Yu, J.-M. Shieh, and C.-L. Pan, "Frequency-dependent complex conductivities and dielectric responses of indium tin oxide thin films from the visible to the far-infrared," IEEE J. Quantum Electron. 46(12), 1746–1754 (2010).

[6] E. C. Garnett and P. Yang, "Silicon nanowire radial p-n junction solar cells," J. Am. Chem. Soc. 130, 9224–9225 (2008).

[7] M. D. Kelzenberg, S. W. Boettcher, J. A. Petykiewicz, D. B. Turner-Evans, M. C. Putnam, E. L. Warren, J. M. Spurgeon, R. M. Briggs, N. S. Lewis, and H. A. Atwater, "Enhanced absorption and carrier collection in Si wire arrays for photovoltaic applications," Nature Materials 9, 239–244 (2010).

[8] R. F. Bunshah and R. S. Juntz, "Influence of condensation temperature on microstructure and tensile properties of titanium sheet produced by high-rate physical vapor deposition process," Metallurgical Transactions 4, 21 (1973).

[9] H. Yumoto, J. Hatano, T. Watanabe, K. Fujikawa, and H. Sato, "Properties and surface morphology of indium tin oxide films prepared by electron shower method," Jpn. J. Appl. Phys. 32(part 1) 1204–1209 (1993).

[10] N. Laman and D. Grischkowsky, "Terahertz conductivity of thin metal films," Appl. Phys. Lett. 93, 051105 (2008).

[11] J. B. Baxter and C. A. Schmuttenmaer, "Conductivity of ZnO nanowires, nanoparticles, and thin films using time-resolved terahertz spectroscopy," J. Phys. Chem. B 110, 25229–25239 (2006).

[12] C. A. Schmuttenmaer, "Using terahertz spectroscopy to study nanomaterials," Terahertz Science and Technology 1, No. 1–8 (2008).

[13] C.-S. Yang, C.-M. Chang, P.-H. Chen, P. Yu, and C.-L. Pan, "Broadband terahertz conductivity and optical transmission of indium-tin-oxide (ITO) nanomaterials," Opt. Express 21(14), 16670–16682 (2013).

[14] C.-S. Yang, C.-H. Chang, M.-H. Lin, P. Yu, O. Wada, and C.-L. Pan, "THz conductivities of indium-tin-oxide nanowhiskers as graded-refractive-index structure," Opt. Express 20(S4), A441-A451 (2012).

[15] C.-S. Yang, C.-J. Lin, R.-P. Pan, C. T. Que, K. Yamamoto, M. Tani, and C.-L. Pan, "The complex refractive indices of the liquid crystal mixture E7 in the terahertz frequency range," J. Opt. Soc. Am. B 27(9), 1866–1873 (2010).

[16] J. Han, Z. Zhu, S. Ray, A. K. Azad, W. Zhang, M. He, S. Li, and Y. Zhao, "Optical and dielectric properties of ZnO tetrapod structures at terahertz frequencies," Appl. Phys. Lett., 89, 031107 (2006).

[17] J. Han, W. Zhang, W. Chen, S. Ray, J. Zhang, M. He, A. K. Azad, and Z. Zhu, "Terahertz dielectric properties and low-frequency phonon resonances of ZnO nanostructures," J. Phys. Chem. C, 111(35), 13000–13006 (2007).

[18] X. Zou, J. Luo, D. Lee, C. Cheng, D. Springer, S. K. Nair, S. A. Cheong, H. J. Fan, and E. E. M. Chia, "Temperature-dependent terahertz conductivity of tin oxide nanowire films," J. Phys. D: Appl. Phys., 45, 465101 (2012).

[19] G. J. Exarhos and X.-D. Zhou, "Discovery-based design of transparent conducting oxide films," Thin Solid Films, 515, 7025–7052 (2007).

[20] N. V. Smith, "Classical generalization of the Drude formula for the optical conductivity," Phys. Rev. B, 64, 155106 (2001).

[21] D. G. Cooke, A. N. MacDonald, A. Hryciw, J. Wang, Q. Li, A. Meldrum, and F. A. Hegmann, "Transient terahertz conductivity in photoexcited silicon nanocrystal films," Phys. Rev. B, 73, 193311 (2006).

[22] T. L. Cocker, L. V. Titova, S. Fourmaux, H.-C. Bandulet, D. Brassard, J.-C. Kieffer, M. A. El Khakani, and F. A. Hegmann, "Terahertz conductivity of the metal-insulator transition in a nanogranular VO_2 film," Appl. Phys. Lett., 97, 221905 (2010).

[23] H. Němec, P. Kužel, and V. Sundström, "Far-infrared response of free charge carriers localized in semiconductor nanoparticles," Phys. Rev. B, 79, 115309 (2009).

[24] H. Němec, P. Kužel, and V. Sundström, "Charge transport in nanostructured materials for solar energy conversion studied by time-resolved terahertz spectroscopy," J. Photoch. Photobio. A, 215, 123–139 (2010).

[25] L. V. Titova, T. L. Cocker, D. G. Cooke, X. Wang, A. Meldrum, and F. A. Hegmann, "Ultrafast percolative transport dynamics in silicon nanocrystal films," Phys. Rev. B, 83, 085403 (2011).

[26] E. Conwell and V. F. Weisskopf, "Theory of impurity scattering in semiconductor," Phys. Rev., 77(3), 388–390 (1950).

[27] J. Ederth, "Electrical transport in nanoparticle thin films of gold and indium tin oxide," Ph.D. dissertation, Dept. Mat. Science, Uppsala University, Uppsala, Sweden, 2003.

[28] J. Gao, R. Chen, D. H. Lin, L. Jiang, J. C. Ye, X. C. Ma, X. D. Chen, Q. H. Xiong, H. D. Sun, and T. Wu, "UV light emitting transparent conducting tin-doped indium oxide (ITO) nanowires," Nanotechnology, 22, 195706 (2011).

[29] C.-L. Pan and C.-S. Yang, "Transparent Conducting Oxides for THz Applications," The 8th International Photonics and OptoElectronics Meetings (POEM 2015), OSA Topical Meeting on Optoelectronic Devices and Integration (ODEI), June 16-19. Wuhan, China, paper OTB.1.

[30] X.-C. Zhang and J. Xu, Introduction to THz Wave Photonics, Springer, New York, US, 2010.

[31] C.-Y. Chen, C.-F. Hsieh, Y.-F. Lin, R.-P. Pan, and C.-L. Pan, "Magnetically tunable room-temperature 2π liquid crystal terahertz phase shifter," Opt. Express 12, 2630–2635 (2004).

[32] C.-F. Hsieh, R.-P. Pan, T.-T. Tang, H.-L. Chen, and C.-L. Pan, "Voltage-controlled liquid-crystal terahertz phase shifter and quarter-wave plate," Opt. Lett. 31, 1112–1114 (2006).

[33] H.-Y. Wu, C.-F. Hsieh, T.-T. Tang, R.-P. Pan, and C.-L. Pan, "Electrically tunable room-temperature 2π liquid crystal terahertz phase shifter," IEEE Photonic Technol. Lett. 18, 1488–1490 (2006).

[34] C.-S. Yang, T.-T. Tang, P.-H. Chen, R.-P. Pan, P. Yu, and C.-L. Pan, "Voltage-controlled liquid-crystal terahertz phase shifter with indium-tin-oxide nanowhiskers as transparent electrodes," Opt. Lett. 39(8), 2511–2513 (2014).

[35] C.-S. Yang, T.-T. Tang, R.-P. Pan, P. Yu, and C.-L. Pan, "Liquid crystal terahertz phase shifters with functional indium-tin-oxide nanostructures for biasing and alignment," Appl. Phys. Lett. 104, 141106 (2014).

Chapter 15

Ultrafast Carrier Dynamics at p-n Junction of $Cu(In,Ga)Se_2$-Based Solar Cells Measured by Optical Pump Terahertz Probe Spectroscopy

Woo-Jung Lee and Yong-Duck Chung

Additional information is available at the end of the chapter

http://dx.doi.org/10.5772/66350

Abstract

Among other materials, the p-type $Cu(In,Ga)Se_2$ (CIGS) alloy has attracted attention as the most efficient absorber in thin-film solar cells. The typical CIGS layer is deposited with a polycrystalline structure containing an amount of native defect states, which serve as carrier traps and recombination centers. These defect states in the CIGS layer can be easily changed after deposition of an n-type buffer layer, due to the formation of p-n junctions. To understand the influence of the p-n junction on these defect states, the behavior of photoexcited carriers, from the CIGS absorber to the buffer layer, is considered to be an important issue and is closely related to solar cell performance. In this study, we performed experiments to investigate the ultrafast carrier dynamics of CIGS-based solar cells, using optical pump terahertz (THz) probe (OPTP) spectroscopy, and demonstrated the correlation between solar cell performance and the behavior of photoexcited carrier dynamics.

Keywords: ultrafast carrier dynamics, $Cu(In,Ga)Se_2$, optical pump terahertz probe spectroscopy, p-n junction, defect states

1. Introduction

1.1. Basic properties of a CIGS solar cell

$Cu(In,Ga)Se_2$ (CIGS) is an attractive material for solar cells because of its beneficial qualities, including a high absorption coefficient for visible light, favorable direct bandgap (Eg), and great power conversion efficiency [1–5]. CIGS is comprised of $CuInSe_2$ (CIS)-$CuGaSe_2$ (CGS) and its *Eg* can be controlled from 1.04 eV of CIS to 1.68 eV of CGS by adding Ga content into the CIS. A CIGS of a ternary compound consists of a group I element (Cu), III elements

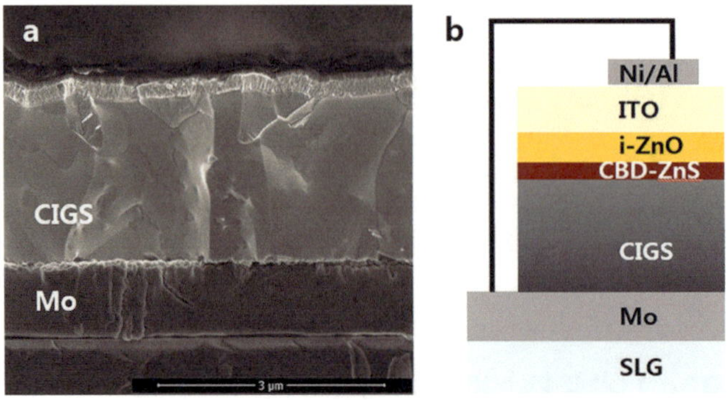

Figure 1. (a) The cross-section image of a CIGS solar cell with a CBD-ZnS buffer layer obtained by scanning electron microscopy and (b) the schematic diagram of a completed CIGS solar cell structure.

(Ga, In), and IV elements (Se), which are tetrahedrally bonded with a chalcopyrite crystal structure. The CIGS is generally deposited as a polycrystalline structure with an amount of native defect states and grain boundaries, which serve as carrier traps and recombination centers [6–8]. Representatively, there are 12 intrinsic defect states in CIGS: 3 vacancies (V_I, V_{III}, and V_{VI}), 3 interstitials (I_i, III_i, and VI_i), and 6 antisite defects (I_{III}, I_{VI}, III_I, III_{VI}, V_I, and VI_{III}). These defect states can work as either acceptors or donors by enthalpy of formation energy. The typical CIGS is considered to be a p-type material because of the amount of negatively charged Cu vacancies (V_{Cu}) with acceptor characteristics [9].

A CIGS solar cell with a chemical bath deposited-ZnS (CBD-ZnS) buffer layer is composed of several stacked layers, as indicated in the scanning electron microscopy (SEM) image of **Figure 1a**. p-Type CIGS layers with a polycrystalline structure approximately 2.2 μm thick were deposited by the conventional coevaporation method of a multistage process on Mo coated soda lime glass (SLG) [10–12]. As the n-type buffer layer, CBD-ZnS with a thickness of ~30 nm was prepared on the CIGS layer by precipitation from an aqueous solution. RF sputtering was then conducted to form an i-ZnO(70 nm)/ITO(150 nm) film as the transparent conducting oxide (TCO) layer. Finally, a current-collecting grid of Ni(50 nm)/Al(3 μm) was deposited by e-beam evaporation, completing the structure of the solar cell described in **Figure 1b**.

In CIGS-based solar cells, understanding the interfacial reaction at the p-n junction between the p-type CIGS absorber and n-type buffer layer is particularly important. This is because the intrinsic defect states in the CIGS, which experience a metastable redistribution of charge carriers after deposition of the buffer layer, have an influence on device performance [13–15]. Moreover, these defect states and the properties of the p-n junction have a decisive effect on the carrier relaxation dynamics in the CIGS layer, dominating the device performance. Thus, to improve the CIGS efficiency in a solar cell, it is essential to understand the intrinsic characteristics of the native defects in the CIGS, and the p-n junction.

1.2. Optical pump-THz probe spectroscopy

Among various measurement tools, optical pump terahertz (THz) probe (OPTP) spectroscopy was utilized in this experiment to verify the effect of the defect states because it is an extremely sensitive tool for investigating scattering mechanisms and dynamic energy transitions of photoexcited carriers in the region of shorter timescales on the order of ~*fs* and *ps* [16–18]. **Figure 2** shows the equipment system of the OPTP spectroscopy used in this study [19]. The detailed explanation for ingredients of the system is described as follows.

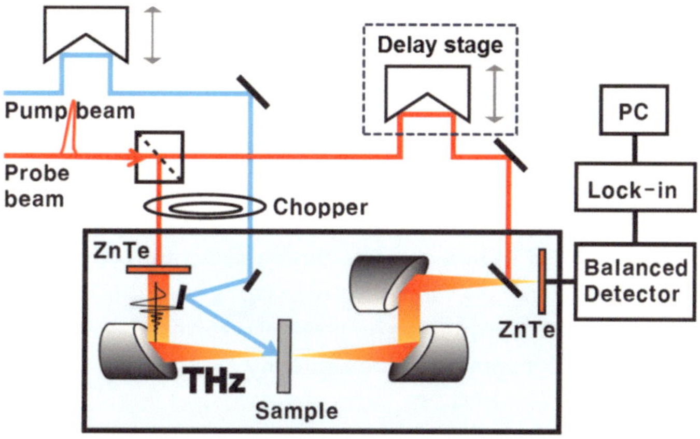

Figure 2. Schematic diagram of OPTP measurement system.

To measure the photoexcited carrier dynamics related with defect states, THz-time domain spectroscopy (in short, THz-TDs) measurement is carried out in advance. A fs pulse laser with 800 nm wavelength and pulse duration of 120 fs is generated from a Ti:sapphire regenerative amplifier system (Micra-Legend Elite, Coherent Inc.), which is split into two beams; one is transformed into THz signal, passing <110> ZnTe crystal with a thickness of 1 mm and the other is maintained as a probe beam with a time delay by moving the delay stage. The former of THz signal is transmitted throughout the sample, and then detected by means of an electro-optic (EO) sampling method of 3 mm thick <110> ZnTe nonlinear crystal, which signal is collected with a lock-in amplifier (Stanford Research System, SR830). For OPTP measurement, the maximum point of the transmitted THz signal is verified and probe beam is positioned at this point by fixing the delay stage. The optical pump beam excites the samples with the time delay and transient change of the sample in the 0.2–2.6 THz frequency range was probed by THz probe pulse via EO sampling. The diameter of the pump beam is 3 mm, which is over two times larger than that of the THz probe beam. Thus, the time evolution of the pump-probe signal can be collected by scanning the time delay of the pump pulses with respect to the THz pulse. All OPTP experiments are carried out at room temperature in confined area by purging dry air.

2. Ultrafast carrier dynamics of CIGS solar cell

2.1. Na effect in CIGS solar cell

In CIGS solar cell, Na supply is known to be one of the methods for enhancing solar cell efficiency. Na atoms are typically supplied by utilizing SLG, which is a preferred substrate material for the industrial manufacturing of rigid CIGS-based modules. It is a generally accepted contention that Na atoms diffuse from the SLG into the CIGS layer through a Mo layer during fabrication of solar cell, beneficially affecting the conversion efficiency of the solar cell [20–24]. On the basis of Na effect, we assume that Na diffusion into the CIGS layer may alter the electronic structure and defect states. However, direct evidence of how Na influences the performance of solar cell has not yet been obtained experimentally. We consider the possibility that electronic structure of abundant defect states existed in the CIGS layer can be changed by Na content, which is a crucial issue in efforts to optimize solar cell design.

In this section, to investigate the formation of defect states in the CIGS layer depending on the Na content, the study of ultrafast carrier dynamics was conducted on the CIGS layers grown on two different substrates, borosilicate (BS) and SLG by measuring OPTP spectroscopy. Carrier dynamics related to defect states can be determined by the relaxation times of photoexcited carriers relative to the scattering rate or carrier-trapping at defect states, ranging from hundreds of *fs* to several *ps*. After p-n junction was formed by depositing a CdS layer as a buffer layer, carrier dynamics were also subsequently analyzed.

2.1.1. The fabrication of CIGS solar cell depending on Na content

CIGS layers of approximately 2.5 μm thick were deposited on Mo coated BS (Na_2O: 4 at.%) and SLG (Na_2O: 14 at.%), respectively. The Ga/III and Cu/III composition ratios of the CIGS layer were about 0.14 and 0.87 in both cases. As a buffer layer, an n-type CdS layer with a thickness of ~70 nm was grown on the CIGS via CBD method. Typical solar cells (ITO/i-ZnO/CdS/CIGS/Mo) were then fabricated on BS and SLG under identical conditions, and their efficiencies were determined to be 8.5 and 10.9%, respectively. The device performance on SLG was superior to that on BS by ~2.4%, which is ascribed to the diffusion of Na atoms from the SLG. To investigate how the substrates affected defect states, PL and OPTP measurements were conducted on CIGS and CdS/CIGS layers directly grown on BS and SLG without a Mo layer. For OPTP measurement, metal layer such as Mo should be removed because THz probe pulse can be easily absorbed by metal layer due to numerous free carriers.

SEM measurement was performed on each CdS/CIGS samples grown on BS and SLG as shown in **Figure 3a, b** and **c, d**, respectively. Approximately 2.5 μm thick CIGS absorber and 70 nm thick CdS buffer layer were similarly deposited on both substrates. We found out that grain size of CIGS grown on SLG is larger than that on BS, which is considered as diffused-Na from the SLG. To verify the existence of the Na content, depth profiles of the elemental constituents in the CdS/CIGS layer grown on BS and SLG were examined by using secondary ion mass spectroscopy (SIMS) as shown in **Figure 4**. A substantial Na content diffused up to the CdS layer in the SLG case, but not the BS. SIMS results clearly demonstrate that the SLG can effectively supply Na atoms into the CIGS and the CdS layer as compared to BS.

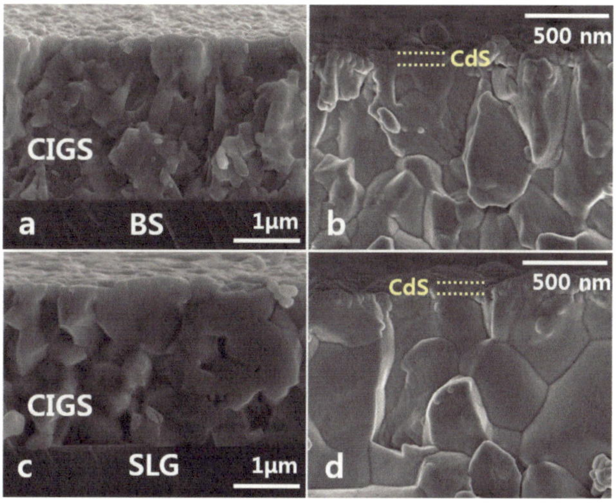

Figure 3. SEM images of CdS/CIGS film grown on BS (a, b) and SLG (c, d).

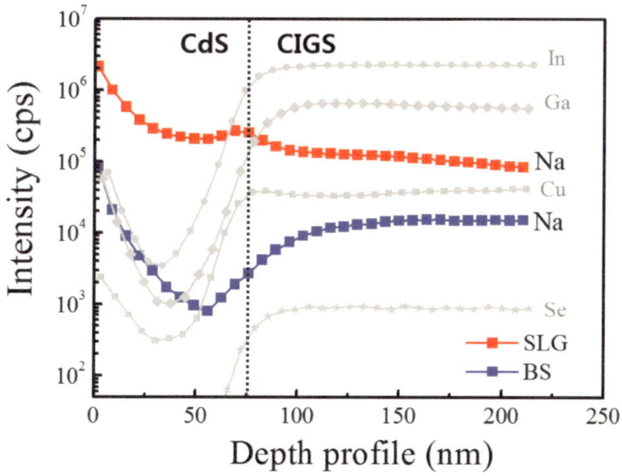

Figure 4. SIMS depth profiles of Na atoms in the CdS/CIGS layers grown on BS and SLG.

2.1.2. OPTP spectroscopy results

To measure the photoexcited carrier dynamics, the measurement of THz-TDs is essential. **Figure 5** shows the THz pulse spectra transmitted through the CIGS films grown on BS and SLG. After penetration of THz pulse through the samples, the intensity of THz pulse was drastically decreased as compared with the reference THz pulse (noted as "air").

Based on the results of THz-TDs spectrum, the measurement of OPTP spectroscopy was conducted on the CIGS and CdS/CIGS layer. In this experiment, pump beam of 400 nm is utilized

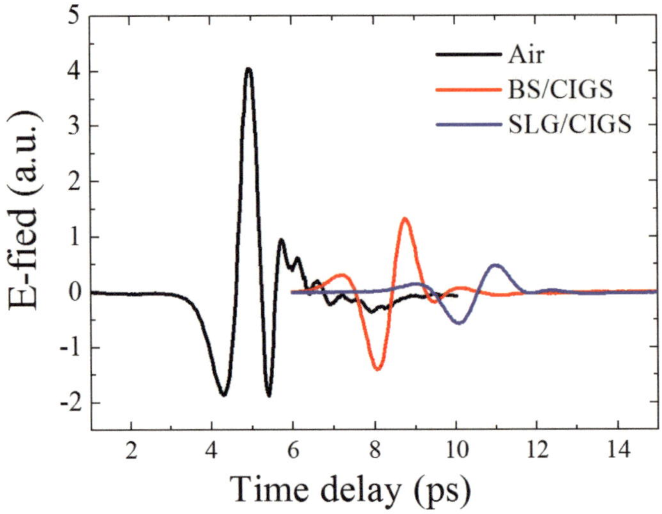

Figure 5. THz pulse spectra transmitted through the CIGS films grown on BS and SLG with a reference THz pulse measured by THz-TDs spectroscopy.

to excite photocarriers from the CdS buffer layer. When an intense *fs* laser pump pulse of 400 nm is injected, charge carriers excite, and the THz probe pulse is transmitted with pump-induced change; that is, the THz probe pulse transmitted through the samples reflects the absorption change induced by the *fs* laser pump pulse of 400 nm, which is expressed with, −ΔT/T. Thus, the near-edge −ΔT/T spectra intensity implies the photoexcited carrier density, and the decay curve corresponds to the carrier lifetime. The −ΔT/T signal is mainly attributed to the carrier relaxation of photoexcited electrons not holes since the effective mass of the electron is smaller than that of a hole in $CuInSe_2$ and $CuGaSe_2$ [9]. The typical −ΔT/T curve, at first, abruptly increases due to the absorption of photoexcited carriers by photon energy above the E_g, and then sequentially decays owing to the intraband relaxation (τ_1) and the recombination process/trapping at defect states (τ_2) [18, 25]. The photocarrier density is drastically decreased

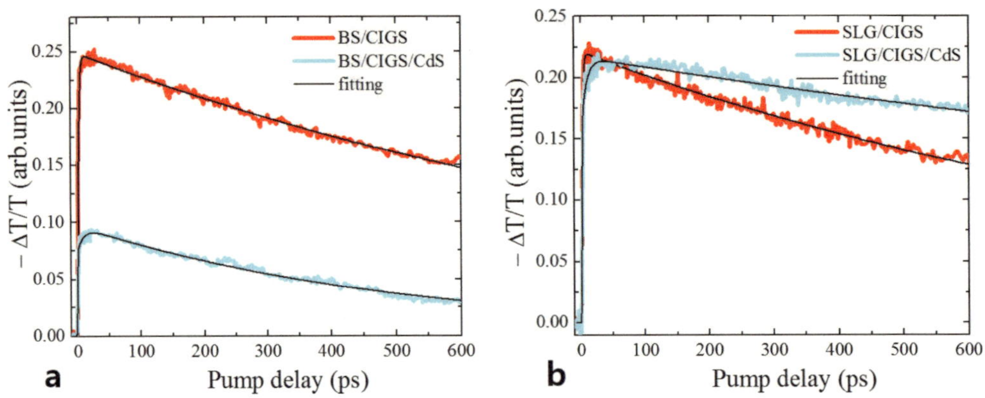

Figure 6. The measured and fitted $-\Delta T/T$ spectra of the CIGS and CdS/CIGS layers grown on BS (a) and SLG (b) produced by the photoinduced change in the THz probe pulse (a 400 nm pump beam was used).

	BS/CIGS	BS/CIGS/CdS	SLG/CIGS	SLG/CIGS/CdS
τ_1	1.7 ps	9.3 ps	1.3 ps	7.6 ps
τ_2	1150 ps	520 ps	1110 ps	2580 ps

Table 1. The carrier lifetimes obtained from the CdS/CIGS and CIGS layer are summarized as a function of substrate.

after the deposition of CdS layer in case of BS (**Figure 6a**) and is almost identical to both CIGS and CdS/CIGS layers in case of SLG (**Figure 6b**), which means that the characteristics of the CIGS layer is significantly changed in contact with the CdS layer. The time evolution of $-\Delta T/T$ can be well fitted with a biexponential function. The lifetimes of τ_1 and τ_2 obtained from the CIGS layer were approximately 1.7 and 1150 ps in BS, and 1.3 and 1110 ps in SLG; this means that a similar relaxation mechanism exists in the two CIGS layers. After deposition of the CdS layer, the lifetimes of τ_1 and τ_2 were noticeably changed, into 9.3 and 520 ps in BS, and 7.6 and 2580 ps in SLG. Interestingly, τ_2 dropped by half in BS, whereas it rose two-fold in SLG. These lifetimes are summarized in **Table 1**.

2.1.3. *Ultrafast carrier dynamics at p-n junction of CdS/CIGS depending on Na content*

To identify the defect states determining the lifetime of the relaxed carriers, PL measurement was performed with an excitation light of 400 nm, an identical value to the pump beam energy of OPTP measurement. PL peak assigned to the donor-acceptor pair (DAP) transition [26–29]. The DAP transition energy is measured of approximately 0.97 and 0.92 eV in CIGS on SLG and BS, respectively. Using the estimated E_g of 1.1 eV, the sum of each donor and acceptor ionization energy ($\Delta E_D + \Delta E_A$) becomes 130 meV in SLG and 180 meV in BS. According to several reports, the enthalpy of the formation energy of defects such as V_{Cu}, V_{Se}^+, and In_{Cu}^+ is very small and even negative [9, 30]. In the CIGS layer on SLG, the PL peak of 0.97 eV stems from the optical transition between the shallow acceptor level of 50 meV formed by V_{Cu} and the donor level of 80 meV created by Se vacancy (V_{Se}) [28]. In CIGS on BS, a new defect level emerges. It is plausible to assign this as an optical transition from antisite In_{Cu} as a donor level of descending order to V_{Cu} [28, 31].

Generally, photocarriers excited by high photon energy comply with Fermi-Dirac distributions through carrier-carrier scattering within a few hundred *fs*, and then lose their excess energy to the lattice, and relax to the bottom of the band by carrier-phonon scattering within several *ps* [18, 25]. In the both cases of CIGS on BS and SLG, τ_1 shows the similar trend regardless of substrate types because τ_1 is determined by the intraband relaxation time to reach the bottom of the conduction band; that is, it means that CIGS layers grown on BS and SLG are almost identical. Moreover, measured lifetimes of τ_1 in our system are similar to other results for CIGS thin film [32, 33]. On the other hand, τ_2 corresponds to an interband relaxation time either by the recombination of the electron-hole pair to restore initial states or by the electron trapping at defect states energetically located within the E_g. The both τ_2 obtained from the CIGS on BS and SLG also indicate similar values. Considering reported lifetimes, τ_2 is ascribed to the relaxation time from conduction band edge to the shallow acceptor energy level localized above the valence band edge within several nanoseconds [33]. After p-n junction is formed, τ_1 becomes longer within several picoseconds in both cases of CdS/CIGS on

BS and SLG. It is due to the CdS band structure with larger E_g of 2.4 eV, bringing about rise of the time for photocarriers to reach the conduction band edge. On the other hand, τ_2 exhibited the opposite trend after the deposition of the CdS layer on CIGS, which implies that different defect states are formed near the p-n junction as a function of substrate type. In BS, τ_2 is about 520 ps, whereas in SLG, it is approximately 2580 ps; that is, in the case of BS, a deeper defect level to trap photocarriers could be formed at the p-n junction, shortening τ_2 value. Considering PL results, the formation of deeper defect states into the E_g is plausible in CIGS on BS because decreased PL intensity in BS is caused by electron trapping at defect states instead of the radiative recombination. We suggest that the Na atoms affect the quality of p-n junctions and defect states located in E_g. According to calculation results, ionized intrinsic In_{Cu}^{2+} can transform into lattice-relaxed, deep defect level "*DX* states" formed as a Frenkel-pair consisting of an In interstitial and a Cu vacancy; that is, $In_{Cu}^{2+} + 2e \rightarrow In_{DX}^{0} = (In_i^{+} - V_{Cu}^{-})$ [34]. Since large quantities of isolated In_{Cu} and isolated V_{Cu} are formed very close to a 1:2 ratio during CIGS growth, most In_{Cu} eventually exists as a neutral defect ($In_{Cu} - 2\,V_{Cu}$), whereas the amount of uncompensated isolated In_{Cu} becomes a donor [35, 36]. The above intrinsic defect complexes can create the defect-localized states in the E_g, namely "*DX* states," in the case of $Cu(In_{1-x}Ga_x)Se_2$ with an $E_g \leq 1.2$ eV and $x < 0.3$ [34]. This has a harmful influence on PV device performance by trapping photocarriers produced by photoexcitation under illumination. These "*DX* states" even cause Fermi level pinning and trapping of photocarriers near the p-n junction, resulting in a decrease of τ_2 in BS. In the case of SLG, the increase of τ_2 is thought to result because the added Na automatically diffuses into the CIGS and CdS layer from the SLG, as demonstrated in SIMS results. The incorporation of less than 1% Na is considered enough to significantly change their electronic properties. Na is known to reduce the defect density inCIGS film by passivation of the Se vacancy or annihilation of the compensating antisite donor defect In_{Cu} by Na [20]. We suggest a possible scenario for the cause of the longer lifetime, τ_2 in SLG: (i) Na eliminates In_{Cu} by substituting for In on Cu sites; that is, forming Na_{Cu} and (ii) Na easily forms Na_{In} antisites defects, resulting in an increase in acceptor density. Thus, there is no "*DX* state" in SLG, which is due to the deficient In_{Cu} being replaced by Na_{Cu} which prevents the creation of defect complexes ($In_{Cu} - 2\,V_{Cu}$), near the p-n junction. In other words, the introduction of Na into the CIGS and CdS layers plays an important role in the PV device; this is because of the removal of "*DX* states" and an increase in acceptors, thereby promoting τ_2 in SLG. The relaxation mechanism of photocarriers at the p-n junction of CdS/CIGS depending on the two substrates of BS and SLG is displayed in a simplified band structure in **Figure 7**. This result clearly indicates that Na atoms exert a positive effect at the p-n junction of PV devices by producing increased photocarrier lifetimes in the case of SLG.

2.2. Variation of buffer layer in CIGS solar cell

Typical CIGS-based solar cells have a buffer layer between the CIGS absorber layer and the transparent ZnO front electrode, which plays an important role in improving the cell performance. Among various buffer materials, Zn-based materials have been frequently studied because of their beneficial properties, for example, good transparency with large direct E_g, less toxicity, and cost effectiveness [37–40]. ZnS film can be grown by various deposition methods such as evaporation, atomic layer deposition, sputtering, and metal organic chemical vapor

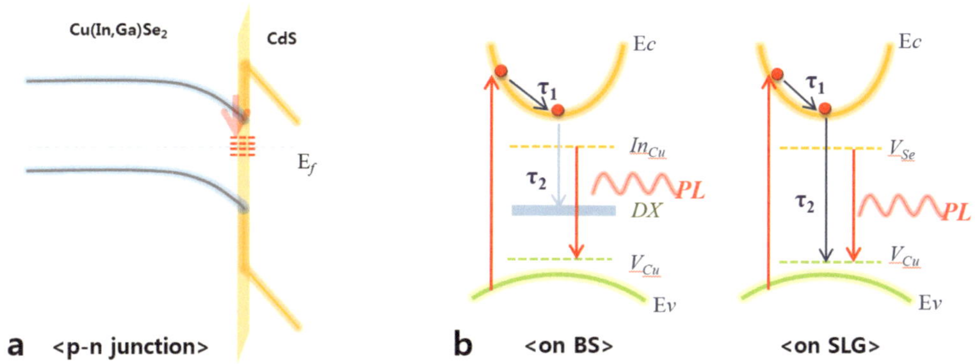

Figure 7. (a) A schematic defect level diagram at the p-n junction of CdS/CIGS. (b) Illustration of the carrier relaxation mechanism of CIGS thin films after deposition of CdS layer depending on substrate type of BS and SLG.

deposition. The efficiency of the CIGS solar cell with a Zn-based buffer layer is comparable to a conventional CIGS solar cell with a CdS buffer layer [41]. Ultimately, the efficiency of solar cells with a Zn-based buffer layer depends on the properties of the p-n junction, which is mainly determined by the buffer layer deposition process.

In this section, we fabricated a CIGS solar cell with Zn-based buffer layers grown by three deposition methods, cracker-Zn(O,S), CBD-Zn(O,S), and sputter-Zn(O,S), based on optimized conditions previously determined by our group [42–45]. To extract the correlation between cell efficiency and the interfacial characteristics between the CIGS and buffer layer, OPTP spectroscopy was utilized to investigate the CIGS, and the CIGS with the buffer layer, with respect to carrier trapping times at defect states.

2.2.1. The fabrication of a CIGS solar cell with a Zn-based buffer layer

A CIGS layer of approximately 2.2 μm thick was deposited on a Mo coated SLG. The Ga/III and Cu/III composition ratios of the CIGS layer were measured to be about 0.3 and 0.96, respectively, by X-ray fluorescence. We prepared two vacuum-based buffer layers of cracker-Zn(O,S) and sputter-Zn(O,S) with thicknesses of ~8 and 70 nm, respectively. We also arranged one chemical-based buffer layer of CBD-Zn(O,S) with a thickness of 30 nm. Since the buffer layers were deposited utilizing the optimized conditions from our group, their thicknesses were different.

Typical solar cells were fabricated with the various buffer layers, and the other deposition conditions were identical. The performance of the solar cell showed that the results depend on the buffer types. The best cell efficiency of approximately 13% was obtained for the CIGS solar cell with CBD-Zn(O,S).

Details of the cell structures and their cell performance are provided in **Figure 8**. For the OPTP measurement, CIGS without a Mo layer and a Zn-based buffer layer was directly deposited on SLG, and the injected and transmitted optical and THz pulses are simply illustrated in **Figure 9**.

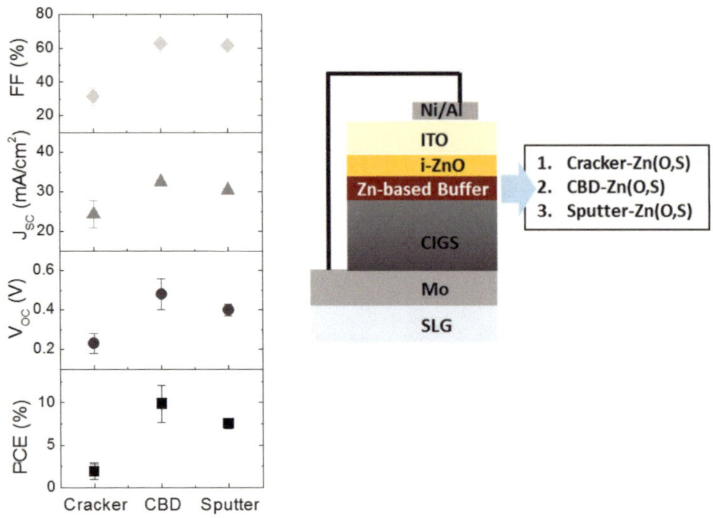

Figure 8. The performance of the CIGS solar cell as a function of the method of deposition of the Zn-based buffer layer (left), and an illustration of the completed CIGS solar cell structure (right).

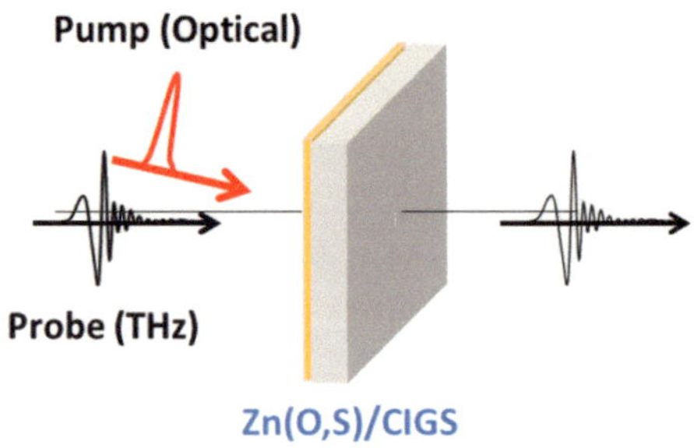

Figure 9. Simplified schematic diagram of the OPTP spectroscopy measurement.

2.2.2. OPTP spectroscopy results

For OPTP measurement, THz-TDs were conducted on the samples by comparing the THz pulse before and after transmission through the sample. **Figure 10a** shows the THz pulse spectra transmitted through the SLG, which is drastically reduced in comparison with the reference THz pulse through "air". After penetration of the THz pulse through the sample of CIGS and the Zn(O,S) buffer layers deposited on the CIGS, the intensity of the THz pulses were found to be similar to each other, as indicated in **Figure 10b**. This means that no photo-carriers were excited by the THz pulse.

Based on the results of the THz-TDs spectra, OPTP spectroscopy measurements were carried out on the CIGS and Zn(O,S)/CIGS layers. In those experiments, two pump beams of 400 and 800 nm were used to investigate optically photoexcited carrier dynamics along the depth distribution.

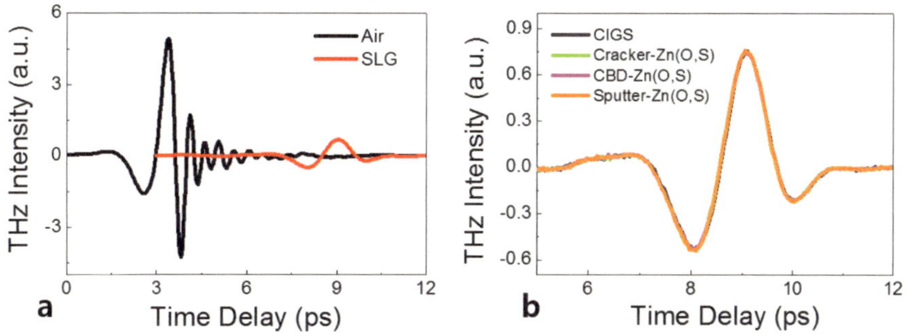

Figure 10. THz pulse spectra transmitted through the SLG and air (as a reference) (a), and CIGS and Zn(O,S) buffer layers/CIGS grown on SLG (b) measured by THz-TDs.

Figure 11a and **b** show the $-\Delta T/T$ spectra and normalized ΔT spectra of the CIGS film as a function of pump beam energy. The spectra intensity at the near-edge implies the photoexcited carrier density, and the decay curve corresponds to the carrier lifetime. In the CIGS film, the photoexcited carrier density and carrier lifetime were drastically increased by the pump beam of 800 nm, as compared with the 400 nm beam, which means that the density of defect states in the CIGS is differentiated along the depth direction.

Considering the penetration depth of the pump beam into the CIGS film, a wavelength of 400 nm should penetrate the surface of the CIGS less than 50 nm, and the 800 nm beam should reach the near-surface of the CIGS at about 150 nm. Thus, we assume that the charge carrier density acting as long-lived photocarriers is higher at the near-surface than the surface of the CIGS.

To extract carrier lifetimes relative to defect states, we fitted the normalized ΔT spectra of the CIGS using the equation (1) that follows:

$$y(t) = [A_e \cdot (\mathrm{erf}\left(\frac{t-t_e}{-\tau_e}\right) + y_e)] \times [A_0 \cdot \exp\left(\frac{t-t_0}{-\tau_0}\right) + A_1 \cdot \exp\left(\frac{t-t_0}{-\tau_1}\right) + A_2 \cdot \exp\left(\frac{t-t_0}{-\tau_2}\right)] \quad (1)$$

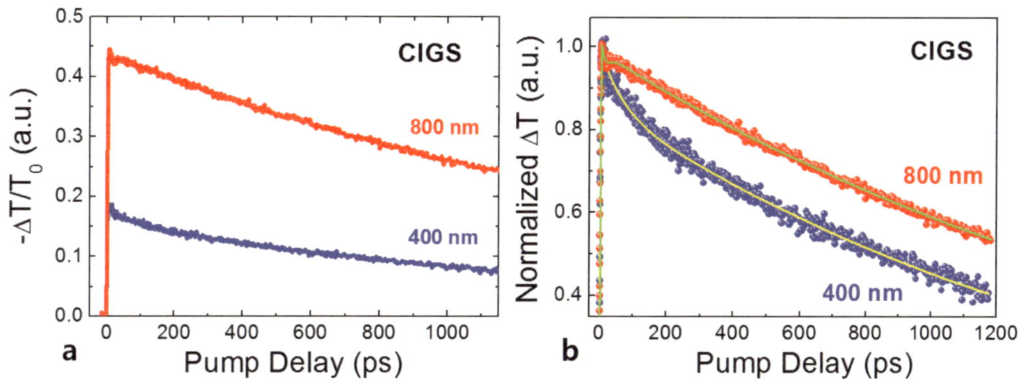

Figure 11. The measured $-\Delta T/T$ spectra of the CIGS (a) and normalized ΔT spectra of the CIGS (b) as a function of pump beam energies of 400 and 800 nm.

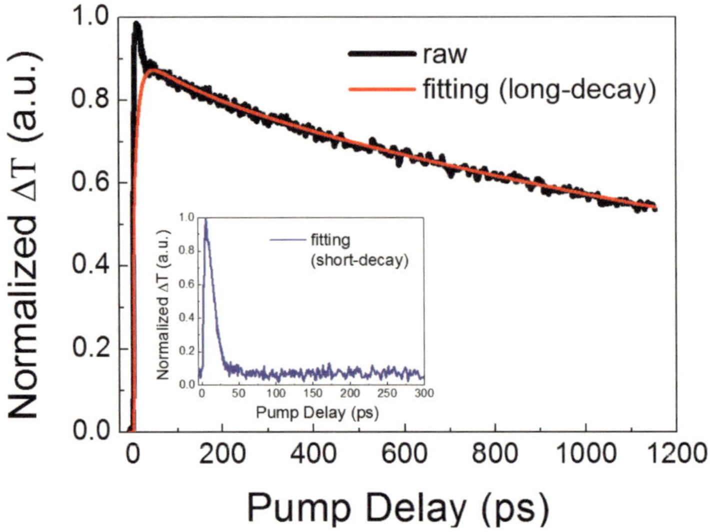

Figure 12. The normalized and fitted ΔT spectra of the CIGS measured at a pump beam energy of 800 nm. This indicates the fitting method, after subtracting the short decay curves caused by the sharp peak.

In this study, a sharp peak appeared at the near-edge of the normalized ΔT spectra with the 800 nm pump beam (**Figure 11b**), which was separately fitted, as indicated in **Figure 12**. To consider only the long decay curve while excluding the very sharp peak, we subtracted the very short decay curves of less than 1 *ps* by fitting, as indicated in the inset of **Figure 12**. From the fitting results, the notable thing to be observed is that the fast carrier lifetime (τ_1) of 61.8 *ps* and the long carrier lifetime (τ_2) of 1552 *ps* existed for the 400 nm pump beam, while τ_1 disappeared and τ_2 became longer, reaching 1930 *ps* for the 800 nm pump beam.

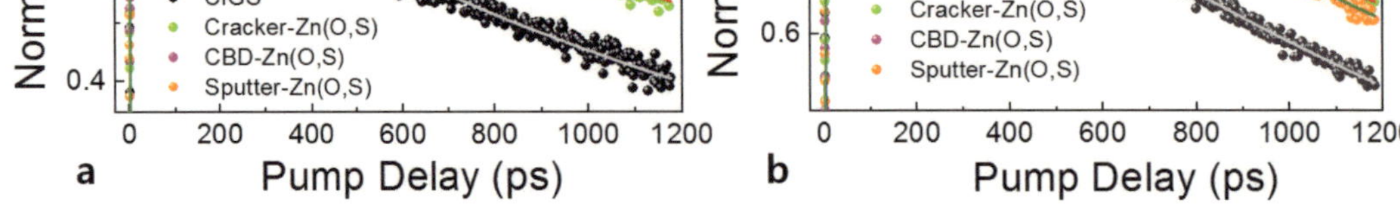

Figure 13. The normalized and fitted ΔT spectra of the pure CIGS film and various Zn(O,S) buffer layers deposited on CIGS film, as a function of pump beam energies of 400 nm (a) and 800 nm (b).

After deposition of the Zn(O,S) buffer layers on the CIGS film, OPTP spectroscopy was also conducted. The photoexcited carrier density and carrier lifetimes at 800 nm were much higher than that at 400 nm, which is a similar tendency to the measured spectra from the CIGS films as shown in **Figure 11a**. After depositing the buffer layer on the CIGS, the decay curves were extended regardless of the deposition method of buffer layer for both the 400 and 800 nm pump beams.

To clearly express the distinctive features of the carrier lifetimes according to buffer types, we fitted the normalized ΔT spectra as presented in **Figure 13a** and **b**. The detailed carrier lifetimes obtained from the fitting results are summarized in **Table 2**. After deposition of the Zn(O,S) buffer layers, two carrier lifetimes of τ_1 and τ_2 existed for the 400 nm pump beam, whereas only the τ_2 value was detected and prolonged for the 800 nm pump beam, as was the case with CIGS. In the case of the sputter-Zn(O,S)/CIGS, only the τ_2 value was found, without τ_1, irrespective of pump beam energy.

Pump beam energy	Lifetimes	CIGS	Cracker	CBD	Sputter
400 nm	τ_1 (ps)	61.8	141	119	-
	τ_2 (ps)	1552	2171	2529	2212
800 nm	τ_1 (ps)	-	-	-	-
	τ_2 (ps)	1930	2611	3325	2500

Table 2. The carrier lifetimes obtained from the pure CIGS film and various Zn(O,S) buffer layers grown on CIGS film are summarized as a function of pump beam energy.

2.2.3. *Ultrafast carrier dynamics of the CIGS/Zn-based buffer layer*

To investigate the carrier dynamics related to the defect states along the depth direction in the CIGS and Zn(O,S)/CIGS, 400 and 800 nm pump beam energies were applied. The 400 nm pump beam is sensitive to the surface of the film, and the 800 nm pump beam reflects from the near-surface of the film. The photocarriers in the CIGS film can be excited by both pump beams, whereas those of the Zn(O,S) film cannot be excited by the 800 nm pump beam due to its low energy of 1.55 eV.

To verify the light absorption of the Zn(O,S) buffer layer at the 800 nm wavelength depending on the deposition method, transmittance spectra were measured for various Zn(O,S) films directly grown on SLG, as indicated in **Figure 14**. After deposition of the Zn(O,S) film, the wavelength of 800 nm mostly penetrated toward the CIGS film.

In the pure CIGS film, two carrier lifetimes (fast: τ_1 of 61.8 *ps* and long: τ_2 of 1552 *ps*) were detected for the 400 nm pump beam, while the fast carrier lifetime τ_1 was dissipated and long carrier lifetime τ_2 was prolonged to 1930 *ps* for the 800 nm pump beam. From these results, we inferred that there were different defect states available to trap photoexcited carriers at the surface and the near-surface, respectively.

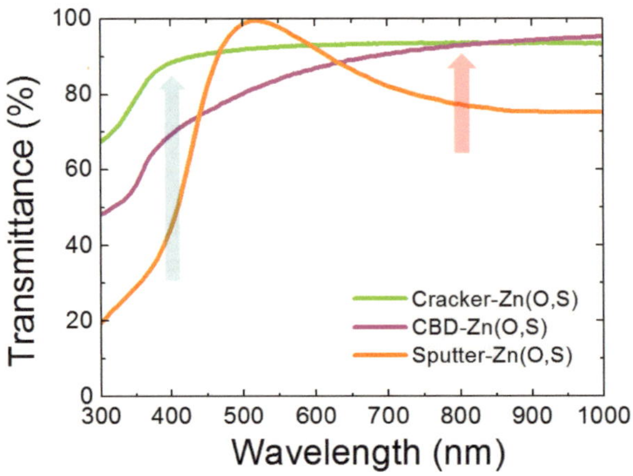

Figure 14. Transmittance spectra of various Zn(O,S) films grown on SLG.

Even though there are several defect levels in the E_g of CIGS, the carrier lifetimes average out to one decay time. Thus, τ_1 and τ_2 should be explained with different photocarrier trapping times, not defect energy levels. A typical CIGS film consists of numerous defect states, caused by several point defects, grain boundaries, and rough surface [7, 9, 34]. With the injection of the 400 nm pump beam into the CIGS film, the rough surface is an important factor, because in a semiconductor, photocarriers can be captured at surface defect states.

Among the several point defects that exist in the CIGS film, Cu vacancies (V_{Cu}) can be easily formed due to their low formation energy, creating a shallow acceptor level [9]. Thus, the V_{Cu} defect level can have a dominant role in the trapping of photoexcited electrons. (In the OPTP experiments, we only considered photoexcited electrons, not holes, due to the higher mobility of photoexcited electrons as compared to photoexcited holes.) Considering that photocarriers are primarily excited in the surface region for the 400 nm, the trapping time of the surface defect states is relatively faster than the bulk defects of V_{Cu}. Based on surface defects and V_{Cu}, we assumed that τ_1 and τ_2 correspond to the trapping times of photoexcited electrons in the surface defect states, and the V_{Cu} defects entirely distributed from surface to bulk, respectively.

With the injection of the 800 nm pump beam into the CIGS film, photoexcited electrons occur at the near-surface, and not the surface. Thus, τ_1 reflecting the CIGS surface is not detected, because the photoexcited electrons are dominantly captured at bulk defect states before being trapped at the surface defect states. The lower the defect density, the more that carrier lifetime is increased. Here, τ_2 is increased as compared to that for the 400 nm pump beam, which means that the bulk defect density, V_{Cu}, decreases along the depth direction, resulting in a rise in τ_2. The carrier dynamics model in the pure CIGS film is illustrated in **Figure 15a**.

After deposition of the Zn(O,S) buffer layers on the CIGS film, both τ_1 and τ_2 were increased for the 400 nm pump beam, excluding the sputter-Zn(O,S) buffer layer. The increase in τ_1 is attributed to the surface curing effect, which reduces surface defect states after the deposition of the buffer layer. Since a thin-buffer layer does not entirely cover the rough surface of

http://dx.doi.org/10.5772/66350

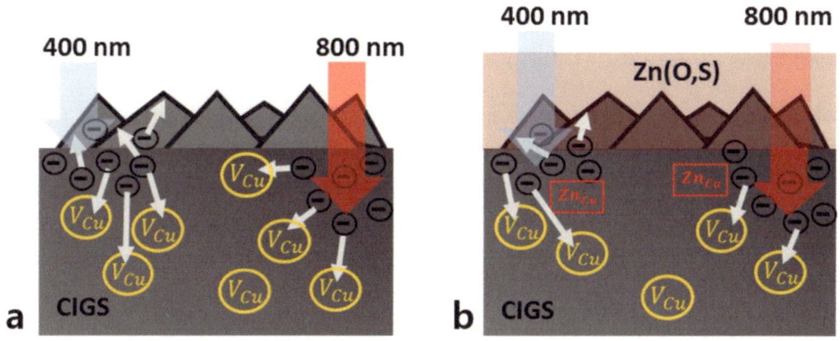

Figure 15. Schematic diagram of photoexcited carrier dynamics in the pure CIGS film (a) and after deposition of Zn(O,S) buffer layers on the CIGS film (b) as a function of pump beam energy.

the CIGS film, τ_1 is slightly increased up to ~140 *ps* and 120 *ps* in cracker-Zn(O,S) and CBD-Zn(O,S) which are about 8 and 30 nm thick, respectively.

In the sputter-Zn(O,S) buffer layer, τ_1 is dissipated because of its thick-film thickness of 70 nm, which relieves the surface scattering effect, as compared with the other buffer layers. In general, Zn atoms of the buffer layer have been known to diffuse toward the CIGS film. The diffused Zn atoms can occupy the V_{Cu} sites, forming a Zn^+ charge state as a donor [46, 47]. Thus, Zn substitution to V_{Cu}, Zn_{Cu}, causes a reduction in the bulk defect states available to trap photoexcited electrons, resulting in a rise of τ_2.

With the 800 nm pump beam, τ_1 also disappeared, like the CIGS film, implying surface defect states were not detected. The τ_2 values also increased, which is ascribed to the combination of increased Zn_{Cu} in the surface region of the CIGS, and decreased bulk defect density, V_{Cu}.

The carrier dynamics model in the Zn(O,S)/CIGS film is illustrated in **Figure 15b**. τ_2 showed the highest values for both pump beams of 400 and 800 nm in the CBD-Zn(O,S) buffer layer grown on CIGS, as compared with the others. From these results, we carefully suggest that solar cell efficiency can be enhanced when carrier lifetime reflecting bulk defect density is prolonged.

3. Conclusion

In summary, we have demonstrated the effectiveness of OPTP measurement for determining the ultrafast carrier dynamics of photocarriers excited from CIGS and buffer layers, depending on substrate type (BS and SLG) and several Zn(O,S) buffer layers (cracker-, CBD-, sputter-Zn(O,S)). By fitting the normalized ΔT spectra, carrier lifetimes were extracted and then carrier behavior was analyzed, related with defect states.

Considering the enthalpy of formation energy, V_{Cu} can be easily formed in CIGS, which has an important influence on carrier lifetimes. In the first study, we found that a deep complex defect level of $(In_{Cu} - 2V_{Cu})$, designated "*DX* states," could be natively formed in the E_g near the p-n junction in the case of BS, resulting in a decrease of τ_2 due to trapping photocarriers.

In contrast, τ_2 was increased in SLG with the addition of a Na supply, which is considered to relieve the defect level located in the E_g, forming the Na_{Cu}, antisite.

In the second study, the behavior of V_{Cu} defect states was determined to play a dominant role. Based on the geometric characteristics of the CIGS layer, we defined the two types of defect states, τ_1 and τ_2, which correspond to the trapping times of photoexcited electrons in the surface defect states at the CIGS surface, and the V_{Cu} defects, which are entirely distributed in the CIGS bulk, respectively. From the fitted τ_1 and τ_2 values, we discovered that the V_{Cu} defect states decrease in the depth direction of the CIGS, and Zn substitution to V_{Cu}, Zn_{Cu}, causes a decrease in bulk defect states available to trap photoexcited electrons, resulting in a rise of τ_2.

Acknowledgements

This work was supported by the "New & Renewable Energy" project of the Korea Institute of Energy Technology Evaluation and Planning (KETEP) grant funded by the Korean government Ministry Of Trade, Industry, & Energy (20153010011990, 20153000000030). The authors also would like to acknowledge the financial support from the R&D Convergence Program of MSIP (Ministry of Science, ICT and Future Planning) and ISTK (Korea Research Council for Industrial Science and Technology) of the Republic of Korea (Grant B551179-12-01-00).

Author details

Woo-Jung Lee and Yong-Duck Chung*

*Address all correspondence to: ydchung@etri.re.kr

Electronics and Telecommunications Research Institute, Daejeon, Korea

References

[1] Andrew M. Gabor, John R. Tuttle, David S. Albin, Miguel A. Contreras, Rommel Noufi, and Allen M. Hermann: High-efficiency $CuIn_xGa_{1-x}Se_2$ solar cells made from $(In_x,Ga_{1-x})_2Se_3$ precursor films. Applied Physics Letters. 1994; **65**: 198–200. DOI: 10.1063/1.112670

[2] Clas Persson, Yu-Jun Zhao, Stephan Lany, and Alex Zunger: n-type doping of $CuInSe_2$ and $CuGaSe_2$. Physical Review B. 2005; **72**: 035211. DOI: 10.1103/PhysRevB.72.035211

[3] Ingrid Repins, Miguel A. Contreras, Brian Egaas, Clay DeHart, John Scharf, Craig L. Perkins, Bobby To, and Rommel Noufi: 19.9%-efficient $ZnO/CdS/CuInGaSe_2$ solar cell with 81.2% fill factor. Progress in Photovoltaics: Research and applications. 2008; **16**: 235–239. DOI: 10.1002/pip.822

[4] Philip Jackson, Dimitrios Hariskos, Erwin Lotter, Stefan Paetel, Roland Wuerz, Richard Menner, Wiltraud Wischmann, and Michael Powalla: New world record efficiency for Cu(In,Ga)Se_2 thin-film solar cells beyond 20%. Progress in Photovoltaics: Research and Applications. 2011; **19**: 894–897. DOI: 10.1002/pip.1078

[5] Philip Jackson, Roland Wuerz, Dimitrios Hariskos, Erwin Lotter, Wolfram Witte, and Michael Powalla: Effects of heavy alkali elements in Cu(In,Ga)Se_2 solar cells with efficiencies up to 22.6%. Physica Status Solidi (RRL)-Rapid Research Letters. 2016; 10: 583–586 DOI: 10.1002/pssr.201600199

[6] B Ohnesorge, R Weigand, G Bacher, A Forchel, W Riedl, and FH Karg: Minority-carrier lifetime and efficiency of Cu(In,Ga)Se_2 solar cells. Applied Physics Letters. 1998; **73**: 1224–1226. DOI: 10.1063/1.122134

[7] Michael Hafemeister, Susanne Siebentritt, Jürgen Albert, Martha Ch Lux-Steiner, and Sascha Sadewasser: Large neutral barrier at grain boundaries in chalcopyrite thin films. Physical Review Letters. 2010; **104**: 196602. DOI: 10.1103/PhysRevLett.104.196602

[8] Harry Mönig, Y Smith, Raquel Caballero, Christian Kaufmann, Iver Lauermann, Martha ch. Lux-Steiner, and Sascha Sadewasser: Direct evidence for a reduced density of deep level defects at grain boundaries of Cu(In,Ga)Se_2 thin films. Physical review letters. 2010; **105**: 116802. DOI: 10.1103/PhysRevLett.105.116802

[9] Shengbai Zhang, Su-Huai Wei, Alex Zunger, and H Katayama-Yoshida: Defect physics of the $CuInSe_2$ chalcopyrite semiconductor. Physical Review B. 1998; **57**: 9642. DOI: 10.1103/PhysRevB.57.9642

[10] Yong-Duck Chung, Dae-Hyung Cho, Won-Seok Han, Nae-Man Park, Kyu-Seok Lee, and Jeha Kim: Incorporation of Cu in Cu(In,Ga)Se_2-based thin-film solar cells. Journal of the Korean Physical Society. 2010; **57**: 1826–1830. DOI: 10.3938/jkps.57.1826

[11] Dae-Hyung Cho, Yong-Duck Chung, Kyu-Seok Lee, Nae-Man Park, Kyung-Hyun Kim, Hae-Won Choi, and Jeha Kim: Influence of growth temperature of transparent conducting oxide layer on Cu(In,Ga)Se_2 thin-film solar cells. Thin Solid Films. 2012; **520**: 2115–2118. DOI: 10.1016/j.tsf.2011.08.083

[12] Jae-Hyung Wi, Woo-Jung Lee, Dae-Hyung Cho, Won Seok Han, Jae Ho Yun, and Yong-Duck Chung: Characteristics of temperature and wavelength dependence of $CuInSe_2$ thin-film solar cell with sputtered Zn(O,S) and CdS buffer layers. physica status solidi (a). 2014; **211**: 2172–2176. DOI: 10.1002/pssa.201431232

[13] Katsumi Kushiya and Osamu Yamase: Stabilization of PN heterojunction between Cu(InGa)Se_2 thin-film absorber and ZnO window with Zn(O,S,OH)x buffer. Japanese Journal of Applied Physics. 2000; **39**: 2577. DOI: 10.1143/JJAP.39.2577

[14] Taizo Kobayashi, Hiroshi Yamaguchi, and Tokio Nakada: Effects of combined heat and light soaking on device performance of Cu(In,Ga)Se_2 solar cells with ZnS(O,OH) buf-

fer layer. Progress in Photovoltaics: Research and Applications. 2014; **22**: 115–121. DOI: 10.1002/pip.2339

[15] Negar Naghavi, Solange Temgoua, Thibaud Hildebrandt, Jean François Guillemoles, and Daniel Lincot: Impact of oxygen concentration during the deposition of window layers on lowering the metastability effects in $Cu(In,Ga)Se_2$/CBD Zn(S,O) based solar cell. Progress in Photovoltaics: Research and Applications. 2015; **23**: 1820–1827. DOI: 10.1002/pip.2626

[16] Rohit P. Prasankumar, Prashanth C. Upadhya, and Antoinette J. Taylor: Ultrafast carrier dynamics in semiconductor nanowires. Physica Status Solidi (b). 2009; **246**: 1973–1995. DOI: 10.1002/pssb.200945128

[17] Priti Tiwana, Patrick Parkinson, Michael B. Johnston, Henry J. Snaith, and Laura M. Herz: Ultrafast terahertz conductivity dynamics in mesoporous TiO_2: influence of dye sensitization and surface treatment in solid-state dye-sensitized solar cells. The Journal of Physical Chemistry C. 2009; **114**: 1365–1371. DOI: 10.1021/jp908760r

[18] Ronald Ulbricht, Euan Hendry, Jie Shan, Tony F. Heinz, and Mischa Bonn: Carrier dynamics in semiconductors studied with time-resolved terahertz spectroscopy. Reviews of Modern Physics. 2011; **83**: 543. DOI: 10.1103/RevModPhys.83.543

[19] Woo-Jung Lee, Dae-Hyung Cho, Jae-Hyung Wi, Won Seok Han, Yong-Duck Chung, Jaehun Park, Jung Min Bae, and Mann-Ho Cho: Na-dependent ultrafast carrier dynamics of CdS/$Cu(In,Ga)Se_2$ measured by optical pump-terahertz probe spectroscopy. The Journal of Physical Chemistry C. 2015; **119**: 20231–20236. DOI: 10.1021/acs.jpcc.5b02282

[20] Leeor Kronik, David Cahen, and Hans Werner Schock: Effects of sodium on polycrystalline $Cu(In,Ga)Se_2$ and its solar cell performance. Advanced Materials. 1998; **10**: 31–36. DOI: 10.1002/(SICI)1521-4095(199801)10:1<31::AID-ADMA31>3.0.CO;2-3

[21] Dominik Rudmann, Antonio F. da Cunha, Marc Kaelin, Fe Kurdesau, Hans Zogg, Ayodhya N. Tiwari, and Gerhard Bilger: Efficiency enhancement of $Cu(In,Ga)Se_2$ solar cells due to post-deposition Na incorporation. Applied Physics Letters. 2004; **84**: 1129–1131. DOI: 10.1063/1.1646758

[22] Shogo Ishizuka, Akimasa Yamada, Muhammad Monirul Islam, Hajime Shibata, Paul Fons, Takeaki Sakurai, Katsuhiro Akimoto, and Shigeru Niki: Na-induced variations in the structural, optical, and electrical properties of $Cu(In,Ga)Se_2$ thin films. Journal of Applied Physics. 2009; **106**: 034908. DOI: 10.1063/1.3190528

[23] Dae-Hyung Cho, Kyu-Seok Lee, Yong-Duck Chung, Ju-Hee Kim, Soo-Jeong Park, and Jeha Kim: Electronic effect of Na on $Cu(In,Ga)Se_2$ solar cells. Applied Physics Letters. 2012; **101**: 023901. DOI: 10.1063/1.4733679

[24] Woo-Jung Lee, Dae-Hyung Cho, Jae-Hyung Wi, Won Seok Han, Jeha Kim, and Yong-Duck Chung: Na effect on flexible $Cu(In,Ga)Se_2$ photovoltaic cell depending on diffusion

barriers (SiOx, i-ZnO) on stainless steel. Materials Chemistry and Physics. 2014; **147**: 783–787. DOI: 10.1016/j.matchemphys.2014.06.021

[25] Andreas Othonos: Probing ultrafast carrier and phonon dynamics in semiconductors. Journal of Applied Physics. 1998; **83**: 1789–1830. DOI: 10.1063/1.367411

[26] Ingo Dirnstorfer, Mt. Wagner, Detlev M. Hofmann, MD Lampert, Franz Karg, and Bruno K. Meyer: Characterization of CuIn(Ga)Se_2 thin films. Physica Status Solidi (a). 1998; **168**: 163–175. DOI: 10.1002/(SICI)1521-396X(199807)168:1<163::AID-PSSA163>3.0.CO;2-T

[27] Rajmund Bacewicz, P. Zuk, and R. Trykozko: Photoluminescence study of ZnO/CdS/Cu(In,Ga)Se_2 solar cells. Optoelectronics Review. 2003; **11**: 277–280.

[28] Sho Shirakata, Katsuhiko Ohkubo, Yasuyuki Ishii, and Tokio Nakada: Effects of CdS buffer layers on photoluminescence properties of Cu(In,Ga)Se_2 solar cells. Solar Energy Materials and Solar Cells. 2009; **93**: 988–992. DOI: 10.1016/j.solmat.2008.11.043

[29] Young Min Shin, Chang Soo Lee, Dong Hyeop Shin, Young Min Ko, Essam A Al-Ammar, Hyuck Sang Kwon, and Byung Tae Ahn: Characterization of Cu(In,Ga)Se_2 solar cells grown on Na-free glass with an NaF layer on a Mo film. ECS Journal of Solid State Science and Technology. 2013; **2**: P248–P252. DOI: 10.1149/2.002306jss

[30] Su-Huai Wei, Shengbai Zhang, and Alex Zunger: Effects of Ga addition to $CuInSe_2$ on its electronic, structural, and defect properties. Applied Physics Letters. 1998; **72**: 3199–3201. DOI: 10.1063/1.121548

[31] S Zotta, Karl Leo, Martin Ruckh, and Hans Warner Schock: Photoluminescence of polycrystalline $CuInSe_2$ thin films. Applied Physics Letters. 1996; **68**: 1144–1146. DOI: 10.1063/1.115704

[32] Shih-Chen Chen, Yu-Kuang Liao, Hsueh-Ju Chen, Chia-Hsiang Chen, Chih-Huang Lai, Yu-Lun Chueh, Hao-Chung Kuo, Kaung-Hsiung Wu, Jenh-Yih Juang, and Shun-Jen Cheng: Ultrafast carrier dynamics in Cu(In,Ga)Se_2 thin films probed by femtosecond pump-probe spectroscopy. Optics Express. 2012; **20**: 12675–12681. DOI: 10.1364/OE.20.012675

[33] Makoto Okano, Yutaro Takabayashi, Takeaki Sakurai, Katsuhiro Akimoto, Hajime Shibata, Shigeru Niki, and Yoshihiko Kanemitsu: Slow intraband relaxation and localization of photogenerated carriers in $CuIn_{1-x}Ga_xSe_2$ thin films: Evidence for the existence of long-lived high-energy carriers. Physical Review B. 2014; **89**: 195203. DOI: 10.1103/PhysRevB.89.195203

[34] Stephan Lany and Alex Zunger: Intrinsic DX Centers in Ternary Chalcopyrite Semiconductors. Physical review letters. 2008; **100**: 016401. DOI: 10.1103/PhysRevLett.100.016401

[35] Yu-Jun Zhao, Clas Persson, Stephan Lany, and Alex Zunger: Why can $CuInSe_2$ be readily equilibrium-doped n-type but the wider-gap $CuGaSe_2$ cannot? Applied Physics Letters. 2004; **85**: 5860–5862. DOI: 10.1063/1.1830074

[36] Stephan Lany, Yu-Jun Zhao, Clas Persson, and Alex Zunger: Halogen n-type doping of chalcopyrite semiconductors. Applied Physics Letters. 2005; **86**: 42109–42109. DOI: 10.1063/1.1854218

[37] Tokio Nakada, Masayuki Mizutani, Y Hagiwara, and Akio Kunioka: High-efficiency Cu(In,Ga)Se_2 thin-film solar cells with a CBD-ZnS buffer layer. Solar Energy Materials and Solar Cells. 2001; **67**: 255–260. DOI: 10.1016/S0927-0248(00)00289-0

[38] Tokio Nakada and Masayuki Mizutani: 18% efficiency Cd-free Cu(In,Ga)Se_2 thin-film solar cells fabricated using chemical bath deposition (CBD)-ZnS buffer layers. Japanese Journal of Applied Physics. 2002; **41**: L165.

[39] Negar Naghavi, Stefani Spiering, Michael Powalla, Bruno Cavana, and Daniel Lincot: High-efficiency copper indium gallium diselenide (CIGS) solar cells with indium sulfide buffer layers deposited by atomic layer chemical vapor deposition (ALCVD). Progress in Photovoltaics: Research and Applications. 2003; **11**: 437–443. DOI: 10.1002/pip.50

[40] Muhammad Monirul Islam, Shogo Ishizuka, Akimasa Yamada, Keiichiro Sakurai, Shigeru Niki, Takeaki Sakurai, and Katsuhiro Akimoto: CIGS solar cell with MBE-grown ZnS buffer layer. Solar Energy Materials and Solar Cells. 2009; **93**: 970–972. DOI: 10.1016/j.solmat.2008.11.047

[41] Theresa Magorian Friedlmeier, Philip Jackson, Andreas Bauer, Dimitrios Hariskos, Oliver Kiowski, Roland Wuerz, and Michael Powalla: Improved photocurrent in Cu (In,Ga)Se_2 solar cells: from 20.8% to 21.7% efficiency with CdS buffer and 21.0% Cd-free. IEEE Journal of Photovoltaics. 2015; **5**: 1487–1491. DOI: 10.1109/JPHOTOV.2015.2458039

[42] Dae-Hyung Cho, Woo-Jung Lee, Sang-Woo Park, Jae-Hyung Wi, Won Seok Han, Jeha Kim, Mann-Ho Cho, Dongseop Kim, and Yong-Duck Chung: Non-toxically enhanced sulfur reaction for formation of chalcogenide thin films using a thermal cracker. Journal of Materials Chemistry A. 2014; **2**: 14593–14599. DOI: 10.1039/c4ta02507e

[43] Jae-Hyung Wi, Tae Gun Kim, Jeong Won Kim, Woo-Jung Lee, Dae-Hyung Cho, Won Seok Han, and Yong-Duck Chung: Photovoltaic performance and interface behaviors of Cu(In,Ga)Se_2 solar cells with a sputtered-Zn(O,S) buffer layer by high-temperature annealing. ACS Applied Materials & Interfaces. 2015; **7**: 17425–17432. DOI: 10.1021/acsami.5b04815

[44] Dae-Hyung Cho, Woo-Jung Lee, Jae-Hyung Wi, Won Seok Han, Tae Gun Kim, Jeong Won Kim, and Yong-Duck Chung: Interface analysis of Cu(In,Ga)Se_2 and ZnS formed using sulfur thermal cracker. ETRI Journal. 2016; **38**: 265–271. DOI: 10.4218/etrij.16.2515.0031

[45] Woo-Jung Lee, Hye-Jung Yu, Jae-Hyung Wi, Dae-Hyung Cho, Won Seok Han, Jisu Yoo, Yeonjin Yi, Jung-Hoon Song, and Yong-Duck Chung: Behavior of photocarriers in the light-induced metastable state in the p-n heterojunction of a Cu(In,Ga)Se_2 solar cell with CBD-ZnS buffer layer. ACS Applied Materials & Interfaces. 2016; **8**: 22151–22158. DOI: 10.1021/acsami.6b05005

[46] Tokio Nakada, Tomoyuki Kume, Takahiro Mise, and Akio Kunioka: Superstrate-type Cu(In,Ga)Se_2 thin film solar cells with ZnO buffer layers. Japanese Journal of Applied Physics. 1998; **37**: L499.

[47] Chang-Soo Lee, Suncheul Kim, Essam A Al-Ammar, HyuckSang Kwon, and Byung Tae Ahn: Effects of Zn diffusion from (Zn,Mg)O buffer to CIGS film on the performance of Cd-Free Cu(In,Ga)Se_2 solar cells. ECS Journal of Solid State Science and Technology. 2014; **3**: Q99–Q103. DOI: 10.1149/2.003406jss